THE UBIQUITOUS ATOM

THE UBIQUITOUS ATOM

EDITED AND WITH AN INTRODUCTION BY

Grace Marmor Spruch

AND *Larry Spruch*

ILLUSTRATED WITH PHOTOGRAPHS AND WITH DIAGRAMS BY RICHARD LIU

CHARLES SCRIBNER'S SONS · NEW YORK

Library of Congress Cataloging in Publication Data

Spruch, Grace Marmor, comp.
The ubiquitous atom.

"Based upon material from booklets in the series Understanding the atom, produced under the aegis of the United States Atomic Energy Commission."
Bibliography: p.
1. Atoms. I. Spruch, Larry, joint comp.
II. United States. Atomic Energy Commission. Understanding the atom. III. Title.
QC777.S67 539.7 74-507
ISBN 0-684-13773-9

1 3 5 7 9 11 13 15 17 19 V|C 20 18 16 14 12 10 8 6 4 2

PRINTED IN THE UNITED STATES OF AMERICA

CONTENTS

ILLUSTRATIONS

DIAGRAMS

THE UBIQUITOUS ATOM

Introduction

SOME years ago, several members of the Institute for Advanced Study, in Princeton, New Jersey, were sitting around talking when one of the nonscientists in the group threw up his hands and confessed he had no idea what a molecule was. J. Robert Oppenheimer, head of the institute at that time and a man known to speak his mind, is said to have responded that if a scientist was unfamiliar with Shakespeare he would be considered an ignoramus, and in our technological era anyone unfamiliar with a molecule, *he* considered an ignoramus.

At one time science was an essential ingredient in the cultural fare of an educated man. Remember the story we all learned as children, about Benjamin Franklin and his kite? His kite flying had the serious scientific intent of testing whether lightning and laboratory electricity were the same phenomenon. Thomas Jefferson, in designing his home at Monticello, incorporated all kinds of scientific devices into the plan, including some of his own invention. Tolstoi's novels contain many scientific references, and several Tolstoi heroes have decided scientific bents. Franklin, Jefferson, and Tolstoi are three of many examples one might cite. They were extraordinary men, of course, but cultured men of the nonextraordinary variety also knew a good deal of the science of their time.

We have been informed ad nauseam that science has undergone enormous changes in this century, particularly in the last 25 years. Almost as enormous, however, have been the changes in attitude toward science.

The end of World War II saw atomic energy lighting the way to peace and technological prosperity for all. And scientists, who had up to then been considered society's eccentrics (when not evil or mad—witness films and comic strips) took on the role of benefactors of mankind. This attitude remained on a slightly downward-slanting plateau in the decade that followed the first atomic explosion, until the launching of the Soviet Sputnik—which launched the space age—boosted the importance of scientists, for it was through them that damaged national pride could be repaired. Funds were poured with such profligacy into projects relating to space that one of us jokingly suggested that Americans had really launched Sputnik and had credited the Russians precisely for the purpose of stimulating the flow of federal funds; the Russians, for their part, had been too embarrassed to deny the credit.

The upsurge in the prestige of science, directly proportional to the funds, lasted about a decade after Sputnik's launch. Then, almost coinciding with our setting a man on the moon, the moonlit period was over. Not only space programs were cut, but scientific programs as well. (The distinction is intentional, as will be made clear as we go on.) Science and scientists were looked upon in another light, as destroyers of nature, as polluters of our planet.

Why should one be concerned about science and attitudes toward it? For reasons that can be lumped in three categories: cultural, social, and aesthetic.

The cultural category takes in what Oppenheimer was talking about when he called his colleague an ignoramus. True, it was easier to be scientifically cultured in Benjamin Franklin's day. Science was not as specialized then. As science became specialized, it was put into the hands of specialists, and left there. Even scientists have become ignoramuses, by Oppenheimer's definition: biologists know very little physics, and physicists know essentially no biology. Specialization is not confined to science. It extends in varying degrees to most areas of life. Few do their own plumbing these days.

There are advantages to specialization, obviously. It frees us from having to think about everything, gives us a finite number of problems to worry about. The plumber worries about the plumbing—which brings us to the social category.

Congress—or its local equivalent—is continually called upon to decide whether a supersonic transport plane should be funded, or nuclear tests permitted, or a power company licensed to build another plant, or cyclamates permitted in soft drinks, or phosphates allowed in detergents (or detergents allowed at all), or whether a costly accelerator should be constructed while poverty remains in our streets. Congress represents us. And if we do not wish to entrust these decisions entirely to scientific advisers, our plumbers in government, then we are obligated to know something about these matters. True, we cannot make judgments that require expertise. But it is possible, on considerably less than expertise, to decide which of several experts makes a better case. Will nuclear tests pollute the atmosphere or won't they? Are there genetic dangers from these tests or not? Should our electricity come from nuclear fuel or from coal? Is there an energy crisis? Is there a shortage of oil?

Some problems can hardly be understood without a great deal of technical knowledge. Others require only a rudimentary knowledge of the facts of science, but they also require a knowledge of what science *is.* It has often been asked: "Should we have sent a man to the moon, or would that money have been better employed to try to cure cancer?" The question is difficult to answer, for it involves a choice between technology and science. Putting a man on the moon was a problem in *technology,* in engineering. All the scientific principles were known—in fact, known since Newton's time. The problem was to develop a technology that would put the scientific principles into practical flight, in particular, to develop engines and fuels that would provide the necessary thrust. The project might take a long or a short time, but sooner or later it would be done. Curing cancer is a problem in *science.* The nature of the disease is not understood. The principles are not known. It is not certain that they will be known even after huge expenditures. Our thinking may not be sufficiently advanced, our minds may not be ready for the crucial ideas.

Some persons have equated curing cancer with making the atomic bomb. When $2 billion was put into the Manhattan Project, the bomb came out. However, the atomic bomb project was similar to that of putting a man on the moon—it was a technological problem.

(An easier question, since it involves a choice within the area of technology, is: "Should we have put a man on the moon, or should it have been a machine?" The cost would have been much less for a machine, for the safety factors need not have been so stringent.)

Confusion of technology with science is so pervasive that even war-oriented technology is called science and written up in the science section of newspapers.

The third category, the aesthetic, relates to the fact that some of the most exciting and stimulating ideas to have emerged in this century have been in science. Quantum theory, relativity—these were revolutions in the realm of thought. Some of the questions continue to be exciting and stimulating because they are not settled. Was the universe created in a big bang some 13 billion years ago, or has it always existed, in a steady state? Other questions have been answered and it is the answers that are fascinating. Does light consist of waves or of particles? Light is both! Are there limits to man's knowledge? Yes. The uncertainty principle sets limits. One question, so simple it seems simple-minded, "Why is it dark at night?" has the mind-boggling answer: "Because the universe is expanding."

A line from Schopenhauer seems to apply: "The task is, not so much to see what no one has seen yet; but to think what nobody has thought yet, about what everybody sees." Or, a bit closer to home, our friend Evelyn's mother says, whenever she learns something surprising: "If I'd have died yesterday I wouldn't have known that."

The last few years have seen a sharp rise in anti-intellectualism. In the theater, plays aim to reach audiences through the gut rather than the more conventional entry points. In music, appeal is mostly through rhythm and volume rather than the more cerebral melody and harmony. And science, until recently the benefactor of man, has become a source of evil.

There is a profound lack of knowledge in both extreme views, that in science lies salvation or damnation. In science lies potentiality. It is the use of science that swings us toward salvation or damnation. Physicist Richard Feynman likes to quote a Buddhist proverb: "To every man is given the key to the gates of heaven; the same key opens the gates of hell."

It is ironic that science, destroyer of established doctrines and articles of faith since almost the time of Copernicus, should now be defending its own doctrines and articles of faith. The kind of attack which comes out of the latest interest in Eastern philosophy and drugs, which maintains that there is more to reality than can be found in scientific description, is difficult to counter with argument. We can only note that it has often occurred in the past. A powerful antiscience movement followed the French Revolution, for example, when it was claimed that Newtonian mechanistic thinking had brought on the revolution.

One reason for today's antiscience attitude, it has been suggested, is that science takes things apart to examine them in their simplest forms, while the trend today is to "put it all together." Thus the "in" field is ecology, which does just that.

Many scientists find themselves, however, on the side of those who attack the doctrine that unlimited research is always good.

The concept of unlimited research is a relatively recent one. Newton and Descartes did not envision an open-ended science in which research would everlastingly expose new frontiers for further exploration. In fact, Descartes claimed that if he were given a "grant," he could solve the 150-odd problems which remained to be solved, and which he proceeded to enumerate. Open-endedness is a modern development.

Even after conceding that there is no limit to possible research, the question arises: "If one can, does that mean one should?" This question plagued some of the developers of the atomic bomb. Hitler made the answer easy then. To give an answer now is harder. And giving an answer should not be left to scientists alone. The scientist's concern with the implications of his science should be shared with the public. Should research on heart transplants be furthered? If so, who should receive the transplants? Everyone? With overpopulation a problem, should anyone be kept alive beyond his time? And what is his time, when medicine has already lengthened it considerably? Shall we proceed with sex determination of offspring? If we do and are successful, who should do the determining, society or the parents? Here, marketplace laws of supply and demand might lead to rather unstable equilibrium. Will research on freezing sperm and on the genetic code lead to genetic manipulation, and possible dictatorial control through such manipulation? The late German born physicist Max Born wrote: "Intellect distinguishes between the possible and the impossible; reason distinguishes between the sensible and the senseless. Even the possible can be senseless."

Ours may be a technological era, but it need not be technology-dominated. Technology may have played too great a role in our lives, but it may also be that we can no longer get along without it. If, as has recently been demanded, we must humanize science, then we must understand science first. A little knowledge may be a dangerous thing, but lack of a little knowledge may be more dangerous.

We have attempted in this book to put together material, ordinarily dispersed in physics books, chemistry books, biology texts, engineering tracts, involving the atom: what it is, what it does, and what can be done with it.

The first part of the book deals with fundamentals, the basic physics and chemistry of the atom. The treatment is largely historical. Social, political, and economic influences on science, although important, have been left out.

Part II deals with the atom and life. Some of the biology and chemistry of life processes are discussed, mainly to provide a background for considering the effects of atomic particles on living things. What are the effects of gamma rays passing through our bodies, and what are the effects on our children's children? Is it dangerous to wear a watch with a radium dial?

Part III spreads out from man to his environment: the cosmic rays that incessantly bombard man from outer space, and the neutrons that rush up at him from the soil; man's attempts to harness particles in medicine and in agriculture, to cure and to learn; the evils that have resulted from man's lack of thought, the evils of pollution that plague us today. Some will undoubtedly turn out to be temporary and give the lie to Marc Antony's

> The evil that men do lives after them,
> The good is oft interred with their bones.

Part IV continues to emphasize the man-made aspect of man's environment: his atomic projects. The story, or again history, starts with the initial production of atomic energy, telling why men of goodwill found it necessary to first use this marvelous new power in a bomb, then goes on to the controlled use of atomic energy in power plants and generators, and in taking man under the polar ice cap in submarines and into outer space. Atomic particles are followed into practical problems in industry, to determine when bottles are full or paper is the right thickness, even into art, to date an ancient vase or detect a forged painting.

Most of the book is readily understandable, and hopefully enjoyable. Some pages, notably those on high-energy physics, are rather difficult, since only enough space was used to state some of the essential ideas. Here too, there surely can be pleasure in knowing that nature can distinguish between right-handedness and left-handedness even without thoroughly understanding the theory or the experiment that proved it. Those who lack technical knowledge of music can derive pleasure from a Beethoven quartet, on another level from that of the person who can appreciate its construction. We take courses in music appreciation without knowing how to perform, art appreciation without knowing how to paint. To fully understand the uncertainty principle takes years of technical training. But everyone can appreciate the principle and its consequences on another level. Shall we call that level "physics appreciation"?

Note on the text: This book is based upon material from booklets in the series Understanding the Atom, produced under the aegis of the United States Atomic Energy Commission. The AEC is to be commended for sponsoring this series of well-written, informative booklets that serve the important function of educating the public.

The Commission has recently been criticized by environmentalists—with much justification, in our opinion—for bias in such areas as nuclear weapons testing and power plants, and for imposing insufficiently strict standards for the use of atomic energy. Whether or not the AEC is biased in these matters, the authors selected to write the different booklets did not, by and large, give the impression of being biased. Where we felt bias did creep in, we invoked editorial prerogative to excise and temper, since the book was prepared independently, with no connection with the AEC. (The booklets are in the public domain.) In any case, the bulk of the contents is unaffected by such considerations.

Material written by authors of the booklets has a reference number in parentheses at the end of a passage from a given booklet. Such passages may not be in consecutive order. For titles and authors of booklets, see Sources following the Appendix. Material written by the editors is interspersed with material from the booklets but distinguished typographically or by enclosure in brackets.

The booklets from which selections were made total about 2500 pages. While we did not wish the resulting abstracts to read like an almanac, we were limited in the amount of connective tissue we could add by considerations of total length. To reduce the choppiness that, of necessity, ensues in a scheme of this sort, footnotes that appeared in the booklets have been incorporated into the text in parentheses, and ellipses have not been added where deletions were made.

PART I

SOME BASIC PHYSICS AND CHEMISTRY

IT is usually easiest to tell—and often easiest to understand—a story that starts with a beginning and goes on to a middle and an end. Let us begin with a bit of history, so that our modern ideas will not appear to have sprung full blown from unsown ground.

In this bit of history we cannot begin to describe adequately the zigzag course of events. An idea conceived by the ancient Greeks may be followed by experiments and theories of the eighteenth century that led to major breakthroughs. Possibly important—if incomplete, and perhaps confused—experiments and theories in between have been left out. Concepts that a child considers obvious today may have taken centuries to develop. One need only consider Leonardo da Vinci's writings—let alone those of the ancient Greeks—on what came to be known as Newton's second law, that force equals mass times acceleration ($F = ma$), to realize how difficult it must have been to grasp the essential ideas, for lack of experimental data or incorrect philosophical guidelines. To use a phrase uttered by the Austrian-born physicist Wolfgang Pauli (1900–58) in another context, the ideas often were so confused they "weren't even wrong."

After the first chapter, various subjects are treated in some depth. Some of the text in this first part is necessarily rather technical, but the bulk of parts II, III, and IV is not based on this more technical material.

1 ◊ Introduction to the Atom

Beyond our solar system and local galaxy of stars, 10 billion other galactic systems swarm to the edges of the universe. Within each of our fingertips, each atom, a million times smaller than the width of a human hair, is, itself, ten thousand times as large as its central nucleus. This enormous range of sizes seems to defy unifying explanations.

Size is not the only obstacle to understanding the nature of matter. The material of the universe assumes a million different forms—gas, liquid, solid, animate, inanimate—thwarting any general classification.

Still, there has always been the feeling—or, at least, a wistful hope—that nature could be explained in terms of simple building blocks, which join together or break apart in simple ways. The Greeks thought that there might be basic atoms, out of which everything else could be formed. Empedocles, in 400 B.C., taught that everything was made from four elements and four essences. His geometric arrangement of these is shown below. Ordinary substances were supposed to be composed of combinations of the pure elements linked by differing essences. Wood might appear to be made of earth, but if you added heat you could see that it also contained fire and air. The system was profound but useless; it was a mystic's scheme, leading only to alchemy and its vain efforts to turn lead into gold.

Yet today we have arrived at a scheme of building blocks that superficially looks almost as sparse as that of the Greeks. In the last 40 years, we have torn apart the atoms that comprise

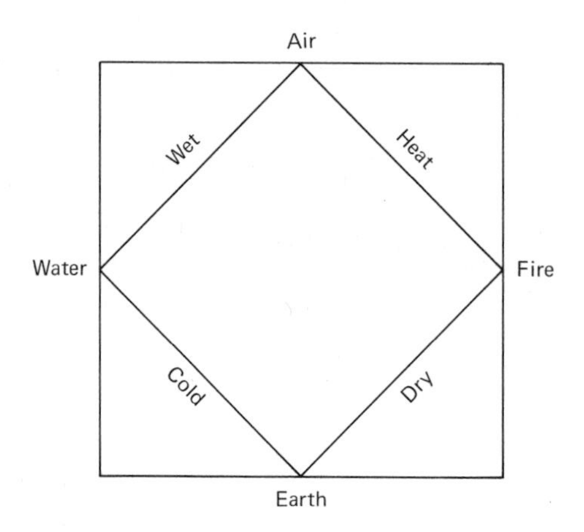

Elements and essences of the Greek universe

the traditional ninety-two elements. We have learned to manipulate the cores of those atoms to realize, at last, the alchemists' dream of converting commonplace substances—not into gold, but into much more valuable radioactive materials. Since World War II, we have been busily tearing apart the protons and neutrons that make up the atomic nucleus. We find that there are several other particles besides protons, neutrons, and electrons, and it has become customary to refer to these as elementary or fundamental particles. If we count all these as particles, we have about as many particles as there are elements. Surely they cannot all be elementary or fundamental. As we shall see, there are ways to classify them into a small number of basic types. (13)

Although some of the ancient Greeks thought matter was composed of atoms, their atomic theory was far from the systematic structure of ideas that constitutes modern atomic theory. The Greeks are credited with fantastic insight because we focus on what agrees with modern ideas and overlook more far-out claims, such as that honey consists of round atoms while the atoms of bitter foods are sharp-edged. Ancient Greek atomic theory is important in its philosophical concepts rather than in its details, for underlying the theory is the idea that the universe is comprehensible. It is comprehensible in terms of simple units. And these simple units are not what is immediately apparent to our senses. Scientific validation of this idea was achieved only in this century. For 2000 years after ancient Greece, the prevailing conception of matter was that it was continuous; that is, matter could be divided and subdivided an unlimited number of times without changing its characteristics.

This concept was humorously expressed by the British mathematician Augustus De Morgan (1806–71) in these lines:

> Great fleas have little fleas upon their backs to bite 'em,
> And little fleas have lesser fleas, and so *ad infinitum.*

De Morgan evidently did not keep up with the latest developments in science, however, because 2 years before his birth, John Dalton (1766–1844), an English schoolteacher, had changed the atomic theory of matter from a philosophical speculation into a firmly established principle. The evidence that convinced Dalton and many other contemporary scientists of the reality of atoms came from quantitative chemical analysis.

Dalton knew that many chemical substances could be separated into two or more simpler substances. Chemicals that could be separated further were called compounds; those that could not were called elements. Careful experiments by Dalton and others showed that whenever two or more elements combined chemically to make a compound the relative amounts of the elements had to be carefully adjusted to fit a definite proportion in order to have no elements left over after the reaction was finished. For example, if hydrogen and oxygen were combined to form water (H_2O), the weight of oxygen had to be eight times the weight of hydrogen; otherwise, either some hydrogen or some oxygen would be left over.

This fundamental truth is now called the law of definite proportions. Another important principle, called the law of multiple proportions, is illustrated by hydrogen peroxide (H_2O_2), which is made up of the same two elements that are found in water. The weight of oxygen in hydrogen peroxide, however, is sixteen times the weight of hydrogen or exactly twice the relative weight found in water.

These principles of chemical combination convinced Dalton that each chemical element consists of small, indivisible units, all just alike, called atoms, and that each chemical compound also has basic units, called molecules, which cannot be divided without reducing the compound into its elements—that is, destroying it as a compound. He visualized a molecule of a compound as formed by the uniting of individual atoms of two or more elements. It was obvious to him that in any molecule of a compound, the weight of each atom of a component element bore a proportionate relationship to the weight of the entire molecule which was equal to the proportion, by weight, of all that element in the compound. And although Dalton had no idea how heavy any individual atom really was, he could tell how many times heavier or lighter it was than an atom of another element.

Incidentally, Dalton mistakenly thought that one atom of oxygen was eight times as heavy as one atom of hydrogen instead of sixteen times as heavy. He assumed a water molecule to be HO instead of H_2O.

Curiosity about the fundamental nature of matter was matched by equally avid curiosity about the fundamental nature of electricity. Before 1850 much had been learned about the behavior of electric charge and electric currents flowing through solids and liquids. Real progress in understanding electric charge, however, had to wait for the development of highly efficient vacuum pumps. (24)

This often happens in science. One tends to think that scientific discovery always precedes the development and application of corresponding technology. But a scientific discovery often cannot be made before some advance in technology. Precise observation of the orbits of the planets had to await precision instruments, and only such precise observation could give rise to the theory of gravitation, which in turn predicted planets as yet unseen, to which future technology may take man (or his machines) and where he may perhaps find rocks that will tell the origin of the universe. And so if there is any ordering it is a kind of spiral: scientific discovery, technology, more science, more technology.

The theory that each element has a fixed combining capacity was proposed by the English chemist Sir Edward Frankland (1825–99) in 1852. This capacity was called the valence of an atom. As most of the elements then known would combine with either oxygen or hydrogen, the valence values were related to the number of atoms of oxygen or hydrogen with which one atom of each element would combine. Two atoms of hydrogen combine with one atom of oxygen to form H_2O, so hydrogen was given a valence of 1, and oxygen a valence of 2. The valence of any other element was then the number of atoms of hydrogen (or twice the number of oxygen atoms) that combined with one atom of that element. In ammonia we have the formula NH_3, so nitrogen has a valence of 3; in carbon dioxide, CO_2, the carbon valence is 4. Valences are always whole numbers. Some elements exhibit more than one valence, and the maximum valence appears to be 8.

In the late 1860s the Russian chemist Dmitri Mendeleev (1834–1907) made an intriguing observation when listing the elements in the order of increasing atomic weights. He found that the first element after hydrogen was lithium with a valence of 1, the second heaviest was beryllium with a valence of 2, the third, boron with a valence of 3, and so on. As he continued he found a sequence of valences that went 1, 2, 3, 4, 3, 2, 1, and then repeated itself. If he arranged the elements in vertical columns next to one another, in the order of increasing atomic weights, he found the elements in each horizontal row across the page had the same valence and strikingly similar chemical properties.

This kind of periodicity, or regular recurrence, had been noted by other scientists, but Mendeleev made a great step forward by leaving gaps in his table where the next known element, in order of weight, did not fit because it had the wrong valence or the wrong properties. He predicted that these gaps would be filled by yet-to-be-discovered elements, and he even went as far as to predict the properties of some of these elements from the position they would occupy in his table. Mendeleev's first draft of his periodic table of the elements and an early version circulated in Russia are reproduced in the photograph section. These were based on the sixty-three elements then known. In later versions of the table the elements are arranged in order across the horizontal rows, and those with similar properties fall in the same vertical column. (See pages 48–49.)

At the time of the setting up of the periodic table the noble gases were still undiscovered. (These are discussed in chapter 3.) There were no gaps left for them, as spaces could be left only where at least one element in a group was already known. When argon was discovered some problem therefore arose as to its place in the periodic system. Its atomic weight suggested it might belong somewhere near potassium. When its lack of chemical reactivity was discovered, Mendeleev proposed that it had zero valence and should come between chlorine and potassium. He suggested that a group of such gases might be found. The valence periodicity then would be 0, 1, 2, 3, 4, 3, 2, 1. This new group led to a complete periodicity of 8, which we shall see is a very significant number.

Both Frankland and Mendeleev based their ideas on their knowledge of chemical properties. The theoretical support for both proposals came with the development of a theory of atomic structure and the electronic theory of valence. (6)

By the 1890s it had become clear that the flow of electricity through a highly evacuated tube consisted of a negative electric charge moving at a very high speed along straight lines between sealed-in electrodes. Since it originated at the negative electrode, or cathode, the invisible stream of charge was named cathode rays.

Although many investigators contributed to knowledge about cathode rays, the experiments of Joseph J. Thomson (1856–1940), a British physicist, are generally considered to have been the most enlightening. Thomson arranged a cathode-ray tube so that the rays could be deflected by magnets and by electrically charged metal plates. By applying certain well-known principles of physics, he was able to confirm that electric charge, like matter, was atomized—the stream of charge consisted of a swarm of very small particles, all alike.

Probably Thomson's most significant result was determining the ratio of the charge of each little particle to its weight. He was able to do this by measuring the magnetic force required to divert a stream of charged particles. This charge-to-weight ratio proved to be nearly two thousand times greater than the already known charge-to-weight ratio for a positively charged hydrogen atom, or ion, which until then was thought to be the lightest constituent of matter. It remained to be determined whether charge or weight caused the difference. Further experimentation showed that the charges were approximately the same in the two cases. It was therefore proven that the weight of the hydrogen atom, lightest of all the atoms, was nearly two thousand times as great as the weight of one of the little negative particles.

The name electron was given to the small negative particles identified by Thomson. Since the electrons had come from the cathode, it was apparent that the atoms in the cathode must contain electrons. Thomson reasoned that electric current in a wire is a stream of electrons passing successively from atom to atom and that the difference between an electrically charged atom and a neutral atom is that the charged one has gained or lost one or more electrons. (24)

In 1895 a German physicist, Wilhelm Konrad Roentgen (1843–1923) noticed that certain crystals became luminescent when they were in the vicinity of a highly evacuated electric-discharge tube. Objects placed between the tube and the crystals screened out some of the invisible radiation that caused this effect, and he observed that the greater the density of the object so placed, the greater the screening effect. He called this new radiation X rays, because *x* was the standard algebraic symbol of an unknown quantity.

A French physicist, Antoine Henri Becquerel (1852–1908), saw that this discovery opened up a new field for research and set to work on some of its ramifications. Becquerel happened to develop a plate that had been in contact with uranium material in a dark drawer. A telltale black spot marking the position of the mineral appeared on the developed plate. His conclusion was that uranium gave off X rays or something similar.

At this point, Pierre Curie (1859–1906), a friend of Becquerel and also a professor of physics in Paris, suggested to one of his graduate students, his young Polish bride, Marie Slodowska (1867–1934), that she study this new phenomenon. She found that both uranium and thorium possessed this property of radioactivity, but also, surprisingly, that some uranium minerals were more radioactive than uranium itself. Through a tedious series of chemical separations, she obtained from pitchblende (a uranium ore) small amounts of two new elements, polonium and radium, and showed that they possessed far greater radioactivity than uranium itself. (31)

Although no one realized it at the time, Becquerel had discovered that atoms of some elements will at random times transform themselves into atoms of a different element by emitting certain extremely high-speed charged particles. Atoms that can do this are said to be radioactive, and it was the radiation from transforming uranium atoms that darkened Becquerel's photographic plate.

The radiation from radioactive materials was found to be of three kinds called alpha rays, beta rays, and gamma rays. Alpha rays were first detected by New Zealand-born Ernest Rutherford (1871–1937), who later identified them as positively charged helium atoms. Becquerel demonstrated that beta rays, like cathode rays, consist of negatively charged electrons. The highly penetrating gamma rays were proved by Rutherford and E. N. da C. Andrade to be electromagnetic radiation similar to X rays.

Rutherford, in collaboration with the English chemist

Frederick Soddy (1877–1956), brought order out of a chaos of puzzling discoveries by establishing the general behavior of radioactive atoms. He determined that certain naturally occurring atoms of high atomic weight can spontaneously emit an alpha or a beta particle and thereby convert themselves into new atoms. These new atoms, being also radioactive, sooner or later convert themselves into still different atoms, and so on. Each time an alpha particle is emitted in this sequence, the new atom is lighter by the weight of the alpha particle, or helium atom. The disintegration process proceeds from stage to stage until at last a stable atom is produced. The end product in this decay process in naturally occurring radioactive elements is lead.

One experiment by Rutherford and his coworkers had a most profound effect on the understanding of atomic structure. What they did was to direct a stream of alpha particles at a thin piece of gold foil. The results were astonishing. Almost all the particles passed straight through the foil without changing direction. Of the few particles that did ricochet in new directions, however, some were deflected at very sharp angles, as the diagram shows.

As a result of this experiment, Rutherford proposed a concept of the atom entirely different from the one which pre-

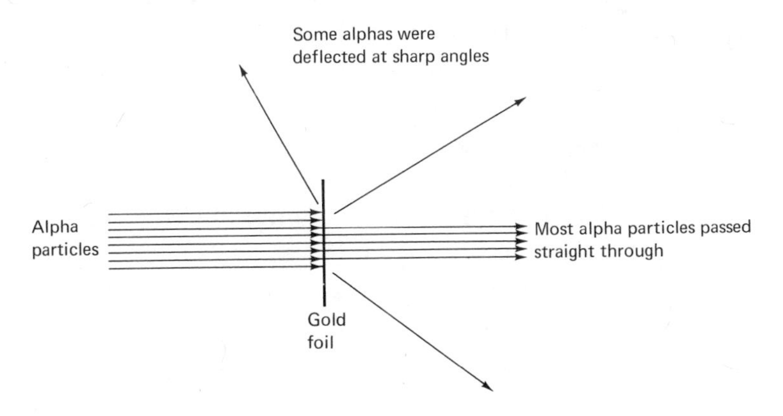

The experiment that led Ernest Rutherford to the concept of the nucleus

vailed at this time. The prevailing notion was one advanced by Thomson which conceived of an atom as a blob of positive electric charge in which were imbedded, in much the same way as plums are in a pudding, enough electrons to neutralize the positive charge. Rutherford's concept, which quickly set aside Thomson's "plum pudding" model, was that an atom has all of its positive charge and virtually all of its mass concentrated in a tiny space at its center. (Collisions with this center, which came to be known thereafter as the nucleus, had been responsible for the sharp changes in direction of some of the alpha particles.) The space surrounding this nucleus is entirely empty except for the presence of a number of electrons (seventy-nine in the case of the gold atom).

To illustrate Rutherford's concept, let us imagine a gold atom magnified so that it is as large as a bale of cotton. The nucleus at the center of this large atom would be the size of a speck of black pepper. If this imaginary bale weighed 500 pounds, the little speck at its center would weigh 499¾ pounds; the surrounding cotton (corresponding to empty space in Rutherford's concept) containing the seventy-nine electrons would weigh but ¼ pound. To express this idea another way, any object such as a gold ring, as dense and solid as it may seem to us, consists almost entirely of nothing.

Rutherford's discovery aroused intense curiosity about the nature and possible structure of this extremely small, but all-important, part of an atom. It was assumed that the positive charge carried by the nucleus must be a whole-number multiple of a small unit equal in size but opposite in sign to the charge of an electron. This conclusion was based on the information that all atoms contain electrons and that an undisturbed atom is electrically neutral. Since it was known that a neutral atom of hydrogen contains just one electron, it appeared that the charge on a hydrogen nucleus must represent the fundamental unit of positive charge, some multiple of which would represent the charge on any other nucleus. Several lines of investigation combined to establish quite firmly that nuclei of atoms occupying adjacent positions on the periodic chart of the elements differed in charge by this fundamental unit. Since the hydrogen nucleus seemed to play such an important role in making up the charges of all other nuclei, it was given the name proton from the Greek *protos,* which means first.

At a historic meeting held in England, in 1913, two apparently unrelated lines of investigation were reported, each of which showed that some atomic nuclei have identical electric charges but different weights.

One report was presented by Frederick Soddy. Soddy knew that the nucleus of a radioactive atom loses both weight and positive charge when it throws out an alpha particle (helium nucleus). On the other hand, when a nucleus emits a beta particle (negative electron), its positive charge increases, but its weight is practically unchanged. Thus Soddy could deduce the weights and nuclear charges of many radioactive products. In several cases the products of two different kinds of radioactivity had the same nuclear charge but different weights. Since it is the positive charge carried by the nucleus of an atom which fixes the number of negative electrons needed to complete the atom, the nuclear charge is really responsible for the exterior appearance, or chemical properties, of the atom.

This conclusion was confirmed by unsuccessful efforts to separate by chemical means different radioactive products having the same nuclear charge but different weights. The products might have had quite different rates of radioactive disintegration, but they appeared to consist of chemically identical atoms of the same chemical element and hence to belong at the same place on the periodic chart of the elements. Soddy suggested that such atoms be called isotopes, from a Greek word meaning same place.

At the same meeting, the English scientist Francis W. Aston (1877–1945), an assistant of Thomson, described what happened when charged atoms, or ions, of neon gas were accelerated in a discharge tube similar to the cathode-ray tube in which Thomson had discovered the electron. The rapidly moving neon ions were deflected by a magnet. Since light objects are more easily deflected than heavy objects, the amount of deflection indicated the weight. By making a comparison with a familiar gas like oxygen, Thomson and Aston were actually able to measure the atomic weight of neon. To their surprise they found two kinds of neon. About nine-tenths of the neon atoms had an atomic weight of 20, and the remainder an atomic weight of 22.

What Thomson and Aston had done was to show that the stable element neon is a mixture of two isotopes. A device that can do what their apparatus did is called a mass spectrograph. Since their time, instruments of this type have shown that more than three-fourths of the stable chemical elements are mixtures of two or more stable isotopes; in fact, there are about three hundred such isotopes in all. The number of known unstable radioactive isotopes (radioisotopes), natural or man-made, is greater than one thousand and is still growing.

During the Middle Ages the desire to find a way to convert a base metal like lead into gold was the outstanding incentive for research in chemistry. (24)

Even Isaac Newton (1642–1727) devoted much of his life to alchemy, an honorable pursuit in his time. But, as John

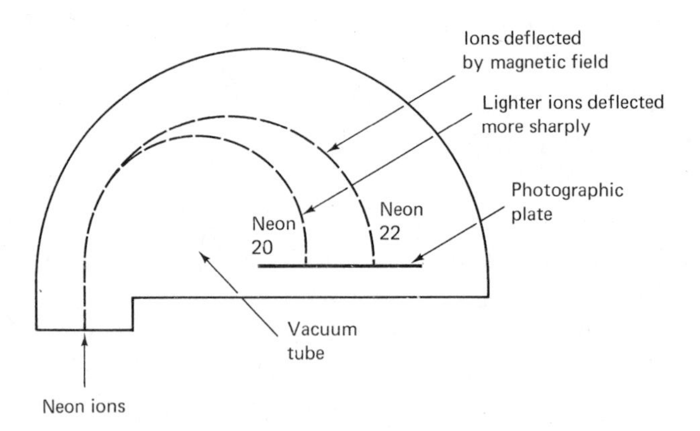

Mass spectrograph as used by Joseph J. Thomson and Francis W. Aston to measure the atomic weight of neon

Maynard Keynes pointed out in his fine essay "Newton the Man" (in James R. Newman, ed., *The World of Mathematics*, New York, Simon and Schuster, 1956, vol. I, p. 277), ". . . Newton was not the first of the age of reason. He was the last of the magicians, the last of the Babylonians and Sumerians, the last great mind which looked out on the visible and intellectual world with the same eyes as those who began to build our intellectual inheritance rather less than 10,000 years ago."

When the important role of the nucleus in determining the chemical properties of an atom became clear and the natural transmutation accompanying radioactivity was understood, the fascinating idea occurred to many people that perhaps man would soon be able to alter the nucleus of a stable atom and thus deliberately convert one element into another. In a historic lecture delivered in Washington, D.C., in April 1914, Rutherford said, "It is possible that the nucleus of an atom may be altered by direct collision of the nucleus with very swift electrons or atoms of helium (i.e., beta or alpha particles) such as are ejected from radioactive matter. . . . Under favorable conditions, these particles must pass very close to the nucleus and may either lead to a disruption of the nucleus or to a combination with it."

World War I began shortly after Rutherford made this statement, and preoccupation with war work stopped his experiments with nuclei. In 1919, however, he published a paper describing what happens when alpha particles pass through nitrogen gas. Very fast protons, or hydrogen nuclei, appear to originate along the paths of the alpha particles. The following is from Rutherford's paper: "If this be the case, we must conclude that the nitrogen atom is disintegrated under the intense forces developed in a close collision with a swift alpha particle, and that the hydrogen atom which is liberated formed a constituent part of the nitrogen nucleus. . . . The results as a whole suggest that, if alpha particles or similar projectiles of still greater energy were available for experiment, we might expect to break down the nuclear structure of many of the lighter atoms."

This prediction has certainly been verified through the use of the atomic artillery provided by extremely powerful particle accelerators, or atom smashers.

It was later shown that, during the nuclear event reported by Rutherford in his 1919 paper, an alpha particle combines with a nitrogen nucleus and that the resulting unstable combination immediately emits a proton and ends up as one of the isotopes of oxygen. This was the first instance of deliberate transmutation of one stable chemical element into another. Since that time practically every known element has been transmuted by bombardment. The dream of the alchemists has been partly fulfilled in that mercury has been changed into gold. We say partly fulfilled because the process is much too expensive to be economically profitable.

During the early 1920s the new technique of bombarding all kinds of matter with alpha particles to see what would happen was widely exploited, and it gradually became clear that in a few instances a peculiar and highly penetrating kind of radiation was produced. In 1932, James Chadwick succeeded in showing that the peculiar radiation must consist of a stream of particles, each weighing about the same as a proton but having no electrical charge.

The name neutron for a possible neutral particle of this type was suggested.

The new particle discovered by Chadwick was destined to play a totally unexpected role, not only in the history of atomic science but also in the fate of nations. It immediately outmoded a previous concept of the nucleus that pictured it as a cluster of protons approximately half of which were neutralized by electrons crowded into the nucleus. A nucleus is now thought of as containing just protons and neutrons. (24)

James Chadwick became Sir James Chadwick because of his contributions to physics. British scientists are not so easy to keep track of when they are transmuted into barons rather than knights and change their names in the process. For example, William Thomson (1824–1907) was made a baron for his important work in thermodynamics and thereupon became Lord Kelvin (after whom the absolute temperature scale is named).

We can see that the old Greek concept of building blocks did not disappear. In the nineteenth century it took the form of the chemical elements. From about 1900 to 1940 the world seemed to be explainable in terms of three building blocks: electrons, protons, and neutrons. In recent years the proton and neutron have themselves been sundered, and a host of new particles have been found. Although these are called elementary particles, it is assumed that they will ultimately be described in terms of a few basic units. The most popular unit that has been postulated is the quark. (The name was taken by American physicist Murray Gell-Mann from a passage in James Joyce's *Finnegans Wake:* "Three quarks for Muster Mark!") If quarks are ever found, scientists will immediately ask whether they can be split. Will there ever be an end? Many scientists hope not, for then physics would lose much of its excitement. But questions are not answered by hopes, and we will return to elementary particles later.

2 ◊ *Inside the Atom*

ASIDE from some minor tampering with chronology in the interest of finishing a story once begun, this history of atomic particles has reached approximately 1930. However, particles aren't everything. What holds the atom together, keeps it in one piece? In considering this aspect of the atom we will have to move freely in time.

Interactions, or Forces

WE know about any object in this world only because of its interactions with other objects. It might seem, in fact, that the world is as complicated as it is because there are so many different expressions of forces, or interactions. There are forces exerted by springs, muscles, wind, expanding gas, gravity, physical contact, magnetism, electricity, and on and on. But in truth we know of only four types of force in all the universe. All others are merely manifestations of the four basic kinds.

GRAVITY

Gravity is the first force we experience. Newton gave a successful description of gravity over 300 years ago. Although the interpretation of this force on a cosmological scale was altered by the general theory of relativity developed by Albert Einstein (1879–1955), Newton's original formula is still satisfactory for most purposes. Even at galactic distances the corrections required by general relativity are small.

Gravity is a simple type of force compared with the other three. Let us examine its features. The basic formula is so familiar that it is easy to forget how remarkable it is.

$$F = G\frac{mM}{r^2}$$

This equation says that the force F between two objects is proportional to the product of their masses m and M and is inversely proportional to the square of the distance r between them. (13)

The value of the gravitational constant G depends upon the units in which force, mass, and length are measured, for example, pounds or kilograms, feet or meters.

Note first of all that the force is always attractive. So far as we know, there is no way to produce gravitational repulsion. Second, the force between the objects is directed along the straight line between them. This seems like a straightforward way to have a force act between objects, but it is not the only way; the electromagnetic force (to be discussed next) can be directed at right angles to the line between interacting objects. Furthermore, the gravitational force does not depend on the velocity or the orientation of the objects as some other forces do.

ELECTROMAGNETISM

We link electricity and magnetism together because they are both part of the same phenomenon. The force between two electric charges depends, among other things, on their relative velocity. If they are at rest with respect to each other, they experience only an electrostatic force. If they are moving, they also exert magnetic forces on each other.

When the charges are at rest, the force can be described with Coulomb's law (named for the French physicist Charles A. Coulomb, 1736–1806):

$$F = K\frac{q_1 q_2}{r^2}$$

At first glance, this formula looks much like Newton's law of gravitation.

Note the dependence on the inverse square of the separation of the charges q_1 and q_2 just as, in the gravitational law, there was an inverse square dependence on the separation of the masses m and M.

In this case, however, the charges can be plus or minus, and the force can thus be attractive or repulsive.

If one electric charge is moving past another, the magnetic force produced on the first one by the second is proportional to the velocities of the charges and can act in a direction perpendicular to the velocity of the first and to the line separating the two charges. Furthermore, if an electric charge accelerates, an entirely different phenomenon appears. Radiant energy is emitted which goes off with the speed of light.

The electromagnetic force is responsible for most of the force we experience in everyday life. It holds the electrons to the atoms and the atoms to each other and therefore is the source of all chemical binding. The force exerted by a spring can be traced directly to the electromagnetic attractions between displaced atoms in the crystal structure of the metal. All biological processes can be described (when they are known) by molecular transformations, which are controlled by electromagnetic forces.

STRONG NUCLEAR FORCE

Both gravity and electromagnetism have been known for some time, but it was not apparent until this century was well along that there was a third type of force—in the nuclei of atoms. In 1933, when the neutron was discovered, the picture of the atomic nucleus was outlined essentially as we know it today. The nucleus contains both protons and neutrons. The protons have positive electric charge and the neutrons have no charge. Nevertheless, they stick very tightly. The problem is: Since like charges repel, why doesn't the electrostatic repulsion of the protons make the whole assembly explode? The gravitational attraction, as we shall see shortly, is far too weak to bind the nucleus together. There must be another force acting between the nuclear particles which is stronger than the repulsive force of electromagnetism.

Thirty years of research have led to a good qualitative description of this strong nuclear force, but not to a complete quantitative theory. We cannot write a formula as was done for gravity and electromagnetism. The most important property of the force is that it is short-range. If the nuclear particles are touching each other [within about 10^{-13} centimeter] the force is very strong. Outside this range the force falls rapidly to zero.

WEAK INTERACTION

Although the Italian-American physicist Enrico Fermi (1901–54) worked out some of the rules for the fourth type of force, the weak interaction, in 1933, its significance was not well understood until the past decade. Some of the most exciting work now being done with the large particle accelerators is concerned with its effects. The weak interaction is responsible for and helps explain the natural radioactive decay of nuclei that emit beta rays. Many of the new particles that have been discovered in the last 15 years also decay because of the weak interaction.

These four—gravity, electromagnetism, the strong nuclear force, and the weak interaction—are all the forces that exist. So far as we know, there are no other ways in which the basic building blocks of matter can interact with one another. We now know how to calculate, at least in principle, the results of any experiment we know how to do involving electromagnetism or gravity. Even with the qualification this is a bold claim. Still, the forces involved are known. There remains the possibility that gravity and electromagnetism are linked in some way which we do not now understand.

In contrast, there is no general theory describing the strong nuclear force, although we know enough about it to transmute elements and derive power from the fission of heavy nuclei.

We have called the nuclear binding force strong and the decay force weak. Let us indicate the reason for these designations by comparing the strengths of the four interactions. If two protons are touching each other, the strong nuclear force can be 100 times greater than the electromagnetic repulsion. The weak interaction is smaller than the strong nuclear force by a factor of 10 trillion [10^{13}—throughout this book, per American usage, a billion is a thousand million = 10^9; 1 trillion = 10^{12}]. And the gravitational attraction, the weakest of all, is smaller than the strong nuclear force by a factor of 10^{39}. (13)

The reason gravity is so important although so much weaker than the other forces is that nuclear forces are strong but very short-ranged, therefore particles are affected only by their immediate neighbors; weak interactions are both weak and short-ranged; and electromagnetic forces, which are much stronger than gravitational forces and are long-ranged, stem from charges. Because we deal most often with neutral systems containing equal numbers of positive and negative charges, the electromagnetic forces tend to cancel out. Gravitational forces are very weak, but long-ranged, and all particles are attracted by all other particles.

Our weight, for example, represents the attraction of 10^{51} nuclear particles in the earth to 10^{29} within our own body. The effects of gravity are negligible in individual nuclear or atomic reactions. At the other end of the scale, gravity dominates the interactions of stars and galaxies.

How do these building blocks of matter (particles) interact through these forces? We can answer this with great success mathematically. Gravitational and electromagnetic effects are thought to be achieved by establishment of gravitational or electromagnetic fields. This concept is satisfactory for many purposes, and it is not at all at odds with another model, which pictures the sources, or charges, of each of the forces as emitting some sort of agents, which impinge upon other, similar sources.

Instead of viewing the influence of one charge on another as a continuous process due to their interaction with each other's fields, we picture a whole series of individual collisions taking place between the charges and their agents. The interchange of these messengers produces the binding force (or, in some cases, the repulsion). This description has been highly successful in explaining the electromagnetic force. In this case the propagating agents are photons, which are electromagnetic radiation (see diagram).

The same type of theory has been worked out for gravity, and the supposed agents have been named gravitons. Because this interaction is so extremely weak, we have not been able to detect gravitons with our present instruments. (13)

Whether or not gravitons exist, there are surely gravitational waves in the universe, but it has been assumed that they are virtually impossible to detect. Undeterred, American physicist Joseph Weber spent over a dozen years building fantastically sensitive apparatus capable of detecting a vibration one one-hundredth the size of the nucleus in a bar 1 meter long. Recently he has claimed to have detected pulses of gravitational waves. Should his claim be confirmed, and should his present estimate of the frequency with which the pulses arrive and their intensity stand up, our current picture of the universe may have to be altered drastically.

The same model was used by the Japanese physicist Hidekei Yukawa in 1935 to explain the strong nuclear force. Be-

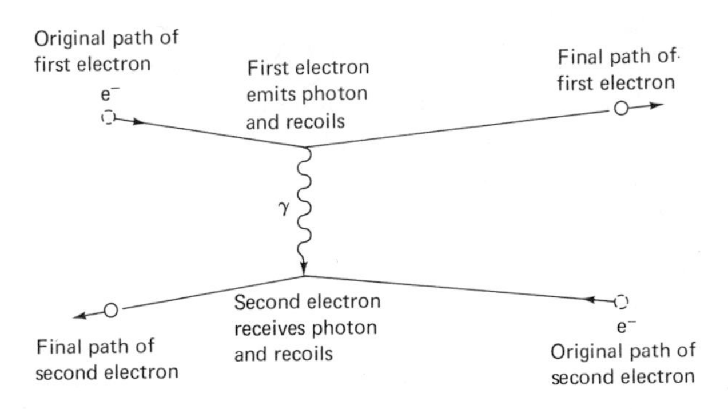

Interaction of two electrically charged particles by way of a photon agent

cause of the strength and the short-range nature of this force, the agent that makes it effective must have mass (the photon has none). This agent of the nuclear force is the meson, described later. Qualitatively, interchange of mesons could provide the nuclear binding force; but it turns out that there are several mesons, and there is much that we do not know about them yet.

Conservation Laws

A productive way to learn how things change is to discover in what ways they remain constant. Even simple events—such as the collision of two balls—are difficult to describe in terms of the forces acting at each instant. If, however, we do not attempt to explain in detail what happens during the collision but look only at the final results, we can make some very general statements. Certain properties of the event always remain constant.

Almost everyone is familiar with some of the conservation laws. (These laws state, in general terms, that matter, energy, momentum, etc., can neither be created nor destroyed.) Many grade-school children know that energy is conserved before they know what energy is. Lest familiarity breed indifference, let us emphasize a few points about energy conservation.

In the first place, when we say that a property is conserved, we are talking about what happens in a particular, limited system which we can isolate either physically or in imagination. The total energy of two colliding balls is not conserved if one of them is fastened to a spring; in that case we have to consider the energy system that includes the spring.

Second, we now realize that mass and energy are interchangeable. Therefore we say that the total mass–energy of an isolated system is conserved. (13)

The concept of conservation of energy is about 130 years old. One of its originators was a medical man, the German Julius Robert von Mayer (1814–78). While serving as a ship's surgeon in the tropics, he noticed that venous blood was brighter red than in cooler climates, and began to speculate about the relationship between heat and work and physiological processes.

Until 1905 matter had been considered as something that has mass or inertia; energy, on the other hand, had been regarded as the ability to do work. It was believed that the two were as different from each other as, say, a square yard is different from an hour. Einstein's special theory of relativity, however, implies that matter and energy are merely two different manifestations of the same fundamental physical reality, and that each may be converted into the other according to the famous equation:

$$E = mc^2$$

where E = quantity of energy
m = quantity of matter
c = speed of light in a vacuum

Some people began to realize during the 1920s that atomic nuclei contain vast stores of energy that might some day revolutionize civilization.

Since any nucleus consists of a certain number of protons and neutrons, it seems logical that the total weight of the nucleus could be determined by adding together the individual weights of the particles in it. When mass spectrographs of sufficiently high accuracy became available, however, it was found that in the case of nuclear weights, the whole was not equal to the sum of its parts. All nuclei [except that of hydrogen, which consists of only one particle, a proton] weigh less than the sum of the weights of the particles in them.

For example, the atomic weight of a proton is 1.00812 and that of a neutron is 1.00893. (These are relative weights based on an internationally accepted scale.) It would seem then that a nucleus of helium containing two protons and two neutrons should have an atomic weight of 2×1.00812 plus 2×1.00893, or 4.0341. Actually the atomic weight of helium as measured by the mass spectrograph is only 4.0039. (See diagram, page 36.)

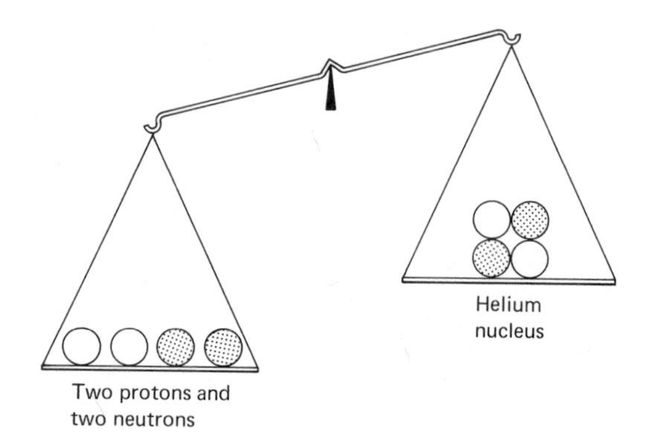

A case where the whole is not equal to the sum of its parts. Two protons and two neutrons are distinctly heavier than a helium nucleus, which also consists of two protons and two neutrons. Energy makes up the difference.

What happens to the missing atomic weight of 0.0302? As postulated in Einstein's formula, it must be converted into energy. The conversion occurs when the protons and neutrons are drawn together into a helium nucleus by the powerful nuclear forces between them.

When the missing atomic weight 0.0302 is multiplied by the square of the velocity of light according to Einstein's theory, it is found to represent a tremendous amount of energy. Indeed, the energy released in forming a helium nucleus from two protons and two neutrons turns out to be 7 million times that released when a carbon atom combines with an oxygen molecule to produce a molecule of carbon dioxide in the familiar process of combustion.

The general behavior of such losses in atomic weight for atoms throughout the periodic table show that, in general, if two light nuclei combine to form a heavier one, the new nucleus does not weigh as much as the sum of the original ones. This behavior continues up to iron, nickel, and cobalt in the periodic table. But if two nuclei heavier than iron are coalesced into a single very heavy nucleus found near the end of the periodic table (such as uranium), the new nucleus weighs more than the sum of the two nuclei that formed it.

Thus, if a very heavy nucleus could be divided into parts, energy would be released, and the sum of the weights of the fragments would be less than that of the original nucleus.

In these two types of nuclear reactions, a small amount of matter would actually vanish! Einstein's special theory of relativity states that the vanished matter would reappear as an enormous quantity of energy. (24)

We might note that Einstein did not receive his Nobel Prize for his major work, the theory of relativity. Alfred Nobel specified in his will that the awards be made for work that benefits mankind. This tends to favor work with implications that can be seen directly over abstract work with long-term consequences. Einstein received his award, therefore, for his work on the photoelectric effect, work which was important but not in the same intellectual league as relativity. Einstein's is not the only case of the award's being made to a major figure for his comparatively minor work.

The total momentum of a system is also conserved. At low velocities, momentum is the product of mass and velocity. (13)

At velocities approaching that of light, Einstein's special theory of relativity must be used and the expression for momentum is more complicated.

Momentum is a fundamental quantity and is often more easily measured than either mass or velocity separately.

Angular (rotational) momentum is an extremely important property of elementary particle systems. On a human scale it is related to the product of ordinary momentum and the length of the radius arm pointing from the moving object to a point around which it is rotating or about which it might rotate. Angular momentum has a definite direction, and both this direction and the magnitude remain constant in an isolated system. Because of this we can balance bicycles while they are moving but not while they are standing still. Because angular momentum is conserved, a figure skater can twirl faster and faster on one toe. She starts her spin with arms outstretched so that the mass-times-velocity-times-radius-arm is large. When she draws her arms in close to her body, the radius arm becomes small and the velocity must increase to compenstate since the total product must remain the same. Many of the particles have intrinsic angular momentum, which we call spin, as well as angular momentum due to their motion around each other.

Many particles behave as if they were spinning on a central axis, but only certain values of this spin appear. In terms of the basic unit of spin, most particles have 0 unit, ½ unit, or 1 unit. The difference between half-integral spin and integral spin (0, 1, 2, etc.) is profound. The half-integral spin particles, such as the electron, proton, and neutron, obey an exclusion law that forbids the existence of more than one of them in the same place (discussed later). The integral spin particles, such as the photon and the meson, obey no such law.

Particle spin must produce observable effects, or the concept would be meaningless. We can best illustrate these effects in the case of the electron. If the electron is actually a spinning electric charge, it should also have the characteristics of a tiny magnet. In a magnetic field it should line itself up in the same direction as that of the field, much as a compass needle will point north-south in the earth's magnetic field. In a magnetic field that changes appreciably over a very small distance, a tiny magnet, when aligned appropriately, would move toward the region of stronger field.

In a famous experiment done in 1924, two German scientists, O. Stern (1888–1969) and W. Gerlach, observed this movement of the electron magnets. The startling result of the experiment was that the electron spin can be either in the direction of a magnetic field or opposed to it, but in no other orientation.

We say that the spin of the electron is quantized, meaning that it can have only certain values. In this respect it resembles electric charge, which is also quantized. One is tempted to think that the electrons can really spin in any direction but that in the magnetic field they just happen to align themselves up or down. But the concept of electron spin is meaningless except in terms of its measurement, and when electron spin is measured it has only one of two values—with the measuring field or opposed to it. (13)

Angular momentum of motion other than spin is also quantized. In the next chapter this will be explained in terms of the periodic table of the elements.

At the end of the 1920s, scientists were disturbed to find that, in the beta decay of radioactive nuclei (see chapter 4), the emitted electron had less energy than the energy given up by the decaying nucleus. Also, angular momentum seemed not to be conserved. The experimental evidence could have been interpreted to mean that the concepts of conservation of energy and angular momentum should be abandoned. But rather than abandon concepts that had proved so very fruitful, Wolfgang Pauli postulated that a particle of very small mass, electrically neutral and therefore extremely hard to detect, was being emitted along with the electron and could account for the discrepancies. Enrico Fermi formulated the theory of beta decay. The diminutive in Italian being *-ino,* Fermi named the new neutral particle the neutrino, the "little neutral one," since the name neutron was already in use. The neutrino was so hard to detect that it was not found experimentally until about 25 years later. In his poem "Cosmic Gall," John Updike describes it as well as anybody.

> Neutrinos, they are very small.
> They have no charge and have no mass
> And do not interact at all.
> The earth is just a silly ball
> To them, through which they simply pass,
> Like dustmaids down a drafty hall
> Or photons through a sheet of glass.
> They snub the most exquisite gas,
> Ignore the most substantial wall,
> Cold-shoulder steel and sounding brass,
> Insult the stallion in his stall,
> And, scorning barriers of class,
> Infiltrate you and me! Like tall
> And painless guillotines, they fall
> Down through our heads into the grass.
> At night, they enter at Nepal

And pierce the lover and his lass
From underneath the bed—you call
It wonderful; I call it crass.

It is a measure of the advanced state of physics that theoretical physicists not only explain experimental data but predict the existence of undiscovered elementary particles—the neutrino is not the only one, as we shall see.

Electric charge is also conserved. This does not imply that charged particles cannot be created or destroyed, but only that in any process, total charge before must equal total charge after.

Conservation laws besides those concerning energy, momentum, angular momentum, and charge will be described later.

3 ◊ *The Natural Elements*

LET us put together the bits and pieces described in the preceding chapters to see how the atom appears today.

The atom came to be considered as being made up of a nucleus, containing most of the mass, and electrons revolving around the nucleus rather like the planets revolve around the sun. Each electron has a single or unit negative charge and the entire atom is electrically neutral, or uncharged, because in the nucleus there are a number of protons (equal to the number of electrons), each of which has a unit positive charge.

The number of protons in a given atom of an element is called the atomic number. In addition to the protons, the nucleus contains uncharged particles called neutrons. The neutrons and protons have about the same mass, and the electrons, by comparison, have negligible mass. An element of atomic mass (A) and atomic number (Z) will have a nucleus consisting of Z protons and $(A - Z)$ neutrons, and this will be surrounded by Z electrons. For example, an atom of lithium with mass (A) of 7 and atomic number (Z) of 3 will have a nucleus consisting of three protons and four neutrons $(A - Z)$, surrounded by three electrons.

The lightest element, hydrogen, has Z equal to 1, and each successively heavier element differs from the one preceding it by an increase of 1 in Z, and has one more proton and one more electron than the next lighter one. Thus, the second heaviest element, helium, has Z equal to 2, and so on. For the heavier elements, such as uranium $(Z = 92)$, one might imagine a chaotic situation with many electrons buzzing all around the nucleus. Fortunately, the electrons are restricted to movement in certain fixed orbits or shells. (6)

Describing electrons as running around in independent orbits is a useful approximation to the actual situation, but only an approximation. A true appreciation requires a detailed knowledge of quantum mechanics (developed in the 1920s). The following can be only a rough picture, a cookbook description of the rudiments of quantum mechanics, given without proof.

The basic premise is that the motion of an electron in an atom is quantized. That is, the electron can only have certain amounts of energy or angular momentum. There is no continuous range. This is as if human beings could only be exactly five, six, or seven feet tall.

The quantized motion of an electron in an atom is characterized by four quantum numbers. The principal quantum number n is a measure of the energy of the electron; the smaller the value of n, the more tightly bound to the nucleus the electron is. Also, n is a measure of the dimensions of the orbit. It is sometimes called the shell number; the larger the shell number, the farther the electron from the nucleus, on the average.

The second quantum number l is a measure of the orbital angular momentum, which as was stated in chapter 2, is quantized. For a given value of n, l can assume only integral values from 0 through $n-1$. A rough description of what l corresponds to physically is the shape of the orbit. In a circular orbit $l=n-1$. In an elliptical orbit l has a lower value, with $l=0$ corresponding to the most eccentric ellipse.

The third quantum number relates to the orientation, or tilt, of the plane of the electron's orbit. The number of possible orientations is $2l+1$. These orientations can also be thought of as the angular momentum analog of the two possible states of the spin, up or down (see chapter 2). The electron is said to have a spin of ½. Note that if we use as the number of allowed spin orientations $2s+1$ (analogous to $2l+1$) and substitute $s=½$, we get $2s+1=2$ for the number of spin states an electron can have, the two being "up" and "down." The fourth quantum number is related to these spin states.

In order to use these four quantum numbers to construct the periodic table of the elements, we need one additional concept: the Pauli exclusion principle, which states that no two electrons in an atom can have the same four quantum numbers. In other words, no two electrons in an atom can be in exactly the same state. Let us apply this to the periodic table.

We start with the simplest case, hydrogen, with a single electron in orbit about the nucleus. That electron will be in the orbit where $n = 1$ (for that corresponds to the lowest energy, and systems tend to seek the state of lowest energy for equilibrium). With $n = 1$, the only possible value for l is 0, since l ranges from 0 to $n - 1$, and $n - 1 = 0$. Then, for the number of orientations of the orbit, $2l + 1 = 0 + 1 = 1$. Three quantum numbers have been accounted for in this first orbit with zero orbital angular momentum and a single orientation of the plane of the orbit. The fourth quantum number has two possibilities, spin up or spin down. Selecting either one, we have hydrogen.

Going on to helium, with atomic number $Z = 2$, we can put two electrons in the $n = 1$ orbit, one with spin up, the other with spin down. All the different states for $n = 1$ have now been exhausted. If a third electron were placed in the $n = 1$ orbit, it would be in the same state as the first or second electron, which is prohibited by the Pauli exclusion principle.

The next atom, lithium, with $Z = 3$ and three electrons in orbit, must have the third electron in an $n = 2$ orbit. The $n = 2$ orbit has lots of openings for electrons. First, l can be 1 or 0. For $l = 1$ we have $2l + 1 = 3$ orbit orientations and for $l = 0$ we have $2l + 1 = 1$ orientation, a total of 4. Each orbit orientation can have two electrons, one with spin up, the other with spin down, giving a grand total of eight possibilities with no two electrons having the same four quantum numbers. These eight possibilities take us from lithium through neon, which has $Z = 10$.

Note that the number of electron possibilities for $n = 1$ is two; for $n = 2$, eight; and more generally, for n, $2n^2$.

Now that the $n = 2$ orbit has no more vacancies, the next atom, sodium ($Z = 11$), must start on the third orbit. Proceeding as above, we get eighteen possibilities ($2n^2 = 18$) for electrons in the third orbit.

The exclusion principle says electrons in an atom must go into different states, but the electrons themselves are identical. If electrons were even infinitesimally different from each other, they could all fit into the first orbit, and our world would be entirely different.

(One religious person, much interested in science, was

deeply disturbed by the exclusion principle. "Since God made electrons," he argued, "*He* must be able to tell them apart!")

The quantum mechanical exclusion principle has no classical analog. In quantum mechanics, very small differences can make for profound changes.

The standard notation in atomic physics was introduced before the theory of the atom was developed. The letters *s, p, d, f* come from spectroscopy and stand for descriptions of spectra: sharp, principal, diffuse, and fundamental. Today this terminology is irrelevant, but it is still used. The letters *s, p, d, f, g,* . . . are used to denote electrons with $l = 0, 1, 2, 3, 4$. . . .

The unit of angular momentum is $h/2\pi$, where h—Planck's constant, named for the German physicist Max Planck (1858–1947)—is a constant of nature which appears again and again in quantum mechanical expressions. It is extremely small, 6.6×10^{-34} joule second. The smallness of h is what makes classical Newtonian physics valid for the macroscopic problems of our everyday world. That is, although the complete theory, quantum theory, contains h, h is so small that it can be neglected in macroscopic situations, and neglecting h essentially reduces quantum mechanics to classical physics. If we lived in a world in which h was much larger, quantum mechanics would have to be used, not only for problems of atomic dimensions, but for ordinary macroscopic situations as well. Classical physics may be thought of as the physics valid in a world where $h = 0$.

The first shell contains two electrons, which we write $1s^2$, the second shell has eight [two $2s$ electrons and six $2p$ electrons], written $2s^2 2p^6$, the third eighteen, written $3s^2 3p^6 3d^{10}$.

The electrons do not necessarily fill the shells and subshells in consecutive order. The first (lightest) eighteen elements' electrons are added regularly, the electrons filling the $1s$, $2s$, $2p$, $3s$, and $3p$ subshells in sequence. However, in the nineteenth element, the new electron does not go into the $3d$ subshell, as might be expected, but into the $4s$ subshell. (Questions of this sort are decided on the basis of energy considerations. It is energetically more favorable to put the nineteenth electron into the $4s$ subshell.) From this point on we can write down the electronic configurations of the succeeding (heavier) elements only if we know the order in which the subshells are filled.

Now we are ready to look at the electronic theory of valence and some of its consequences. About 1920 a number of chemists, most notably the American Gilbert N. Lewis (1875–1946), suggested that the electrons in the outermost shells were responsible for elements' chemical reactions. Compounds (that is, molecules) are formed by the transfer or sharing of electrons, and the number of such electrons provided or obtained by an atom of any element during the combining process is its valence. However, there is a kind of regulation of the number of electrons that can participate in this bonding. Atom tend to adjust their electronic structure to that of the nearest element with a completed outer shell. The adjustment is made by losing, gaining, or sharing electrons with other atoms.

The closed-shell arrangement of electrons happens to be the electronic structure of atoms of the noble gases. Moreover, only the six noble gases have this arrangement of maximum stability. The electronic structure of sodium is $1s^2$, $2s^2$, $2p^6$, $3s^1$; sodium has one electron more than the closed-shell arrangement $1s^2$, $2s^2$, $2p^6$, which is the electronic structure of neon. The closed-shell arrangements are also called cores.

Two atoms with the same number of electrons outside a stable core would tend strongly to adjust their electronic configuration in a similar manner; that is, they would have the same valence and therefore the same chemical properties. This fact is borne out by the fact that elements in the same group in the periodic table have the same outer electronic structures.

The fact that the noble gases have completed outer shells means that they have nothing to gain by losing, gaining, or sharing electrons. They already have the stable electronic structures that other elements are striving to attain. This means that they should have zero valence and should not form chemical compounds. Thus, the observed experimental fact that the gases were inert was supported by theory. This startling agreement between experiment and theory was successful in discouraging attempts to make chemical compounds with the noble gases for a period of almost 40 years.

The six gases helium, neon, argon, krypton, xenon, and radon were the confirmed bachelors among the known elements. All the other elements would enter into chemical combination with one or another of their kind, irrespective of whether they were solids, gases, or liquids in their normal state. Not so helium, neon, argon, krypton, xenon, and radon. They were chemically aloof and would have nothing to do with other elements, or even with one another.

This behavior earned them a unique position in the periodic table of the elements, and they were called names like the inert gases or the noble gases by reason of their apparent reluctance to mingle with the common herd of elements. They were also labeled the rare gases, although helium and argon are not really rare. Xenon, however, is the rarest stable element on earth. (6)

The names of some of the helium group gases (the term used today) reflect characteristics attributed to them. Argon is the Greek word for inactive, idle. When three more gases were isolated they were named neon (Greek *neos,* new), krypton (Greek *kryptos,* hidden), and xenon (Greek *xenon,* stranger). Radon is named for its association with radium, and helium derives from the Greek *helios,* the sun, because the element was discovered in the sun (that is, in an analysis of light from the sun) before it was found on earth. Helium, incidentally, is the second most abundant element in the universe, making up about one-quarter of the total mass. Hydrogen makes up most of the remaining three-quarters, and all the other elements combined account for about 1 percent.

Helium is so light that it is continually escaping from the earth's atmosphere into interstellar space. The present concentration of helium in the atmosphere therefore probably represents a steady-state concentration, that is, the amount being released from the earth's crust is equal to the amount escaping from the atmosphere into space. The constant escape explains why there is so little to be found in our air.

This helium in the earth is continually being formed by radioactive decay. All radioactive materials that decay by emitting alpha particles produce helium, since an alpha particle is nothing more than a helium nucleus with a positive charge. Most of the helium in the earth's crust comes from the decay of uranium and thorium.

Radon is obtained from the radioactive decay of radium. One gram of radium produces about 0.0001 milliliter of radon per day. (We should keep in mind, however, that 1 gram of radium is a very large amount in terms of the total available. It is doubtful whether there are more than 100 grams of pure radium available in the Western world today.)

Many of the uses of these gases are outgrowths of their inertness. The greater abundances, and hence lower costs, of helium and argon result in their use as inert atmospheres in which to weld and fabricate metals. The electrical and other properties of the noble gases make most of them ideal gases for filling numerous types of electronic tubes and in lasers. For this, the gases may be used singly or mixed with one or more of the others. Perhaps the best known use is in the familiar "neon" advertising signs. The glow produced by neon alone is red. The other gases produce less brilliant colors: helium (pale pink), argon (blue), krypton (pale blue), and xenon, (blue-green). (6)

A modern version of the periodic table of the elements, with the electronic structures indicated, is shown on pages 48–49, and an alphabetical listing of elements and symbols on pages 50–51. Continuing with some of the more complex elements, the electronic structure of iron is shown in the diagram on page 52.

Iron has a $1s^2$, $2s^2$, $2p^6$, $3s^2$, $3p^6$, and $4s^2$ configuration; its sublevels all have the maximum permissible number of electrons except the *3d*, which could contain four more. We should note that the *4s* sublevel is completed before the *3d* level is filled. This schedule of filling is the general rule, not an exception, for the elements that have an atomic number of 21 or greater.

Let us proceed from iron to elements with higher atomic numbers. If we continue to add electrons to the *3d* sublevel, remembering only four more are needed to fill it, we will reach zinc. The next electrons are added to the *4p* sublevel, which when filled constitutes the rare gas krypton.

PERIODIC TABLE

GROUP / PERIOD	Ia	IIa	IIIa	IVa	Va	VIa	VIIa	VII	
1	1 1.00797 **H** $1s$								
2	3 6.939 **Li** $2s$	4 9.0122 **Be** $2s^2$							
3	11 22.9898 **Na** $3s$	12 24.312 **Mg** $3s^2$							
4	19 39.102 **K** $4s$	20 40.08 **Ca** $4s^2$	21 44.956 **Sc** $3d^1 4s^2$	22 47.90 **Ti** $3d^2 4s^2$	23 50.942 **V** $3d^3 4s^2$	24 51.996 **Cr** $3d^5 4s^1$	25 54.938 **Mn** $3d^5 4s^2$	26 55.847 **Fe** $3d^6 4s^2$	27 58.93 **Co** $3d^7 4s^2$
5	37 85.47 **Rb** $5s$	38 87.62 **Sr** $5s^2$	39 88.905 **Y** $4d^1 5s^2$	40 91.22 **Zr** $4d^2 5s^2$	41 92.906 **Nb** $4d^4 5s^1$	42 95.94 **Mo** $4d^5 5s^1$	43 (99) **Tc** $4d^5 5s^2$	44 101.07 **Ru** $4d^7 5s^1$	45 102.90 **Rh** $4d^8 5s^1$
6	55 132.905 **Cs** $6s$	56 137.34 **Ba** $6s^2$	57 138.91 **La*** $5d^1 6s^2$	72 178.49 **Hf** $5d^2 6s^2$	73 180.948 **Ta** $5d^3 6s^2$	74 183.85 **W** $5d^4 6s^2$	75 186.2 **Re** $5d^5 6s^2$	76 190.2 **Os** $5d^6 6s^2$	77 192.2 **Ir** $5d^9 6s^0$
7	87 (223) **Fr** $6s^2 6p^6 7s^1$	88 (226) **Ra** $6s^2 6p^6 7s^2$	89 (227) **Ac**** $6s^2 6p^6 6d^1 7s^2$						

Key to Table

Atomic number

Atomic weight

20 40.08 **Ca** $4s^2$

Electronic structure

* Lanthanides (Rare Earths)	58 140.12 **Ce** $4f^2 6s^2$	59 140.907 **Pr** $4f^3 6s^2$	60 144.24 **Nd** $4f^4 6s^2$	61 (147) **Pm** $4f^5 6s^2$	62 150.3 **Sm** $4f^6 6s^2$
** Actinides	90 232.038 **Th** $6d^2 7s^2$	91 (231) **Pa** $5f^2 6d^1 7s^2$	92 238.04 **U** $5f^3 6d^1 7s^2$	93 (237) **Np** $5f^4 6d^1 7s^2$	94 (24 **Pu** $5f^6 7s^2$

OF THE ELEMENTS

	Ib	IIb	IIIb	IVb	Vb	VIb	VIIb	0
								2 4.0026 **He** $1s^2$
			5 10.811 **B** $2s^22p^1$	6 12.0111 **C** $2s^22p^2$	7 14.0067 **N** $2s^22p^3$	8 15.9994 **O** $2s^22p^4$	9 18.9984 **F** $2s^22p^5$	10 20.183 **Ne** $2s^22p^6$
			13 26.9815 **Al** $3s^23p^1$	14 28.086 **Si** $3s^23p^2$	15 30.9738 **P** $3s^23p^3$	16 32.064 **S** $3s^23p^4$	17 35.453 **Cl** $3s^23p^5$	18 39.948 **Ar** $3s^23p^6$
28 58.71 **Ni** $3d^84s^2$	29 63.54 **Cu** $3d^{10}4s^1$	30 65.37 **Zn** $3d^{10}4s^2$	31 69.72 **Ga** $4s^24p^1$	32 72.59 **Ge** $4s^24p^2$	33 74.922 **As** $4s^24p^3$	34 78.96 **Se** $4s^24p^4$	35 79.909 **Br** $4s^24p^5$	36 83.80 **Kr** $4s^24p^6$
46 106.4 **Pd** $4d^{10}$	47 107.870 **Ag** $4d^{10}5s^1$	48 112.40 **Cd** $4d^{10}5s^2$	49 114.82 **In** $5s^25p^1$	50 118.69 **Sn** $5s^25p^2$	51 121.75 **Sb** $5s^25p^3$	52 127.60 **Te** $5s^25p^4$	53 126.904 **I** $5s^25p^5$	54 131.30 **Xe** $5s^25p^6$
78 195.09 **Pt** $5d^96s^1$	79 196.967 **Au** $5d^{10}6s^1$	80 200.59 **Hg** $5d^{10}6s^2$	81 204.37 **Tl** $5d^{10}6s^26p^1$	82 207.19 **Pb** $6s^26p^2$	83 208.980 **Bi** $6s^26p^3$	84 (210) **Po** $6s^26p^4$	85 (210) **At** $6s^26p^5$	86 (222) **Rn** $6s^26p^6$

63 151.96 **Eu** $4f^76s^2$	64 157.25 **Gd** $4f^75d^16s^2$	65 158.924 **Tb** $4f^96s^2$	66 162.50 **Dy** $4f^{10}6s^2$	67 164.930 **Ho** $4f^{11}6s^2$	68 167.26 **Er** $4f^{12}6s^2$	69 168.934 **Tm** $4f^{13}6s^2$	70 173.04 **Yb** $4f^{14}6s^2$	71 174.97 **Lu** $4f^{14}5d^16s^2$
95 (243) **Am** $5f^77s^2$	96 (247) **Cm** $5f^76d^17s^2$	97 (247) **Bk** $5f^76d^27s^2$	98 (251) **Cf** $5f^96d^17s^2$	99 (254) **Es** $5f^{11}6s^26d^67s^2$	100 (253) **Fm** $5f^{12}6s^26p^67s^2$	101 (256) **Md** $5f^{13}6s^26p^67s^2$	102 (254) **No**	103 (257) **Lw**

ALPHABETICAL LIST OF

ELEMENT	SYMBOL	ELEMENT	SYMBOL
Actinium	Ac	Erbium	Er
Aluminum	Al	Europium	Eu
Americium	Am	Fermium	Fm
Antimony	Sb	Fluorine	F
Argon	Ar	Francium	Fr
Arsenic	As	Gadolinium	Gd
Astatine	At	Gallium	Ga
Barium	Ba	Germanium	Ge
Berkelium	Bk	Gold	Au
Beryllium	Be	Hafnium	Hf
Bismuth	Bi	Helium	He
Boron	B	Holmium	Ho
Bromine	Br	Hydrogen	H
Cadmium	Cd	Indium	In
Calcium	Ca	Iodine	I
Californium	Cf	Iridium	Ir
Carbon	C	Iron	Fe
Cerium	Ce	Krypton	Kr
Cesium	Cs	Lanthanum	La
Chlorine	Cl	Lawrencium	Lw
Chromium	Cr	Lead	Pb
Cobalt	Co	Lithium	Li
Copper	Cu	Lutetium	Lu
Curium	Cm	Magnesium	Mg
Dysprosium	Dy	Manganese	Mn
Einsteinium	Es	Mendelevium	Md

ELEMENTS AND SYMBOLS

ELEMENT	SYMBOL	ELEMENT	SYMBOL
Mercury	Hg	Samarium	Sm
Molybdenum	Mo	Scandium	Sc
Neodymium	Nd	Selenium	Se
Neon	Ne	Silicon	Si
Neptunium	Np	Silver	Ag
Nickel	Ni	Sodium	Na
Niobium	Nb	Strontium	Sr
Nitrogen	N	Sulfur	S
Nobelium	No	Tantalum	Ta
Osmium	Os	Technetium	Tc
Oxygen	O	Tellurium	Te
Palladium	Pd	Terbium	Tb
Phosphorus	P	Thallium	Tl
Platinum	Pt	Thorium	Th
Plutonium	Pu	Thulium	Tm
Polonium	Po	Tin	Sn
Potassium	K	Titanium	Ti
Praseodymium	Pr	Tungsten	W
Promethium	Pm	Uranium	U
Protactinium	Pa	Vanadium	V
Radium	Ra	Xenon	Xe
Radon	Rn	Ytterbium	Yb
Rhenium	Re	Yttrium	Y
Rhodium	Rh	Zinc	Zn
Rubidium	Rb	Zirconium	Zr
Ruthenium	Ru		

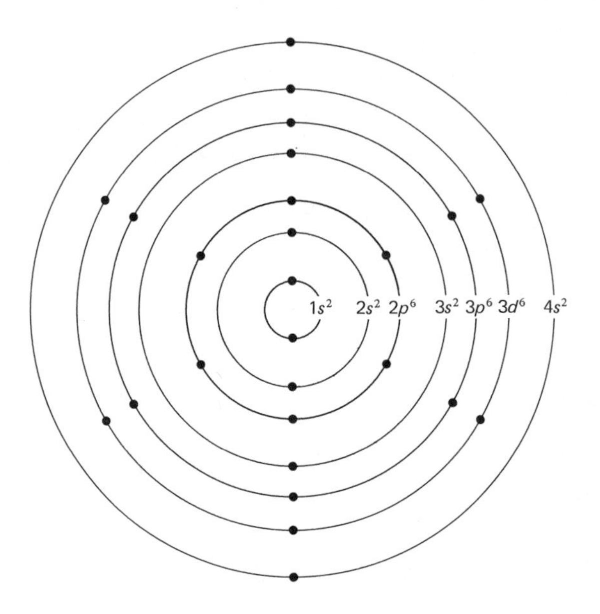

Electronic structure of iron

The next row of elements is built up in the same manner: The first two electrons go into the 5*s* sublevel, before the 4*d* and 5*p* sublevels are filled, at which point we have reached the end of the fifth period, with xenon. The next two electrons go into the 6*s* sublevel (barium), and it looks as if the same process is about to repeat itself for a third time. This is, indeed, true for the next electron added—but thereafter the periodic table is never the same again!

After barium, the next added electron goes into the 5*d* sublevel, giving us lanthanum—the first of the rare earths.

The rare earths make up a fascinating family of elements which for many years were something of a scientific mystery because:

They have such nearly identical chemical properties they are difficult to tell apart.

They make up about one-sixth of all naturally occurring elements, but the entire group occupies only one position in the periodic table.

They are not naturally radioactive, but often are found in the same minerals with radioactive thorium or uranium.

Lanthanum has a $1s^2$, $2s^2$, $2p^6$, $3s^2$, $3p^6$, $3d^{10}$, $4s^2$, $4p^6$, $4d^{10}$, $5s^2$, $5d^1$, $6s^2$ configuration with no $4f$ electrons. The next electron that is added does not go into the $5d$ sublevel, as might be expected from the pattern of the fourth and fifth periods, but goes in the $4f$ sublevel.

The next thirteen electrons also go one by one into the $4f$ sublevel, which then is completely filled at lutetium. For all fifteen rare earths, the outermost electrons are the same: $6s^2$ and $5d^1$. Since the chemical and most of the physical properties of any element depend most strongly on these outer electrons ($6s^2$ and $5d^1$), we see why the rare earths are trivalent (the outermost $6s$ and $5d$ subshells give up three electrons), why they are so similar to one another, and why $4f$ electrons do not enter into chemical bonding. Moreover, we can appreciate the difficulties the discoverers had in separating, isolating, and identifying the individual rare earths.

It is these $4f$ electrons which give each element its particular personality and thus distinguish one rare earth from another. For example, two of the outstanding differences among these very similar metals is the occurrence of ferromagnetism in gadolinium and of superconductivity in lanthanum. At room temperature gadolinium metal is attracted to a magnet because it is ferromagnetic, but the other rare-earth metals are not. At low temperatures lanthanum is the only rare-earth metal that loses all its resistance to electrical flow. At about $-267°$ C it becomes a perfect conductor of electricity, that is, a superconductor.

The term rare earths has its origin with the early discovery of these elements. The word rare arises from the fact that these elements were discovered in scarce minerals. The word earth comes from the facts that they were first isolated from their ores in the chemical form of oxides and that the old chemical terminology for oxide is earth.

The name rare earths is actually a misnomer for these ele-

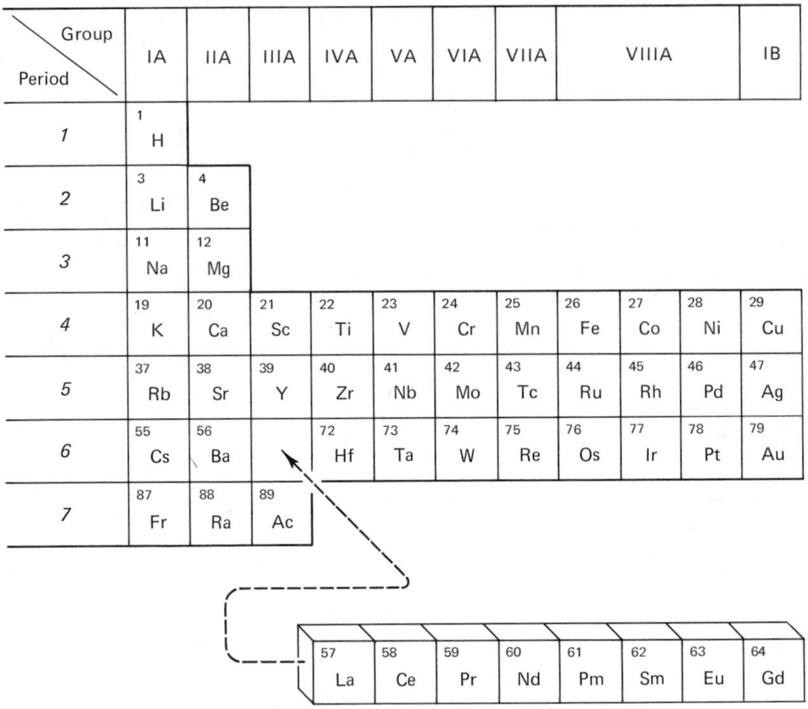

Period \ Group	IA	IIA	IIIA	IVA	VA	VIA	VIIA	VIIIA			IB
1	1 H										
2	3 Li	4 Be									
3	11 Na	12 Mg									
4	19 K	20 Ca	21 Sc	22 Ti	23 V	24 Cr	25 Mn	26 Fe	27 Co	28 Ni	29 Cu
5	37 Rb	38 Sr	39 Y	40 Zr	41 Nb	42 Mo	43 Tc	44 Ru	45 Rh	46 Pd	47 Ag
6	55 Cs	56 Ba		72 Hf	73 Ta	74 W	75 Re	76 Os	77 Ir	78 Pt	79 Au
7	87 Fr	88 Ra	89 Ac								

57 La	58 Ce	59 Pr	60 Nd	61 Pm	62 Sm	63 Eu	64 Gd

All fifteen of the rare earths belong in the sixth row, third column, of the periodic table.

ments are neither rare nor earths. They are metals, and they are quite abundant. Cerium, which is the most abundant, ranks twenty-eighth in the abundances of the naturally occurring elements. The least abundant naturally occurring rare earth, thulium, is more plentiful than cadmium, gold, iodine, mercury, platinum, or silver. Indeed, 25% of the elements are scarcer than thulium. (32)

In the periodic table as shown above the rare earths pop out of the sixth row of elements.

Modern man's first acquaintance with rare earths dates back to 1787, when Lieutenant C. A. Arrhenius of the Swedish army stumbled onto a unique black mineral while examining ores in Ytterby, Sweden, but the chemistry of the rare earths was

IIB	IIIB	IVB	VB	VIB	VIIB	VIIIB
						2 He
	5 B	6 C	7 N	8 O	9 F	10 Ne
	13 Al	14 Si	15 P	16 S	17 Cl	18 Ar
30 Zn	31 Ga	32 Ge	33 As	34 Se	35 Br	36 Kr
48 Cd	49 In	50 Sn	51 Sb	52 Te	53 I	54 Xe
80 Hg	81 Tl	82 Pb	83 Bi	84 Po	85 At	86 Rn

65	66	67	68	69	70	71
Tb	Dy	Ho	Er	Tm	Yb	Lu

hopelessly snarled until quantum theory came along. Then not only did matters get straightened out, but the existence of a rare earth with $Z = 61$ was predicted.

How the rare earths got their names may be of interest from a linguistic standpoint, or even more as a comedy of errors.

Lanthanum was derived from the Greek word *lanthanein,* which means to be hidden or concealed, because it was found in a mixture of oxides which for 36 years had been thought to be pure cerium.

Praseodymium was discovered to be one of the two major components of didymium. The name is derived from the Greek words *prasios* and *didymos,* meaning green twin. Didymium was thought to be a pure element by its discoverer, who derived the name from the Greek word *didymos,* which means twin, because it was discovered along with lanthanum.

Neodymium was discovered to be the other major component of didymium. The name is derived from the Greek words *neos* and *didymos,* which mean new twin.

Terbium was originally called erbium. Later workers confused erbium and terbium, and today, because of common usage, element 65 is known as terbium. The name terbium is derived from the town of Ytterby, Sweden.

Erbium was discovered by the Swedish chemist Carl G. Mosander (1797–1858) in 1843, who originally called it terbium. Because later workers by accident called "Mosander's terbium" erbium, element 68 is now known as erbium. The name erbium is also derived from the village of Ytterby.

Ytterbium obviously, is named for the town of Ytterby.

Dysprosium was named from the Greek word *dysprositos,* which means hard to get at, because it was one of the last rare-earth elements discovered.

Cerium was named after the newly sighted asteroid Ceres in 1803.

Promethium was discovered in 1947 in the fission products of uranium. The name, from Greek mythology, was derived from Prometheus, who stole fire from heaven and gave it to man.

Europium was named after the continent Europe.

Thulium was named for Thule, the ancient name of Scandinavia.

Holmium is derived from the Latinized word for the city of Stockholm, *Holmia.*

Lutetium was named after the ancient name of Paris, Lutetia.

Gadolinium was named after the Finnish chemist Johan Gadolin.

Samarium was named after a 19th century Russian mine official, Colonel M. von Samarski.

Most of us do not realize that the rare earths play important roles in everyday life. For example, the eyeglasses you may be using to read this probably were ground and polished by using cerium dioxide, CeO_2. Most of the cigarette lighters in use today contain a flint composed of a cerium-iron alloy. Lanthanum oxide, La_2O_3, is added to camera lenses to reduce the chromatic aberration (the spreading of colors as they pass through the lens). The steel, iron, copper, aluminum, and magnesium in our automobiles, cooking utensils, home appliances, metal furniture, and tools contain small amounts of rare earths to improve their properties. The light from carbon-arc searchlights and motion-picture projectors is stronger, steadier, and more uniform because of rare earths in the core of the electrodes. The color of beautiful stained-glass windows and glass vases often is produced by one or more of the highly colored rare earth oxides dissolved in the glass.

One of the most exciting developments in rare-earth technology came late in 1964 when europium was first used in color television tubes. The red color emitted by the europium-activated phosphor is about four times brighter than the phosphor previously used. This made it possible to increase the brightness of color television images, because the intensity of the red was now more nearly equal to that of the other principal colors, blue and green.

Another recent advance has been made in using the laser properties of several of the rare earths. At least one commercial neodymium laser is available.

Besides these uses the rare earths are used as catalysts, paint driers, glass decolorizers, ceramic opacifiers, activators for fluorescent lighting, and reagent chemicals. They are valuable, also, in photosensitive glass, electronic equipment (such as vacuum tubes, capacitors, masers, and ferrites), nausea preventives, sunglasses, and glass blowers' and welders' goggles.

High purity rare earths are produced commercially in pound quantities. [The following prices are as of 1966; they are included to show comparative values.] Cerium and lanthanum metals cost about $30 per pound. The most expensive rare earth metals (europium, terbium, thulium, and lutetium) cost about $1000 to $1500 per pound. This compares to about $5 per pound for high-purity lead, mercury, and zinc; $50 to $150 per pound for silver, titanium, and chromium; and $500 to $1200 per pound for gold, platinum, and palladium. (32)

4 ◊ *The Unnatural Elements*

CHEMICAL elements are often called the building blocks of nature. Everything in our physical world—the rocks and minerals of the earth's surface, the waters of rivers and seas, the gases of the atmosphere, the leaves of plants, and the flesh of animals—is constructed of a few dozen chemical elements. It is rather certain that the elements [other than hydrogen and, perhaps, helium] found in our solar system were produced in a series of nuclear fires burning inside ancient stars.

Many unstable forms of the elements were produced. These have been transmuted to the stable ones we are familiar with in everyday life by the spontaneous and continuous changes in the constitution of their nuclei which we call radioactivity. Of the first eighty-two elements in the periodic system, there are two which are missing in nature because they exist in no stable forms. These are elements 43 and 61, technetium [the name comes from the fact that it was the first element to be produced artificially] and promethium. The elements above lead (element 82) have a pronounced nuclear instability because of the large number of protons in their nuclei; it is a lucky cosmic accident that any of these elements have survived since the cosmic event of element synthesis. The heavy elements uranium and thorium, elements 92 and 90, respectively, are unstable, but their rates of spontaneous radioactive change are so low that not all the original stock has been converted to lighter elements. In the course of their radioactive decay, uranium and thorium produce other radioactive elements, such as radium and polonium. None of the primordial stock of these elements remain, but, since they are continually replenished by the decay of uranium and thorium, we are able to find equilibrium amounts of them in natural ores such as pitchblende.

From earliest times men have dreamed of the transmutation of one element into another, particularly of the transmutation of the base metals into the noble coinage metals, gold and silver. Enormous effort went into the chemical experiments of the medieval alchemists. We know now that their efforts were foredoomed to failure because there was no way their chemical treatments could affect the tiny nuclei of the atoms, wherein the identity of each element resides.

To alter the composition of the atomic nucleus, it is necessary to change the number of neutrons and protons in it. This may occur spontaneously by the process known as radioactivity.

For example, it sometimes happens that the neutron-to-proton ratio is higher than the value most favored by the special forces acting between the constituents of the nucleus. In such a case the nucleus seeks to achieve greater stability by converting one of its neutrons into a proton by beta decay (see chapter 2), expressed in this equation:

$$\text{neutron} \rightarrow \text{proton} + \text{electron} + \text{neutrino}$$

In order to maintain electrical neutrality, an electron (beta particle) is created. This electron is ejected from the nucleus with great kinetic energy [energy associated with motion]. A neutrino is also created and ejected.

A second type of radioactive process is alpha decay. In this process the nucleus spontaneously ejects a cluster of nuclear particles. This cluster is an alpha particle and consists of two neutrons and two protons bound together. The loss of two protons (two units of charge) converts the nucleus of one element into one which is two places lower in atomic number. Consider the radioactive decay of radium by alpha decay:

$${}^{226}_{88}\text{Ra} \rightarrow {}^{222}_{86}\text{Rn} + \alpha$$

This is a shorthand way of stating that a nucleus of radium (element 88) with 226 nuclear constituents has disintegrated to form a nucleus of radon (element 86) with 222 nuclear constituents and an alpha particle (cluster of two neutrons and two protons).

The rates at which radioactive processes occur vary enormously. One species may disintegrate completely in a fraction of a second. A collection of nuclei of another species (uranium 238, for example) may disintegrate only partly in billions of years. Each species (that is, each type of nucleus with a specific number of neutrons and protons) has a specific decay rate (see diagram). Radioactive materials disintegrate at a rate proportional to the amount of material present at the time of disintegration. This gives rise to a logarithmic decay curve characterized by a half-life, or the time required for half of the material to disintegrate. (37)

At the end of 1 half-life, half the material remains. At the end of 2 half-lives, half the remainder, or one-fourth of the

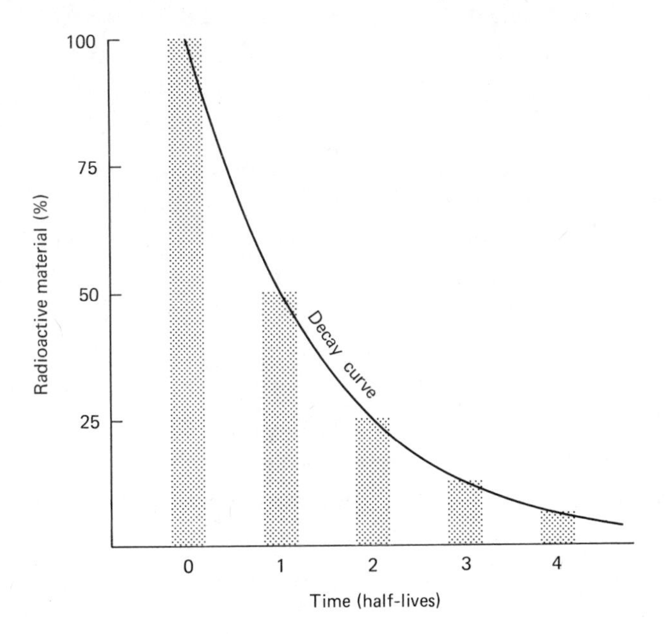

Radioactive decay curve

original material, remains, and so on. The half-life is independent of the history of the material. (Contrast, for example, human beings: if, of a group of infants, half live to age 60, the number of additional years that will elapse until half of *them* remain alive will be a lot less than 60.)

Aside from spontaneous changes, there are artificial changes in nuclear composition which may be caused by adding neutrons or protons to the nucleus. Experimental methods are known by which we may add a single neutron or proton or a group of neutrons and protons.

It is not an easy matter to transform one chemical element into another by adding protons to its nucleus. We may try to do this by directing a stream of protons, or of more complex nuclear projectiles, at a target composed of the element we wish to change. But protons are charged particles and hence repel each other. So the projectile protons (or more complex projectiles) must have sufficient kinetic energy to overcome the electrical repulsion of the protons in the nucleus. This barrier is described by Coulomb's law, which states that the repulsion of like particles varies as the product of their charges and inversely as the square of the distance between them. Because nuclei have extremely small dimensions, the inverse square law requires that the Coulombic repulsion increase to an enormous value at the distance of approach required for contact between projectile and target. The uranium nucleus (for example) has a nuclear radius of about 10^{-12} centimeter (one-millionth of one-millionth of 1 centimeter).

Let us consider the quantitative magnitude of the barrier. The physicists' unit for energy on the atomic scale is the electron volt (see chapter 5, after heading "Identification Properties"). The Coulomb barrier is measured in millions of electron volts, usually expressed by the abbreviation MeV. In the case of the uranium nucleus, the barrier to the approach of protons or deuterons is 12 MeV and to the approach of helium nuclei is 23 MeV. To make this magnitude more understandable, let us consider some energy values from our general experience.

A railroad locomotive traveling at 100 miles an hour has an enormous amount of kinetic energy; yet the energy of a single iron atom in the steel of the speeding locomotive is only 0.006 electron volt. (37)

Kinetic energy equals $\frac{1}{2}mv^2$, one-half the mass times the velocity squared. Although the velocities of the locomotive and the iron atom are the same, the mass of the locomotive is vastly greater, giving it a proportionally greater kinetic energy.

A bullet traveling at a muzzle velocity of 3000 feet per second has a great deal of energy, but each atom of lead in the bullet has only 1 electron volt of energy. The most violent chemical explosions raise the energy of the explosion products to only a few electron volts. From these examples we see that the upper limit of energy required for events in the large-scale world is a million or more times less than that required to force charged matter past the electrical defenses of the nucleus.

This is the fundamental fact that explains the importance to physics of the invention of such accelerating machines as the cyclotron. The purpose of such accelerators is to raise the kinetic energy of nuclear projectiles to an enormous value so they can penetrate the nuclei of target atoms and bring about nuclear reactions.

Soon after Sir James Chadwick discovered the neutron in 1932, it was learned that a copious source of neutrons was available from the reaction between light elements and the alpha rays emitted by polonium and radium. It was also found that neutrons are remarkably effective in causing nuclear transformations. Being electrically neutral, they are unaffected by the Coulomb barrier, and even at low velocity they can penetrate into the nuclear interior of atoms placed in their path. Once inside a nucleus they are acted upon by strong nuclear forces and bound into the nucleus. The immediate result of this absorption is simply to raise the nuclear mass by one unit without changing the proton number. But in most cases the changed neutron-to-proton ratio is not a stable one, and the neutron is transformed into a proton by beta decay, which we discussed earlier. The net result of neutron absorption and beta decay is to convert the original element to an element lying one place higher in the periodic chart of the elements.

Reactions of this kind were intensively studied in the 1930s, and many dozens of new radioactive species of the known elements were produced and identified.

It was natural that scientists should investigate the effects of neutron irradiation of uranium, the heaviest known element, since, on the basis of past experience, it seemed likely that a new element of atomic number 93 would be formed. The prospect of the synthesis of an element not present in nature was exciting and intriguing. In the mid-1930s Enrico Fermi and his co-workers in Rome started a series of experiments with uranium. They expected uranium to capture a neutron and that this would produce a new uranium isotope by the reaction:

$$^{238}\text{U} + \text{neutron} \rightarrow {}^{239}\text{U}$$

Furthermore, they expected the new isotope to undergo radioactive decay to produce an isotope of element 93:

$$^{239}\text{U} \rightarrow {}^{239}93 + \text{electron} + \text{neutrino}$$

This sequence of events does indeed occur in uranium samples bombarded with slow neutrons, but the element 93 product was incorrectly identified in these experiments. The reason for this is that, in bombarding uranium with neutrons, Fermi and his group had unknowingly caused the rare isotope of uranium, ^{235}U, to undergo nuclear fission. When ^{235}U undergoes fission it produces two radioactive elements of medium weight. In each fission event the splitting is slightly different. A group of fissioning uranium nuclei gives rise to many fission products, including many dozens of radioactive forms of many medium weight elements. This complex mixture of radioactive fission products obscured the radiations of the new element 93. Fermi and his coworkers did not know that fission had occurred, for no physicist at that time had dreamed that such a violent rearrangement of nuclear matter could occur. Naturally this led to considerable confusion. In the next few years several of the fission products were isolated by chemical analysis and incorrectly claimed to be new elements lying beyond uranium in the periodic chart. Elements as high as number 96 were reported. These interpretations were recognized as incorrect when the crucial 1938 experiments of the German chemists Otto Hahn (1879–1968) and Fritz Strassmann revealed the reality of nuclear fission.

Thus it was that the first attempt to produce a synthetic transuranium element led to one of the most striking and important discoveries of all scientific history. (37)

When Fermi was awarded the Nobel prize for having produced transuranic elements, the prize was once again "incorrectly" awarded. However, much did come from this experiment, and since Fermi had done so much other work that warranted the Nobel prize, the prize was given to the correct man, if not for the correct experiment.

When fission was discovered in 1938, every nuclear physicist in the world hastened to perform some experiment that would reveal some new information about this astonishing phenomenon. At Berkeley, California, the American physicist Edwin M. McMillan thought it might be useful to measure how far the energetic fragments of uranium fission would travel in matter. His experimental technique was quite interesting as it shows that not all important experiments need be performed with elaborate apparatus. He took ordinary cigarette papers, the kind that people use who prefer to roll their own, and stacked them up to form a little book. On the top sheet he put a thin layer of uranium oxide and then exposed the whole stack to neutrons. The fission fragments flew into the cigarette papers, stopping at various depths. Then he took the papers apart and measured the radioactivity in each one by means of a Geiger counter. So far so good. The results were about as expected. There was, however, an unexpected result that turned out to be of greater significance. The first piece of paper with the uranium layer on it also contained a radioactivity which did not travel with the rest and which did not emit the same radiations.

In trying to think why this should be, McMillan returned to the original thoughts of Fermi. If some of the neutrons did get absorbed in the uranium without causing fission, then some new heavy uranium isotope might be produced, and this might decay to form an isotope of element 93. This isotope would not fly out of the uranium oxide target.

At this point Philip H. Abelson joined McMillan in an intensive study of the new radioactivity with the hope of identifying it as element 93. McMillan and Abelson were successful in this but only after doing some careful rethinking of the location of the heaviest elements in the periodic chart. This led them to the conclusion that their element 93 might resemble uranium in its chemical properties.

In naming their synthetic transuranium element, these researchers used an astronomical analogy. The element uranium had been named after the planet Uranus. The next planet beyond Uranus is Neptune so McMillan and Abelson chose the name neptunium (symbol Np) for the first element beyond uranium.

The isotope of neptunium which they discovered has a half-life of 2.2 days, and only tiny invisible amounts of it could be prepared.

The work on neptunium made it seem highly likely that an additional element could be prepared, so they turned attention immediately to its synthesis. The first identification of an isotope of element 94 was made early in 1941. The discoverers suggested the name plutonium (symbol Pu) for this element, since Pluto is the second planet beyond Uranus, just as plutonium is the second element beyond uranium.

Following the discovery of elements 93 and 94, nine more elements (americium, curium, berkelium, californium, einsteinium, fermium, mendelevium, nobelium, and lawrencium) were synthesized and identified. The nuclear chemists had to make use of detailed theoretical knowledge and good scientific intuition in the fields of nuclear structure, radioactivity, and radiochemistry in order to predict the properties of the unknown element and in order to design properly an experiment for its discovery. Each series of experiments was, in its way, a tour de force of experimental techniques covering target preparation, accelerator operation, radiochemical separations, and the design and use of advanced detectors for the measurement of radiations.

Several of the synthetic elements were first prepared by the transmutation of a target made up of atoms of one of the transuranium elements of lower atomic number; for example, element 95 was first made by neutron irradiation of plutonium. The rather large (microgram to milligram) quantities of the target element required in these experiments were synthesized in nuclear reactors, as discussed below, some years after the first synthesis of a few atoms of the element. This necessity to build upon the work of the past explains why the synthetic elements were not all discovered at once, but in a regular sequence.

Element 95 was first prepared by a neutron capture plus beta-decay sequence. In this case a plutonium target absorbed two neutrons successively before undergoing beta decay.

Elements 96, 97, 98, and 101 were first prepared by bombardments of target elements with energetic alpha particles.

Elements 99 and 100 were synthesized in an unexpected and dramatic fashion. An atomic explosion test produced an enormous quantity of neutrons at the moment of detonation. Many of these neutrons were absorbed by uranium metal in the device. Because of the huge neutron flux, the uranium nuclei absorbed not just one but several neutrons in quick succession. Some uranium nuclei absorbed fifteen or more neutrons, producing such superheavy nuclei as ^{253}U, ^{254}U, ^{255}U, etc. These nuclei had much too high a ratio of neutrons to protons for stability and therefore rapidly underwent beta decay to produce neptunium isotopes. These also were unstable and rapidly decayed to plutonium. Multiple beta decay continued by the following sequence as the nuclei sought to obtain a more favorable neutron-proton balance:

$$ {}_{92}U \xrightarrow{\beta^-} {}_{93}Np \xrightarrow{\beta^-} {}_{94}Pu \xrightarrow{\beta^-} {}_{95}Am \xrightarrow{\beta^-} {}_{96}Cm \xrightarrow{\beta^-} {}_{97}Bk \xrightarrow{\beta^-} {}_{98}Cf \xrightarrow{\beta^-} 99 \xrightarrow{\beta^-} 100 $$

Part of the dust and debris from the explosion was collected and examined by nuclear chemists. When many unknown heavy isotopes of plutonium, americium, and curium were found, a search was made for the possible presence of elements 99 or greater. This search resulted in the discovery of elements 99 and 100, later named einsteinium and fermium after the famous physicists.

Elements 102 and 103 were first synthesized by reactions of transmutation induced by rather unusual nuclear projectiles. The nuclei of the elements carbon and boron (containing six and five protons, respectively) were used instead of neutrons, deuterons, or helium ions. With such complex projectiles the atomic number of the target element can be raised by six or five units in one step. Special accelerators have been constructed for the acceleration of such complex nuclei to the required energy. Most attempts to produce new transuranium elements are being made with their assistance. The Coulomb barrier to the approach of carbon and boron nuclei, and similar projectiles, is much greater than that operating on helium ions so these accelerators must raise the energies of these projectiles accordingly. (37)

The production of lawrencium, ${}^{257}_{103}Lw$, for example, which has 103 protons and $257 - 103$, or 154, neutrons, was achieved by the bombardment of californium by boron:

$$ {}^{11}_{5}B + {}^{250}_{98}Cf \rightarrow {}^{257}_{103}Lw + 4 \text{ neutrons} $$

After the cosmic inspiration that operated in the naming of the first two transuranic elements one cannot help noticing a provincialism that crept in with element 97, named after the city of Berkeley, and with element 98, californium. We have seen that there is a precedent for provincialism, however, in the naming of the rare earths. Who can claim that Berkeley, California, is less deserving than Ytterby, Sweden? If there is europium, why not americium? And as for naming elements after persons, if a Russian mine official could be immortalized, why not Curie, Einstein, Fermi, and Mendeleev?

The synthetic elements represent a 10% extension of the roster of known elements. Their placement in the periodic table of the elements is a matter of great interest and importance. The periodic classification, as we have seen, groups the elements according to families with common properties. This grouping is based on the systematic way the electrons surrounding the atomic nucleus are placed in specific planetlike orbits with precisely quantized energy. These orbits are occupied in a regular sequence, and the completion of certain sequences (referred to as the filling of a shell of electrons) produces particular stability. Elements at some points in the periodic table have some of the outer shells or orbits occupied before all available inner shells are completely filled. When this happens there occurs further on in the table a group of elements, called transition elements, in which the unfilled inner shells are filled in with electrons, one by one. In these elements the outer electrons, which determine the majority of the chemical properties, are not disturbed; hence members of each group of transition elements have very similar chemical characteristics.

The natural elements thorium, protactinium, and uranium plus the eleven synthetic elements constitute a special group of transition elements which are given the family name of actinide elements. In these, the 5*f* inner shell of electrons is gradually filled in while, throughout the group, there is no addition of electrons to the outermost shells.

This type of transition-element family had been encountered in the natural elements many years earlier in the case of the rare-earth group of elements, the fourteen elements beginning with element 58, which have the family name of lanthanide elements.

In many of their chemical and physical properties, the

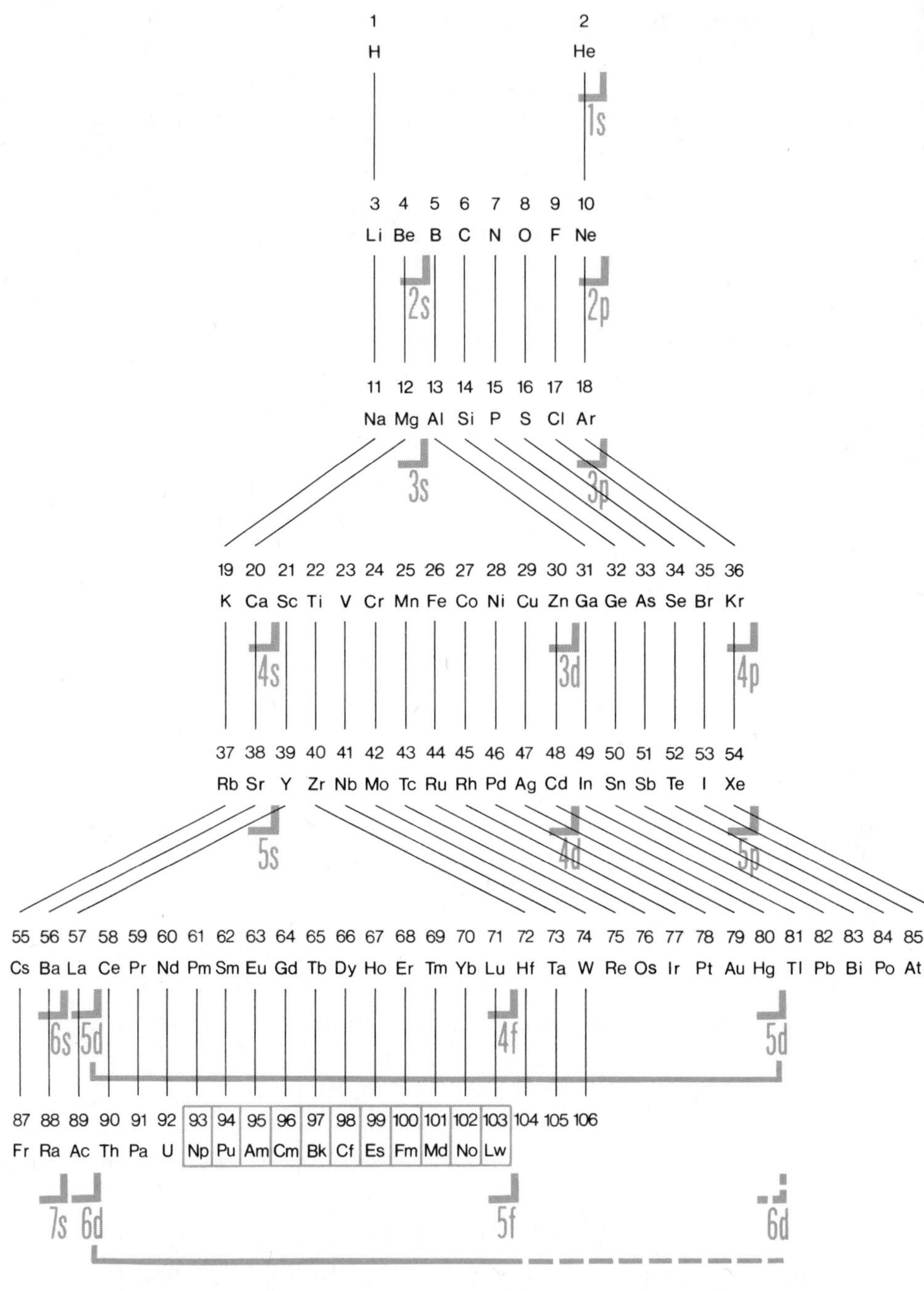

An alternate form of the periodic table of the elements. Elements of similar chemical properties are connected by the lines running from top to bottom. The atomic number of each element is given above the element symbol. The synthetic transuranium elements are shown in shaded squares in the last horizontal line. In each row there appear brackets with the labels 1*s*, 2*s*, 2*p*, and so on. Each of these brackets indicates the filling of an electronic subshell of electrons in the succession of the elements. The characteristic feature of the transuranium elements is that they belong to a series of elements (the actinide elements) in which the 5*f* subshell is being filled. The maximum number of electrons (with spin up or down) in a state of given principle quantum number $n = 1,2,3, \ldots$ and specified angular momentum l (where $l = 0,1,2,$ and 3 are denoted by *s,p,d,* and *f*, respectively) is $2(2l + 1)$. Thus there are at most two electrons in an *s* state, six in a *p* state, ten in a *d* state and fourteen in an *f* state.

actinide and lanthanide elements are very similar. Much interesting chemical research has developed from a detailed intercomparison of the properties of both series of elements.

Following the first identification of one isotopic form of a synthetic element, attempts are made to synthesize and identify as many isotopic species as possible. Since all nuclear species of these elements are radioactive, the experiments have as their purpose the determination of the nuclear decay properties. A summary of data on more than ninety species reveals the steady decrease in the nuclear stability of these elements, a decrease which can be shown by listing the half-life of the most stable forms of each element.

Element 103 should complete the actinide series of elements. Elements 104 through 118 should occupy places in the periodic table similar to the elements 72 through 86 (hafnium through randon). That is, element 104 should resemble hafnium, element 105 tantalum, etc. It would be a fascinating study to discover whether this prediction is true. The nuclear stability of these elements will be low and perhaps the lifetimes of the most stable nuclei will be so short that a detailed study of the chemical properties of these elements will be out of the question. However, this can be settled only by additional work which may reveal that nature has some surprises for us in the synthetic elements of the future. (37)

Several elements in the region 104 through 118 have been synthesized, but the situation is blurred by claims and counterclaims. Some physicists speculate that there are "islands of stability." The increasing instability of the transuranic elements as they increase in atomic number does not continue indefinitely, these physicists believe; there may be stable regions around higher atomic numbers.

Detailed information on the nuclear radiations of about ninety species has been invaluable for the development and testing of theories of the structure of nuclei and of the qualities of the forces at work in the nuclear interior. To cite but one interesting development, it is now clearly established that the nuclei of very heavy elements are not spherical, as was previously believed, but are of spheroidal shape. This has many consequences for the nuclear energy states.

Heavy nuclei with their large content of positively charged protons are particularly susceptible to alpha decay. They are also the only group of nuclei which readily undergo nuclear fission. The heaviest of the synthetic elements fission spontaneously at an appreciable rate. This is a particularly pure form of nuclear fission, and one that is especially easy to investigate. Its study provides much information for a deeper understanding of the important phenomenon of nuclear fission.

A knowledge of the transuranium elements and of the nuclear reactions by which they are made and transformed is of great importance to cosmologists, the theorists who seek to explain the origin, the energy-producing reactions, and the evolution of the stars. Knowledge of the heavy elements is essential in developing theories for the syntheses of elements in various types of stars, the peculiar features of the relative abundance of the elements and their isotopes in our solar system, the evolution of stars in which dense fluxes of neutrons are present, and other matters. (37)

5 ◊ *Elementary Particles*

This study of the atom began with electrons and protons. With the subject of radioactivity, the alpha particle and gamma ray were introduced. Along the way to complex atoms, neutrons and neutrinos were fitted into the scheme. At the beginning of this part the statement was made that there are about as many elementary particles as chemical elements. Before describing this plethora of particles and the ways of classifying them into a small number of basic types, it is time to mention some concepts that characterize the modern atom.

Some concepts in this chapter are difficult to absorb. It is not just that they are abstruse; they are new and therefore unfamiliar. And to some extent the nonmathematical treatment given here does not do them justice. But much of the difficulty is that the concepts do not yet have a proper theoretical basis.

When a theory acquires a proper foundation, it becomes simpler. Disparities fall into place. Electromagnetic theory, for example, is now embodied in four simple, elegant equations. When first promulgated, electromagnetic theory was a network of complications. High-energy physics, the physics of elementary particles and their interactions, is still at a developing stage.

High-energy physics does not yet have much relation to the rest of science. While it is essential, in order to understand chemistry or biology, to know nonrelativistic quantum mechanics (to be discussed in the first part of the chapter), it is not essential to know high-energy physics, which has had no influence on genetics or chemistry, and certainly not on technology. It is possible, therefore, to read the rest of the book without knowing anything about elementary particles. On the other hand, high-energy physics is one of the most exciting areas of physics, with intensive exploration and rapid developments in both theory and experiment. Ultimate applications cannot at this moment be foreseen.

Perhaps this chapter should be approached aesthetically, like an esoteric film, sitting back and getting what one can out of it, without being frustrated when individual sections seem obscure. And, pursuing the film analogy, there is a temptation to classify our subject as science fiction, for it deals with a stranger-than-fiction world with dimensions of 10^{-14} centimeter and times of 10^{-23} second.

Wave-Particle Duality

THE familiar concept of the structure of an atom is shown in the diagram of the Bohr atom. Today mathematical models are regarded as more nearly correct since we know that an electron's position, for example, cannot be measured accurately enough to justify picturing it in orbit around the nucleus. The mathematical description predicts the probability of finding the electron at different distances from the nucleus, and these predictions agree exactly with experiment. This vagueness in pic-

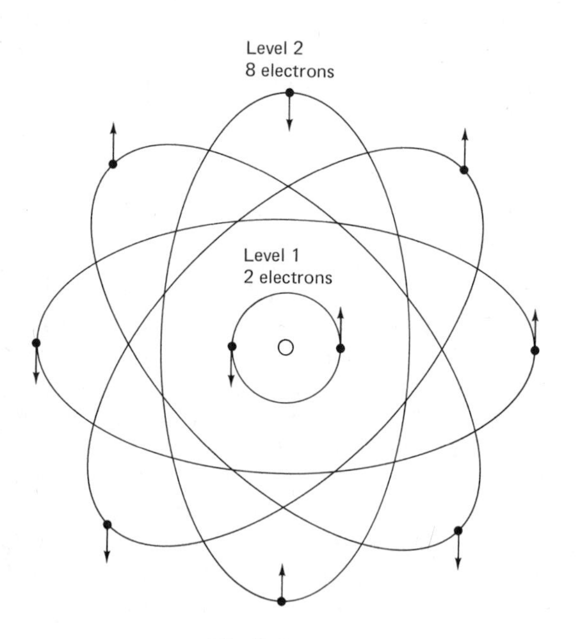

The Bohr atom

turing the electron orbits is typical of our difficulty in describing behavior of all the particles.

It is impossible to measure the position of a very small object without disturbing it with our measuring tool and thereby altering its velocity slightly. Similarly the energy of a moving system during a very short time interval is indefinite by an unknown amount. These tiny variations are insignificant in our everyday world, but on the atomic scale of time and distance they produce unexpected results. These relations between measurements of position and velocity, and between energy and time, are summed up in the Heisenberg uncertainty principle.

$$\Delta x \times \Delta p_x \geqq \hbar \approx 1 \times 10^{-34} \text{ joule-second}$$
$$\Delta E \times \Delta t \geqq \hbar \approx 1 \times 10^{-34} \text{ joule-second}$$

The first equation states that the uncertainty in the position times the uncertainty in the momentum must be greater than or equal to a very small constant (Planck's constant). Either can be measured with great precision, but only at the expense of the precision in measurement of the other. The second equation states that the minimum uncertainty in the energy is ΔE in a measurement of the energy that takes place in a time interval Δt. The symbol $\hbar$ denotes $h/2\pi$.

The equations describing the motion of a particle tell us the probability that a particle will have certain values of position and velocity. Experimental findings confirm that there is a spread, or probability distribution, in the measured values. Particles with large momentum act very much like hard bullets, and so we expect the equations that describe their actions to be similar to those for bullets' actions. Their position at a given time should be quite definite. On the other hand, when low-momentum particles interact, their subsequent positions are more uncertain. The mathematics describes this uncertainty in terms of probability. Solutions to the equations for low-momentum particles are very similar to the solutions of equations describing wave motion.

It is not claimed here that the particle *is* a wave or a hard particle; it merely behaves in this or that fashion under particu-

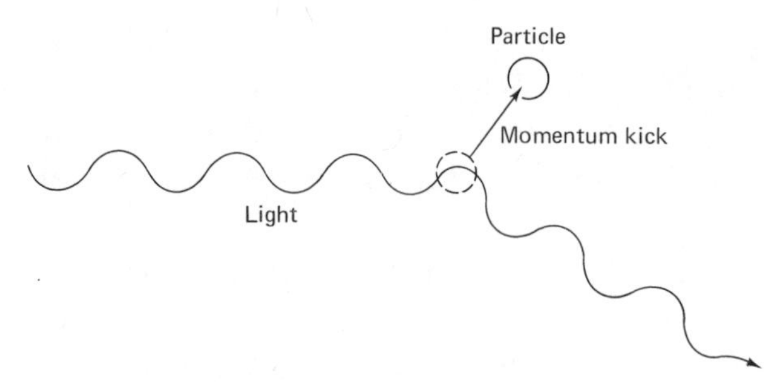

Object disturbed by measurement with a probe

lar conditions. Nor should it be imagined that the particle moves with a wiggly, wavelike motion. It is the probability function which has a wave character, and this function can only predict the chances of finding particular values of position or velocity for a given particle at a particular time.

A demonstration of the wave properties of particles is illustrated in the photograph section. This interference pattern was produced by electrons shot through a thin foil. No electron, of course, really moved along a wavy path, but the wavelike distribution of electrons is described by mathematics of waves, and the alternate bright and dark lines in the photograph are similar to effects of interference phenomena in water waves or in light. (13)

How does light originate? One way is the following.

The atoms so far described have been in the lowest possible energy state. This is the normal state. For example, in the hydrogen atom the single electron will be in the shell corresponding to lowest energy, the orbit of smallest radius. If energy is put into the atom, the electron can be excited into an orbit of larger radius, but the electron will jump down eventually to its lowest energy state, ridding itself of the excitation energy acquired by emitting a quantum, or clump, of energy in the form of electromagnetic radiation (also called a photon). The energy in the quantum is the difference in energy between the excited and the normal state (if the electron jumps directly down to the normal state). Light is emitted by an excited atom in becoming de-excited.

Light, X rays, gamma rays, radio waves, are all electromagnetic radiation (see chart of the electromagnetic spectrum, page 78). Electromagnetic radiation (or any wave) is characterized by a frequency (in cycles per second) and a wavelength, the product of the two being the velocity of the wave, which is c, the velocity of light (the same c that appears in Einstein's $E = mc^2$) for all electromagnetic radiation whether light or radio waves. The frequency of light determines its color; the frequency of radio waves determines the position of a radio station on your radio dial: for example, WQXR operates at a frequency of 1560 kilocycles.

All the remarks about particle versus wave apply to the photon, the natural unit of electromagnetic radiation, as well as to the other particles. For long radio wavelengths the best description of the photon is the wave model. For high-energy X rays the particle model is usually best. For visible light the wave model gives simple explanations of effects where the wavelength is about the same size as the object with which the light is interacting. For example, if light goes through a very small hole, it will spread out like a wave. On the other hand, if the object with which the light interacts is large compared with the wavelength, it is easier to describe the effect by making use of a particle model.

An example of the usefulness of both models can be seen in the shadow cast by parallel light rays (from a distant source) passing a knife edge. At first glance, the shadow is observed to have a very sharp boundary, as would be expected with streams of particles which either pass the knife edge or are blocked. Close examination of the shadow boundary, however, reveals that over very small distances the edge is indistinct and consists of close bands of light and shadow characteristic of wave action. There is no duality problem here, however. For some purposes light is best described by a particle model. Light is not a wave or a particle, however. Light is light. (13)

The originator of the uncertainty principle, the German physicist Werner Heisenberg, is one of the two founders of modern quantum mechanics (the other is the Austrian physicist Erwin Schrödinger 1887–1961). The uncertainty principle is

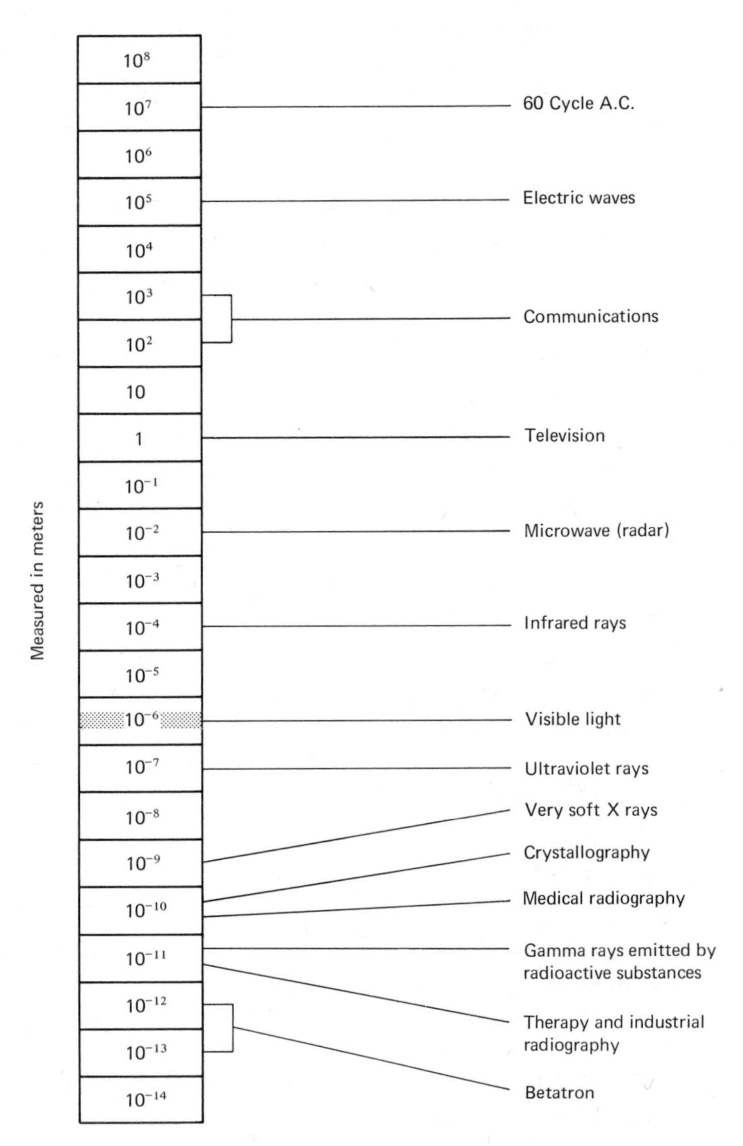

The electromagnetic spectrum, with some uses of various portions indicated

the core of quantum mechanics, both in concept and computation. And the departure of quantum mechanics from Newtonian mechanics is profound.

The influence of Newton's mechanics was much broader than on science alone; it altered the outlook of an era. Using Newton's laws of mechanics, if you know the position and velocity of every particle in the universe at any instant, you can specify their positions and velocities at all later times. The theory, in other words, is completely deterministic. (This does not mean that you can actually sit down and compute the positions and velocities. First of all, it would be a more than formidable mathematical task. Second, you could not simply "decide" to do it, unless the positions and velocities of the particles that constitute you had been so arranged previously that they would lead to your making the computation.)

The uncertainty principle introduces a limitation to the strict determinism of Newtonian mechanics. No one can simultaneously determine the exact position and the exact velocity of a body: the act of determining exact position makes it impossible to determine exact velocity and vice versa. The limitation cannot be overcome by improving measuring instruments; it is fundamental to nature. The uncertainty principle does not introduce free will, it merely rules out strict determinism.

If these remarks seem to overemphasize the impact of science on society, the reader need only recall the denunciations of the uncertainty principle made by the government and press of the Soviet Union during the Stalin period. The Soviets found the intrinsic obstacles to complete determinism that are embodied in the uncertainty principle extremely disturbing. Their state philosophy demanded a strict determinism, for only then could individuals and society be made to take a specified direction. In texts written by Soviet scientists during the Stalin period, there were introductory statements that the authors were not of the "bourgeois Copenhagen school"—that is, Niels Bohr's institute in Copenhagen, from which quantum mechanics or, to be exact, its interpretation, can be said to have emerged. After the disclaimer, however, the Soviet scientist would employ the uncertainty principle just as any scientist from the West would.

Quantum mechanics is associated in many minds with uncertainty and unpredictability; however, there is another view. Victor Weisskopf, in his excellent book *Knowledge and Wonder* (Garden City, N.Y.: Doubleday, Anchor Books, rev. ed., 1966), has pointed out that classical physics cannot account for specific properties of things, that steam is always steam wherever it is found, rock is always rock. Classical physics cannot explain definiteness because classical physics allows continuous gradations of all properties: a body can have a little more of something, a little less of something else, leading to an unlimited range of possibilities. Quantum theory changes this. Things are quantized. Not any orbit or any energy is possible, but only certain ones. It now makes sense to say that two iron atoms are exactly alike; the specified orbits are the same in any two iron atoms in the universe.

Relativistic Effects

ANOTHER feature of particle behavior is that very often the particles are traveling almost as fast as the speed of light. As they approach this critical speed limit, relativistic effects (those explained by the special theory of relativity) assume major importance. The most familiar of these effects is that energy and mass become interchangeable; that is,

$$E = mc^2$$

Particles with mass can be created out of surplus energy of other particles, and can decay back, losing their mass. An example of such creation is shown in the lefthand diagram on page 81. Here a high-energy X ray, or gamma ray, turns all its energy into the mass and kinetic energy of two electrons, one positive and one negative. The inverse effect can be seen in the righthand diagram. Here an antiproton, which will be described later, collided with an ordinary proton. The mass of both was completely annihilated and turned into the mass and kinetic energy of new particles, mesons, which in turn decayed yielding almost all their mass to the energy of yet lighter particles. (See also photograph section, following page 211.)

Another relativistic effect is change in the duration of time for objects traveling close to the speed of light. While you are reading this page, many particles called muons will go through your body. They are created near the top of the atmosphere by

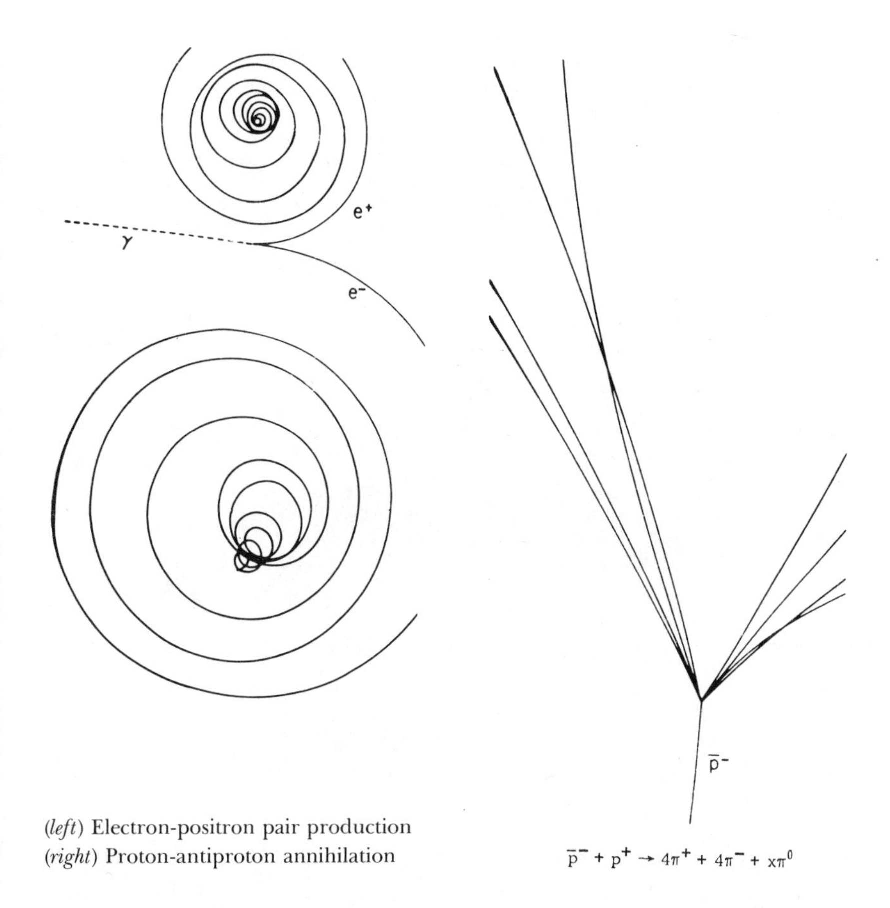

(*left*) Electron-positron pair production
(*right*) Proton-antiproton annihilation

cosmic ray protons (see diagram, page 82). When they are at rest, relative to us, they decay in a few millionths of a second. With that lifetime they should travel only a few hundred yards, even at the speed of light. The fact that many live to reach the earth shows that, at speeds close to the speed of light, time slows down for them.

Identification Properties

We have said that there are only a few properties possessed by fundamental particles. Some of these are obvious enough. The

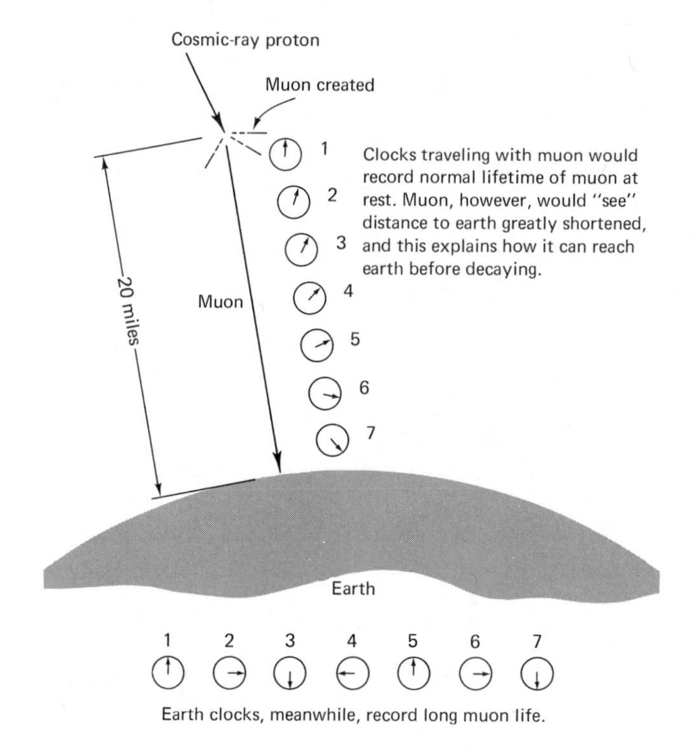

Relativistic effect on time

particles have mass, although in a few cases the mass is zero—which is as reasonable a number as any other. The mass of particles is so small on the scale of pounds or grams that it is usually given in terms of the energy it would take to create the particle. Nature's unit of energy on the atomic scale is an amount about the size of the electron volt. This is the amount of energy needed to lift an electron (or any other particle with the electron's electrical charge) through a potential difference of 1 volt.

We call the electron volt nature's unit of energy because all chemical interactions on the molecular level involve about 1 electron volt of energy. The photons of visible light, which can instigate such chemical changes, possess an energy of a couple of electron volts. Every time there is one molecular interaction in a flashlight battery, a single electron is lifted through a potential of 1½ volts, using up 1½ electron volts of chemical energy. On the human scale, this much energy is extremely small. It takes about 10^{19} such exchanges every second to power a flashlight bulb.

In the business of producing high-energy particles by accelerators, energies are usually expressed in millions of electron volts (MeV). For instance, a dentist's X-ray machine can accelerate electrons (to produce its X rays) to an energy of 0.1 MeV. The mass of an electron pair can be created with an expenditure of only 1.0 MeV. (13)

To give a macroscopic example, if an ant crawls up on a cube of sugar, the energy expended is about 10^{14} electron volts. Why so much? Because an ant is composed of about 10^{17} particles; the electron volt is the energy characteristic of *one* particle.

We have already discussed charge and spin as properties of particles, and the conservation laws associated with them. Other properties must be introduced now in order to explain the classification scheme of the particles.

The properties of mass, electric charge, and rotational spin are familiar to us in everyday life. Some other properties of particles cannot be described in familiar terms. One of these has been given the name isotopic spin since its mathematics is much like that for rotational spin. (13)

Just as there were $2l + 1$ states for angular momentum l, and $2s + 1$ states for spin s, there are $2I + 1$ states for isotopic spin I.

However, no real spin is involved.

The chart of the particles on pages 88–89 shows that most of the particles are members of groups arranged horizontally. Within each group the members have almost the same mass. The pi-meson, for instance, is a triplet group containing π^+, π^-, and π^0. The proton and neutron are in many ways the same particle except for their electric charge effects. We consider them a doublet. The lambda zero, Λ^0, is a singlet. According to our present views, all members of multiplet groups have the same interaction with the strong nuclear force but have different electromagnetic interactions, which are thought to be responsible for the slight mass differences among members of a multiplet.

One way of looking at this is to picture the members of a multiplet as being the same particle until the electromagnetic interaction is "turned on." Then the individual members assume different energy states, made apparent by their mass differences. This is very much like the description of what happens to spinning particles when a magnetic field is turned on. They assume different energy states—with the field, against it, or perpendicular to it.

We use the spin terminology, in fact, to describe the electric charge effect. Since the proton-neutron particle, usually called the nucleon, is a doublet, we assign to it an isotopic spin, $I = ½$. When this isotopic spin is pointing "up," we say that the value of I in the up direction is I_3 (I-sub-three) $= +½$. This represents the proton. When I is pointing "down," $I_3 = -½$ and represents the neutron.

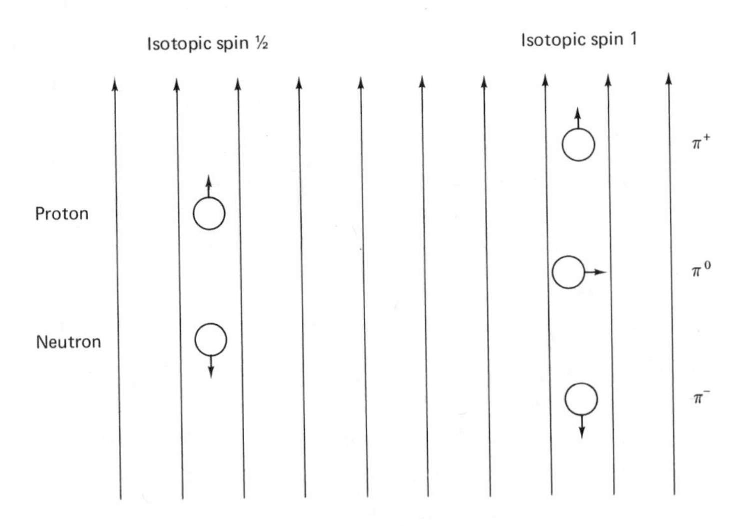

Isotopic-spin multiplets

The pi-meson must have isotopic spin $I = 1$ so that it can assume three different orientations: $I_3 = +1$, $I_3 = 0$, $I_3 = -1$. The lambda zero, with only one possibility, must have isotopic spin of $I = 0$. These orientations of isotopic spin are illustrated on page 84; the numerical value of isotopic spin for each particle is given in the particle chart on pages 88–89.

Another attribute that has no everyday counterpart is called strangeness. It became necessary to assign strangeness values to some particles to account for a strange feature of their production and subsequent decay. These strange particles—the K-mesons, lambda zero, sigma, xi, and omega—are produced in high-energy collisions (in the billion-electron-volt range) but are remarkably stable; that is, they have relatively long lifetimes. There appears to be no reason why a lambda zero, for example, should not fall apart into a pi-meson and a proton in the same time in which it was made, the time it takes light to cross a proton diameter—about 10^{-23} second. Instead, it waits an eternity (on the nuclear time scale) and takes 10^{-10} second to decay. One way to explain this situation is to invent a special conservation law that describes what the particles do.

The conserved attribute is called strangeness, and we assign strangeness values to the particles in a way that will account for what we observe. For instance, the strange particles are always produced in pairs. The lambda zero and the K zero are produced rapidly [in a collision involving other particles] but decay slowly. We assign strangeness of $S = -1$ to the lambda and $S = +1$ to the K. During the production process, therefore, strangeness is conserved (-1 plus $+1$). Once the lambda and the K are separated from each other, they cannot decay without violating the strangeness conservation law because their decay products, protons and mesons, have strangeness numbers of zero. The strangeness conservation law is violated eventually by a decay that goes from $S = -1$ or $+1$ to $S = 0$, but it is a slow process. (The particular numbers of strangeness S shown in the chart were assigned to the particles that were known in 1957. With this scheme, predictions were then made that other particles would be found and that certain reactions would occur and others would not. All these predictions have been fulfilled.) (13)

In 1932 the American physicist Carl Anderson discovered the positive electron—now known as the positron. This positive electron is a mirror image of the negative electron. It has the same mass, the same spin, but the opposite charge of the negative electron. The electron and positron form a particle-antiparticle pair. If an electron and positron come together they can annihilate each other, and out of this annihilation two photons emerge.

$$e^{+} + e^{-} \rightarrow \gamma + \gamma$$

The discovery of the positron had been anticipated by theory. In this case it was the English physicist Paul A. M. Dirac who, in the late 1920s, had predicted its existence. Dirac formed a theory of the electron that united relativity and quantum mechanics and contained the positron as an unavoidable consequence. Since the rest of the theory was in such excellent agreement with experiment, Dirac took the position that the positron had to exist.

When Fermi began working on the neutrino theory it was inevitable that he apply Dirac's ideas to the neutrino. He was led to the idea that if Pauli's neutrino existed then Pauli's antineutrino must also exist. Since neither particle had been directly observed (and neither was until 1953), physicists in the 1930s and 1940s had a queasy feeling about the whole business. This feeling was compounded when many theorists began asking themselves: What is the physical distinction between a neutrino and an antineutrino? For charged particles one of the distinctions between a particle and an antiparticle is very simple. If the particle has a positive charge then the antiparticle must have a negative charge. (Which one is called the particle and which the antiparticle is really irrelevant, but physicists tend to call the most familiar species the particle. Thus, the electron, which was known long before the positron, is called the particle while the positron is called the antiparticle. Similarly, we call the positively charged proton the particle and the negatively charged mirror image of the proton discovered in 1955 the antiproton.) However, if a particle is electrically neutral how does it differ from its antiparticle? (8)

All particles have antiparticles. These have all been produced and are as real as the particles themselves. The track left by the antielectron—the positron—is shown in the lefthand diagram on page 81, and most of the tracks shown in the right-hand diagram on the same page were made by antiprotons. The antiparticles are as stable, or as unstable, as their counterparts, but in our section of the universe they are in the minority. Whenever an antiparticle meets its corresponding particle, the two annihilate each other, their mass turning into other forms of energy.

A universe consisting of atoms with antiprotons and antineutrons in their nuclei and with positrons in their outer shells would be perfectly stable. Why our local situation exists is not known. Perhaps some of the distant galaxies are composed of antimatter.

The electric charge and strangeness number of an antiparticle are opposite to those of its particle. Thus the antiproton is negative, and the positron (antielectron) is positive. The antineutrons are necessarily neutral, but when produced by the high-energy accelerators they meet and annihilate ordinary neutrons just as expected. A few of the particles—η^0, π^0, γ—serve as their own antiparticles. (13)

The Language of the Microworld

It is important to understand the names of the major features of the microworld, as used in the particle chart on pages 88–89 and throughout the text. Some of the nomenclature follows for ready reference. Other terms are explained elsewhere in this chapter.

PARTICLES STABLE AGAINST

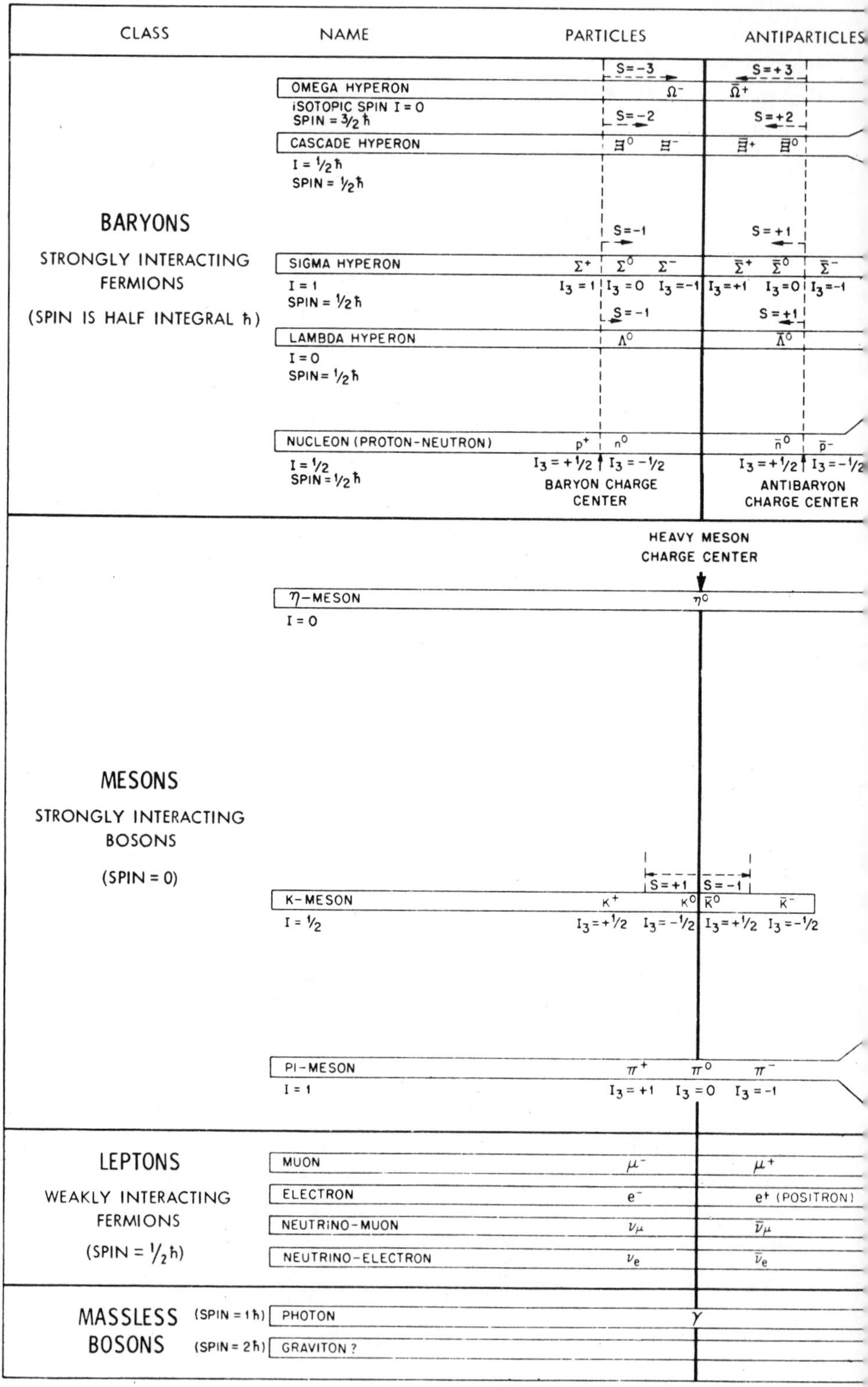

CLASS	NAME	PARTICLES	ANTIPARTICLES
BARYONS STRONGLY INTERACTING FERMIONS (SPIN IS HALF INTEGRAL $\hbar$)	OMEGA HYPERON ISOTOPIC SPIN I = 0 SPIN = $3/2\,\hbar$	$S=-3$ Ω^-	$S=+3$ $\bar{\Omega}^+$
	CASCADE HYPERON $I = 1/2\,\hbar$ SPIN = $1/2\,\hbar$	$S=-2$ Ξ^0 Ξ^-	$S=+2$ $\bar{\Xi}^+$ $\bar{\Xi}^0$
	SIGMA HYPERON I = 1 SPIN = $1/2\,\hbar$	$S=-1$ Σ^+ ($I_3 = 1$) Σ^0 ($I_3 = 0$) Σ^- ($I_3 = -1$)	$S=+1$ $\bar{\Sigma}^+$ ($I_3 = +1$) $\bar{\Sigma}^0$ ($I_3 = 0$) $\bar{\Sigma}^-$ ($I_3 = -1$)
	LAMBDA HYPERON I = 0 SPIN = $1/2\,\hbar$	$S=-1$ Λ^0	$S=+1$ $\bar{\Lambda}^0$
	NUCLEON (PROTON-NEUTRON) I = $1/2$ SPIN = $1/2\,\hbar$	p^+ ($I_3 = +1/2$) n^0 ($I_3 = -1/2$) BARYON CHARGE CENTER	$\bar{n}^0$ ($I_3 = +1/2$) $\bar{p}^-$ ($I_3 = -1/2$) ANTIBARYON CHARGE CENTER
MESONS STRONGLY INTERACTING BOSONS (SPIN = 0)	η-MESON I = 0	HEAVY MESON CHARGE CENTER η^0	
	K-MESON I = $1/2$	$S=+1$ K^+ ($I_3 = +1/2$) K^0 ($I_3 = -1/2$)	$S=-1$ $\bar{K}^0$ ($I_3 = +1/2$) $\bar{K}^-$ ($I_3 = -1/2$)
	PI-MESON I = 1	π^+ ($I_3 = +1$) π^0 ($I_3 = 0$)	π^- ($I_3 = -1$)
LEPTONS WEAKLY INTERACTING FERMIONS (SPIN = $1/2\,\hbar$)	MUON	μ^-	μ^+
	ELECTRON	e^-	e^+ (POSITRON)
	NEUTRINO-MUON	ν_μ	$\bar{\nu}_\mu$
	NEUTRINO-ELECTRON	ν_e	$\bar{\nu}_e$
MASSLESS BOSONS	(SPIN = $1\hbar$) PHOTON	γ	
	(SPIN = $2\hbar$) GRAVITON ?		

STRONG NUCLEAR DECAY

	REST MASS IN Mev	HALF-LIFE IN SECONDS	DECAY SCHEMES	
	1676	$\sim 10^{-10}$	$\Omega^- \rightarrow \Lambda^0 + K^-$	
	$\Xi^{\pm} \approx 1320$	0.9×10^{-10}	$\Xi^- \rightarrow \Lambda^0 + \pi^-$	
	$\Xi^0 \approx 1310$	1.0×10^{-10}	$\Xi^0 \rightarrow \Lambda^0 + \pi^0$	
			$\Sigma^+ \rightarrow p^+ + \pi^0$	(50%)
		0.6×10^{-10}	$\Sigma^+ \rightarrow n^0 + \pi^+$	(50%)
	≈ 1190	$< 10^{-12}$	$\Sigma^0 \rightarrow \Lambda^0 + \gamma$	
		1.2×10^{-10}	$\Sigma^- \rightarrow n^0 + \pi^-$	
	1115	1.7×10^{-10}	$\Lambda^0 \rightarrow p^+ + \pi^-$	(67%)
			$\Lambda^0 \rightarrow n^0 + \pi^0$	(33%)
n	939.5	0.7×10^3	$n^0 \rightarrow p^+ + e^- + \bar{\nu}$	
p	938.2	STABLE		
	548	$< 10^{-16}$	$\eta^0 \rightarrow \pi^+ + \pi^- + \pi^0$	
			$K^+ \rightarrow \pi^0 + e^+ + \nu$	(5%)
			$K^+ \rightarrow \pi^0 + \mu^+ + \nu$	(5%)
			$K^+ \rightarrow \mu^+ + \nu$	(64%)
			$K^+ \rightarrow \pi^+ + \pi^0$	(19%)
			$K^+ \rightarrow 2\pi^+ + \pi^-$	(6%)
$(\bar{K}^- = K^+)$	494	0.8×10^{-8}	$K^+ \rightarrow \pi^+ + 2\pi^0$	(2%)
			$K_1^0 \rightarrow \pi^+ + \pi^-$	(≈34%)
K^0 K_1^0		0.7×10^{-10}	$K_1^0 \rightarrow 2\pi^0$	(≈16%)
	498			
K_2^0		4×10^{-8}	$K_2^0 \rightarrow \pi^+ + \pi^- + \pi^0$	(7%)
			$K_2^0 \rightarrow 3\pi^0$	(7%)
			$K_2^0 \rightarrow \pi^{\pm} + \mu^{\mp} + \bar{\nu}$	(19%)
			$K_2^0 \rightarrow \pi^{\pm} + e^{\mp} + \nu$	(24%)
π^-	140	1.8×10^{-8}	$\pi^- \rightarrow \mu^- + \bar{\nu}$	
π^0	135	0.7×10^{-16}	$\pi^0 \rightarrow \gamma + \gamma$ ($\pi^0 \rightarrow \gamma + e^+ + e^-$	1%)
π^+	140	1.8×10^{-8}	$\pi^- \rightarrow \mu^+ + \nu$ ($\pi^+ \rightarrow e^+ + \nu$	0.01%)
	105.7	1.5×10^{-6}	$\mu^- \rightarrow e^- + \nu + \bar{\nu}$	
	0.51	STABLE		
	0	The neutrinos associated with $\mu^{\pm}$ are different from those with $e^{\pm}$		
	0			
	0	STABLE		
	0	STABLE	NOT DETECTED	

In the chart, groups of particles have been classified as bosons and fermions, and also as baryons, mesons, and leptons. The word hyperon is applied to several particles. Let us see what these and other terms refer to.

All elementary particles are classed in two groups, fermions and bosons. Two bosons can exist in the same state at the same time, but two fermions cannot. [To be precise, bosons do not obey the exclusion principle, while fermions do. Two bosons can have the same set of quantum numbers.]

A baryon is one of a class of heavy particles which includes protons, neutrons, hyperons, and cascade particles (or cascade hyperons). All free baryons heavier than a proton decay into a proton plus other end products.

A meson is a medium-mass particle which has mass greater than that of an electron and less than that of a nucleon (proton or neutron). There are several kinds of mesons: K-mesons, η (eta)-mesons, and π (pi)-mesons (or pions).

A lepton is one of a class of light particles which have no strong nuclear interactions and which include electrons, muons, and two kinds of neutrinos.

These three classes embrace many of the most important particles.

A hyperon is one of a class of baryons heavier than nucleons (the constituents of atomic nuclei). There are several kinds of hyperons: Ω (omega) hyperons, cascade hyperons or Ξ (xi) hyperons, Σ (sigma) hyperons, and Λ (lambda) hyperons. In other words, the different hyperons are four of the subclasses of baryons. There also are three subclasses of mesons, four subclasses of leptons, and two kinds of massless bosons.

Other particle terms, such as lambda zero (Λ^0), refer to specific particles and their charges. Descriptive terms like multiplet refer to mesons and baryons which occur in families of states differing only in electric charge.

Greek letters are used as symbols for many of the particles.

A bar over a particle symbol indicates that it is an antiparticle. A plus, zero, or minus superscript after a symbol indicates its charge. (13)

It is time to introduce the additional conservation laws promised earlier.

There appear to be two conservation laws associated with two families of particles. The particle chart groups together electrons, muons, and neutrinos into a class of light particles called leptons. The nucleons and heavier particles are called baryons. The total number of leptons in the universe remains constant and the total number of baryons remains constant. To be sure, antiprotons and antielectrons can be produced, but only if ordinary baryons or leptons are also produced in the same process. When the positron—the antielectron—is annihilated, it takes with it an electron.

The conservation laws listed so far apply to all types of interactions. There are, however, some conservation laws that do not hold for one or more of the four types of forces. For instance, the total strangeness value of a particle system remains constant when the strong nuclear and the electromagnetic interactions are in control; but the weak interaction can, after a delay, violate the law. The total isotopic spin cannot change in strong nuclear interaction, but it can in both electromagnetic and weak interactions.

One other limited conservation law must be mentioned because until a few years ago no one thought that it was limited. (13)

Parity is one of those difficult abstract ideas which a popular exposition like this one can only hint at the flavor of. Parity conservation or parity symmetry refers to the equivalence of physical events using right- and left-handed coordinate systems. The diagram shows a right- and a left-handed system. The latter has been deliberately oriented to illustrate the fact that right-handed and left-handed systems are related to each other by reflecting all the axes. The first two diagrams on page

92 illustrate the distinction between a reflection and a rotation. In the first we have reflected the arrow about the zero point.

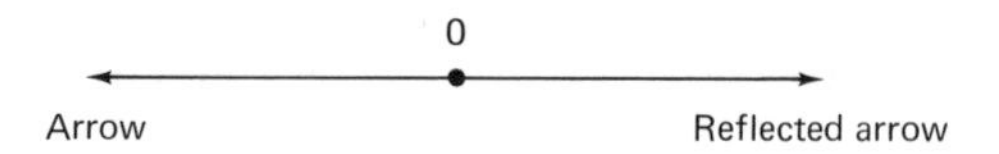

In the second we have rotated the arrow by ninety degrees.

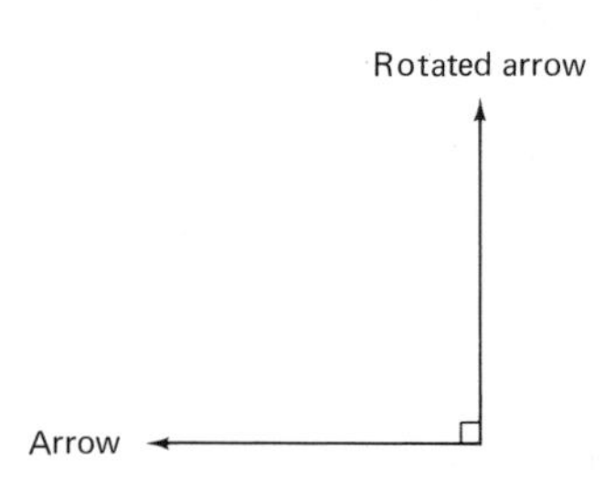

In this case if we continue the rotation by another ninety degrees we produce a configuration that is identical to the reflection. This is possible in two dimensions. But in three dimensions it is not. Thus the configurations are related by reflection of the arrows but cannot be related by any rotation or combination of rotations. No rigid rotation will transform one system

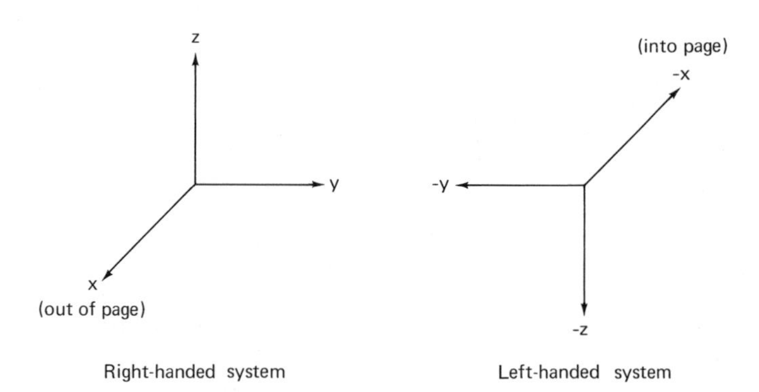

into the other. (Try it with the thumb and two fingers on the right hand and see if you can rotate them so that they look like the reflected configuration made out of the same fingers of the left hand. You cannot.) (8)

In the mathematical description of particles and their interactions, it is possible to describe particles so that, if all the coordinates (x, y, and z) are reversed and made negative, the resulting description will either be identical to or have exactly the negative value of the previous description. This mathematical property is called parity, and it is either even or odd, depending on whether the resulting description remains positive or turns negative. (For example, the sine function of an angle has odd parity; the cosine function has even parity.)

It was believed until recently that there was an absolute conservation law of parity. If the description of a system had even parity before an interaction, it must have even parity after. The implication of this for the physical world was that any process could work just as well in a mirror world as in a real world. Of course, we have our local human peculiarities of driving on the right side of the road and having our clock hands turn clockwise; but it was thought that all particle processes were completely independent of our definitions of right-left or clockwise-counterclockwise and that nature had no preference.

In 1958 the Chinese-born American physicists, Tsung-Dao Lee and Chen Ning Yang, pointed out that, although nature did not favor clockwise over counterclockwise in the everyday world controlled by the strong nuclear and electromagnetic forces, no such evidence existed for the world of the weak interaction. Their suggestion was immediately tested with beta decay, which takes place through the weak interaction, and with some of the new strange particles. True to expectations, the electrons of beta decay come out preferably in one direction. The decay of strange particles is such that one of the decay products usually comes out on only one particular side of the plane defined by the particle motions (see diagram, page 94). In the weak interactions, then, parity is not conserved, and nature does distinguish between right and left, up and down, and clockwise and counterclockwise.

There is a catch to this distinction, however. For antiparticles the natural preferences are reversed. Anticobalt emitting

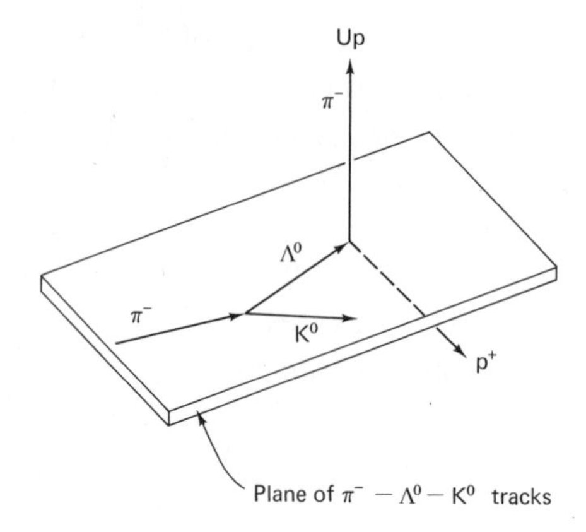

Violation of parity in weak interaction decay of Λ^0. The π^- heads preferably in one direction out of the production plane, thus defining "up." (The π^- at the lower left is involved in the production of the Λ^0 and K^0.)

antielectrons would shoot them preferentially in the opposite direction to that preferred by cobalt. For some reason the direction sense of space seems to be tied up with particle-antiparticle nature.

There may be other conservation laws that we are not aware of. Each law is associated with one of the attributes of the particles or with the properties of our description of them. Since the present system of laws and attributes does not explain all the observed processes, we can surely look forward to other surprises.

We have considered the interactions among the particles and some of the peculiar features of the microworld. Now we shall meet the actors themselves. The organizational scheme for some of the particles shown in the particle chart on pages 88–89 is comparable to Mendeleev's periodic table of the elements. In 1957 Murray Gell-Mann proposed that the particles could be arranged according to a tabular scheme and that more particles would be found with the required characteristics to complete the chart.

During the last few years, Gell-Mann's chart has served the same function that Mendeleev's table did years ago. Undiscovered particles and their characteristics were predicted, sought after, and subsequently found. Gell-Mann's prophecy has now been fulfilled. In August 1963 the last predicted member of the chart—the anti-xi—was found.

Mendeleev's table contained no explanation for the observed periodicity in chemical behavior. Instead, it served as a great proving ground for all the atomic theories that were attempted during the following 50 years. (The quantum theory development of the original Bohr atom successfully explained the periodic table many years after its conception.) In like manner, we now have organization schemes for the particles, but we still have no theory to explain why the particular schemes work and why particles have their particular values of mass and other attributes. All of the particles in the chart are stable against decay in nuclear times of 10^{-23} second. Most of them are actually unstable but decay more slowly through the electromagnetic interaction and much more slowly through the weak interaction.

Two other particles, not included in the original scheme, are also stable against nuclear decay and have been included on the chart. These are the Ω baryon and the η^0 meson.

To make sense of the chart, first locate the familiar particles—the electron, proton, and neutron. Since the particles increase in mass from the bottom to the top of the table, the electron is near the bottom with a creation mass-energy of 0.51 million electron volts (MeV). The neutron at 939.5 MeV is just slightly heavier than the proton. The difference between them is enough, however, to make the neutron unstable when it is by itself and not in a nucleus. It decays into a stable proton and two lighter particles.

Notice also that the photon, the agent of the electromagnetic field, is listed as a particle with mass zero. In the form of a radio wave, a chunk of electromagnetic radiation may not appear very much like a particle, but at the much higher frequency and energy of an X ray it behaves very much like one.

The chart has two major divisions vertically—particles and

antiparticles. All of both kinds have been produced, and the antimatter is every bit as real as the matter. Three particles—the photon, the π^0, and the η^0,—are placed on the dividing line. In effect, they act as their own antiparticles. They can do this because they are members of a class of particles, the bosons, which we must now distinguish.

There are four horizontal divisions, each with a name derived from the Greek.

The baryons (heavy particles) consist of particles like the proton. The leptons (light particles) may be thought of as the electron group. Within each group the particle number must remain constant. A new particle can be created out of energy but only if its antiparticle from the same group is created at the same time. (The particles of a group are assigned positive particle numbers while the antiparticles are assigned negative numbers.) No baryon can be created from energy unless an antibaryon is created at the same time. When an antibaryon is annihilated, it must take with it a baryon. The same is true for the lepton family. When a radioactive nucleus emits an electron (a beta ray), an antilepton must also be shot out—in this case an antineutrino.

The members of the other two classes, called bosons, obey no such conservation principle. There can be as many mesons or photons generated as there is energy available.

One interpretation of this difference is that the baryons and leptons are the sources of the interactions, or forces, and that particles of the other two classes are the agents of these forces. We have already noted that the photon is the agent of the electromagnetic field and that Yukawa predicted the existence of an intermediate-weight particle—the meson—as the agent of the strong nuclear force. It was later discovered that there are several kinds of mesons—the π, K, and η.

There is no agent of the weak interaction on the chart, but the prediction is that it should have a mass larger than that of the K-mesons and, necessarily, have mesonlike characteristics.

One is tempted to think of the heavier baryons as being

compounds of the proton (or neutron) and mesons. However, at this point we are not even sure that protons or neutrons are fundamental particles. The stability of the proton may be almost accidental, and nuclear reactions may depend on the properties of lambda or sigma particles. As far as the nuclear forces are concerned, the proton and the neutron are the same particle with different states of electric charge. In some theories it is assumed that all of the baryons are essentially the same particle with different isotopic spins. Remember that there is no complete theory to explain these particles yet and many strange ideas are still being explored.

The lepton family presents even more puzzles. There seems to be no excuse for the existence of the heavy electron called the muon. In almost every way it behaves like the electron except that its mass is two hundred times greater. The muon shows up in the decay of the pi-meson. When the pi-meson decays, a neutral particle must be emitted in the direction opposite that of the muon in order to conserve momentum.

Since the same kind of thing happens in beta decay, it was assumed until 1962 that the neutral particle was the neutrino. In an experiment at the then largest particle accelerator in the world, the alternating gradient synchrotron at Brookhaven, it was proved that the neutrino accompanying the muon is different from the neutrino accompanying the electron. Thus the strange difference between the electron and the muon becomes even more curious and, so far, is unexplainable.

Both kinds of neutrinos have been detected in their inverse process of colliding with material, in spite of the fact that their rate of interaction with matter is extremely small. (In going through the whole earth, only 1 low energy neutrino in 10 billion would suffer a collision.) Since the neutrinos have no electric charge, they are not subject to the electromagnetic force and can be affected only by the weak interaction. None of the leptons is subject to the strong nuclear force.

Even before all the stable and semistable particles of the

chart were discovered, a new type of event had been observed. In certain types of high-energy collisions, a particle was found which almost immediately decayed to a proton and a pi-meson. The lifetime of this temporary particle was only a little longer than the basic time of strong nuclear interactions: 10^{-23} second. Such a short time cannot be measured by observing the particle's flight in a track chamber. The path would not be much longer than the diameter of a nucleus.

The existence of the particle and its lifetime could be detected only by analyzing the tracks of its decay products. These showed that the proton and the pi-meson had moved as a single particle before breaking up. Similarly, when pi-mesons were shot through a hydrogen target at a particular energy, a large number were momentarily captured and then released again in all directions.

Since 1962 many such short-lived particles have been found. They appear to be particles composed of baryons and mesons and also combinations of only mesons. An immediate question is: Are these in any sense true particles, or are they just compounds of the more elementary particles? When the first proton–pi-meson combination was observed, most physicists thought of it as a resonance or excited state of the proton. Similar excited states of the atom had been measured early in this century. We do not think of the excited hydrogen atom just before it emits a photon as being a separate type of atom. On the other hand, it seems arbitrary to say that a particle which decays through the electromagnetic or the weak interaction is a true particle and that, if it decays through the strong nuclear interaction, it is a compound or an excited state.

Present theories view these interactions not so much as forces as routes by which particles can go from an initial state to a final one. For example, in the decay of the neutron to a proton, an electron, and an antineutrino, the route is the narrow one of the weak interaction (see diagram). It is a two-way route since, if the energy balance were satisfied, a proton, an electron, and an antineutrino could produce a neutron. [Though it is not at all obvious, it is also true that, if the energy balance were sat-

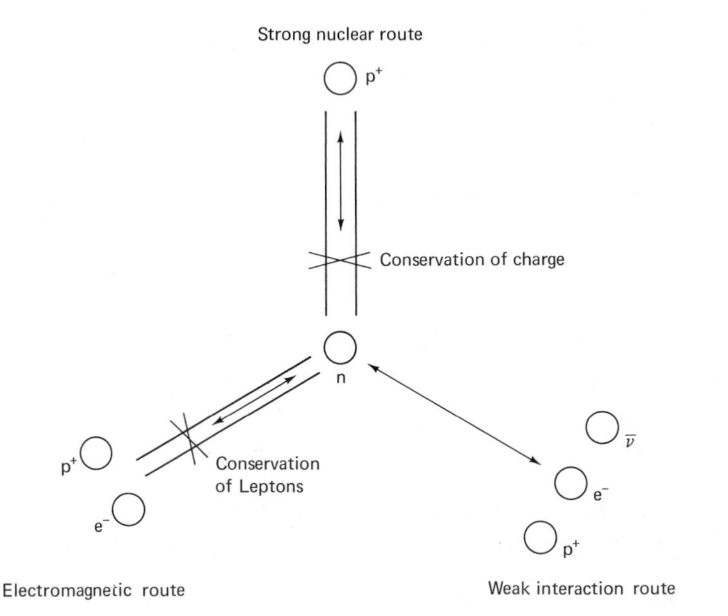

Particles will transform into certain combinations if linking routes are available and not forbidden and if energy permits.

isfied, a proton and an electron could produce a neutron and a neutrino.]

Which way the reaction goes depends on the available energy. In general, any particle will try to decay to a lighter combination. If the neutron could decay to a proton alone, the broad route of the strong nuclear interaction would make the process go immediately. This is forbidden, however, since electric charge would not be conserved. The electromagnetic route is narrower than the strong nuclear by a factor of one hundred but is still 10^{11} times wider than the weak. By the electromagnetic route the neutron could decay into a proton and an electron and so conserve electric charge. However, if this occurred, leptons would not be conserved since a single electron would be created where none existed before. The electromagnetic route could not lead to a proton, an electron, and an antineutrino because the neutrinos are not affected by the electromagnetic force. Only the narrow route of weak interaction satisfies all the conservation laws and connects the necessary decay products.

The peculiar feature of the strong nuclear interaction is that, since it so strongly connects many different combinations of baryons and mesons, any combination acts as if it were partially all the other combinations. A pi-meson, for example, can apparently exist momentarily as a baryon-antibaryon. Such temporary states can be found which will satisfy the conservation of electric charge, strangeness, spin, and everything except energy. Remember, however, that the Heisenberg uncertainty relationship allows energy fluctuations during very brief times. (13)

> Here $\Delta E \sim \hbar/\Delta t \sim 10^{-34}/\Delta t$, where ΔE is the smallest possible uncertainty in the energy if the measurement time is Δt. If a temporary state exists for a sufficiently short time, ΔE will be larger than the energy difference between the pi-meson state and the temporary state. We will then be unable to measure the energy sufficiently accurately to prove that energy is not conserved.

This strange possibility is very useful in explaining many particle events. For example, the neutral pi-meson has no electromagnetic interaction and yet decays to two photons, which are the agents of the electromagnetic field. According to our model, this can occur because about 1% of the time the neutral pi-meson is a proton-antiproton pair. These annihilate each other, giving off two photons. Energy is conserved in the final state.

Under these circumstances the distinction between source and agent of the strong nuclear interaction begins to fade. All that is important is the interaction itself which in some way gives rise to more or less stable resonances we call particles. Attempts are being made to build theories that will explain why particles exist with certain characteristics of mass, spin, etc., and not others. Although it may be possible to tie together the strongly interacting particles—the baryons and mesons—there is no indication that the leptons and photons can be explained in a similar way. Obviously many mysteries remain. What is the nature of the weak interaction agent? Why are there two types of electrons—heavy and light—each with its own neutrino? What other particles may exist? (13)

6 ◊ *The Tools of Exploration*

Accelerators

THE exploration in the microworld requires, paradoxically, giant machines and instruments. The reason is clear when you consider that most of the particles must be created if they are to be examined. This means that their mass-energy must be supplied.

In practice, protons or electrons are accelerated to very high energy and then shot at a target of a chosen element. At energies above 100 million electron volts the nature of the material in the target is usually unimportant. The bombarding particles effectively collide with the individual protons and neutrons in the nuclei. In an individual collision of a proton with a proton, some of the kinetic energy can be used to create the mass of the other particles (see diagram, page 102).

Whether we picture the particle as already existing within the proton or neutron and being knocked out during the collision or as being created during the process is a moot question. The point is that without sufficient energy being provided it cannot be freed.

The energy needed to create some particles, such as the antibaryons, is over 2 billion electron volts (2 BeV). The bombarding particle must have an energy much greater than that, and such energies can be provided only by very large accelerators. Brookhaven's alternating gradient synchrotron (AGS) is almost half a mile in circumference and can accelerate protons to 33 BeV.

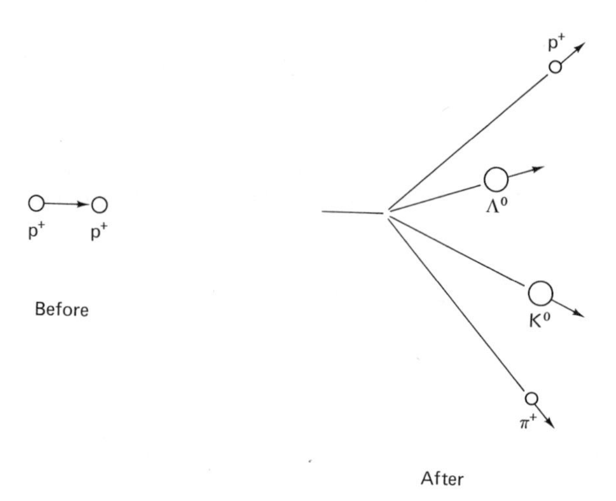

Collision of high-energy proton with stationary proton converts some kinetic energy into mass of π^+ and K^0.

The accelerators that produce energies over 1 BeV accelerate either protons or electrons, depending on the design of the machine. Most of these are circular machines, although there are several linear accelerators for electrons. In all these machines a group of particles is shot in at low energy and then is given repetitive electrical shoves as it passes through accelerating stations in the vacuum tank. Each group may contain as many as 10^{12} particles.

In the circular accelerators, such as the AGS or Berkeley's bevatron, the particles follow a racetrack path which leads them past the same accelerating stations over and over again. The magnetic field in the channel constrains them to follow the circular orbit and also provides the necessary focusing to keep the particles from oscillating too far from the center of the group. This magnetic field is weak at first since the particles have low momentum. As the particle energy increases, the magnetic field must increase, and also the accelerating frequency at the stations must become faster to keep pace with the more rapidly passing particles. After a million or so revolutions, the particles are directed out of the machine or onto a target. The whole cycle usually takes several seconds. The plan view of Brookhaven's 3-BeV cosmotron shows how the different components are arranged. (13)

BeV has been giving way to GeV, where G stands for giga-, from the Greek word for giant. The switch is a measure of the internationalism of high-energy physics, where these large numbers come into play. The pint may be "a pound the world around" but the billion is not one thousand millions, as in this book, the world around. It is one million millions in Britain, for example.

The internationalism characteristic of science in general is reinforced in the field of high-energy physics by the vast sums

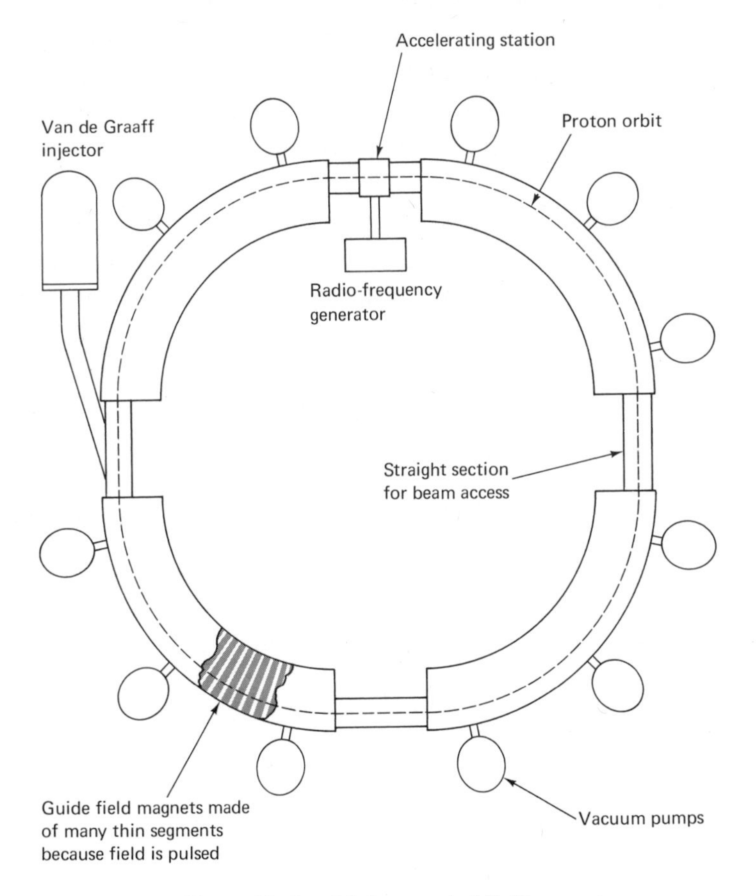

Brookhaven National Laboratory's 3-BeV cosmotron

which must be spent on accelerators in order to achieve energies measured in GeV. The expenditures cannot be met by a nation which is not a superpower, so such nations pool resources, as in CERN, the multination European nuclear research center based in Geneva, Switzerland.

Detectors

SINCE particles cannot be seen directly, we can only search for the debris left behind as they plow through matter. This is like estimating a tornado's winds by studying its path of destruction. A high-speed particle leaves a trail of ions behind it, the number per centimeter of path depending upon the particle species and its velocity. The operation of most radiation instruments depends upon these short-lived tracks of sundered atoms and the phenomena that take place when they recombine.

The primitive electroscope (see diagram) measures the cumulative ionization created by all particles passing through it. As ions and electrons in the filler gas drift toward the oppo-

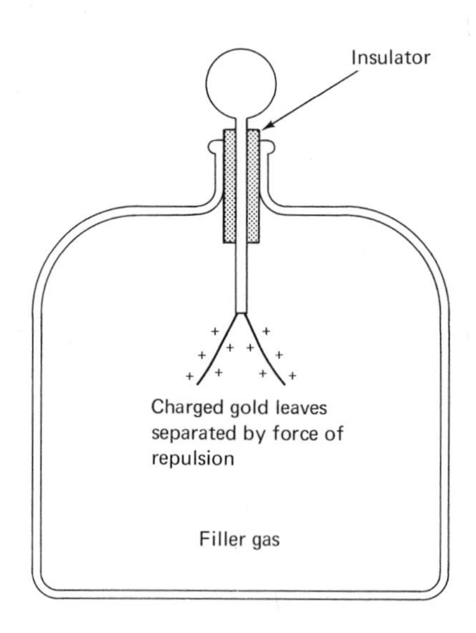

Classic gold-leaf electroscope

sitely charged gold leaves, the leaves are neutralized and subsequently collapse when the force of electrostatic repulsion weakens. The rate of collapse is proportional to the rate ions are created within the electroscope.

Modern ionization chambers also measure the rate of ionization within the chamber walls. In a volley-ball-sized ionization chamber, a central collector rod and surrounding walls supplant the electroscope's gold leaves. The chamber discharges as electrons and ions are attracted to the charged rod and walls. As the pivoted central rod edges toward its discharge position, it trips a switch that sends a pulse of electric current to a recording circuit and also recharges the chamber. Note that an ionization chamber does not reveal particle numbers, energies, directions, or species by itself.

The ionization chamber is frequently associated with a Geiger-Müller (GM) counter that totals the number of ionizing particles in the vicinity. Imagine the ionization chamber pulled out into the shape of a cylinder. When a charged particle deposits a trail of ions and electrons in a GM tube, the voltage applied between the central wire and the walls accelerates the electrons to such high speeds that they collide with neutral

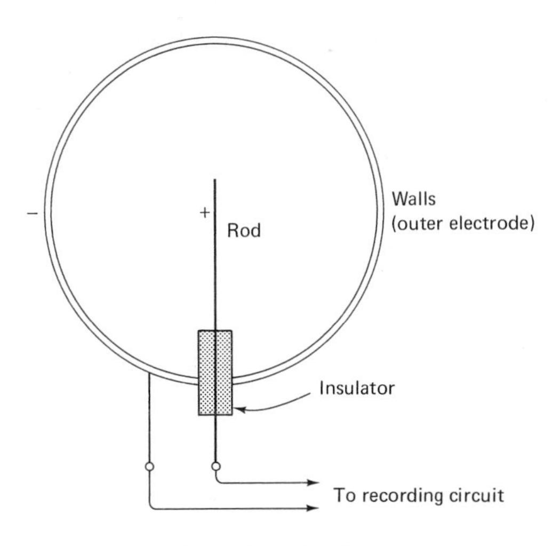

Ionization chamber

atoms in the filler gas and ionize them. These secondary electrons also go on to cause more ionization. On the process goes—avalanche fashion—until the whole tube is filled with ionized gas. The passage of a single particle thus discharges the GM tube and creates an electrical pulse that is fed to the recording circuits. A comparison of the total rate of energy deposited in the ionization chamber with the number of particles in the vicinity yields the average particle energy.

A scintillator crystal gives the experimenter a somewhat better estimate of particle energy. Each particle, passing through a crystal made, say, of thallium-activated cesium iodide, CsI(Tl), leaves the usual trail of ions and electrons. But when these ions and electrons recombine, a flash of light (a scintillation) is emitted that is proportional in intensity to the amount of energy deposited in the crystal. The flash is usually measured by an adjacent photomultiplier tube. If the crystal is thick enough to stop the incident particle, the magnitude of the flash is proportional to the particle's entire supply of energy.

Both scintillators and GM counters are often arranged linearly to help determine particle direction. In such a telescope, a particle is recorded only when it triggers the linear elements in what is termed a coincidence. Telescopes are thus sensitive to particle direction-of-arrival. The combination of thin scintilla-

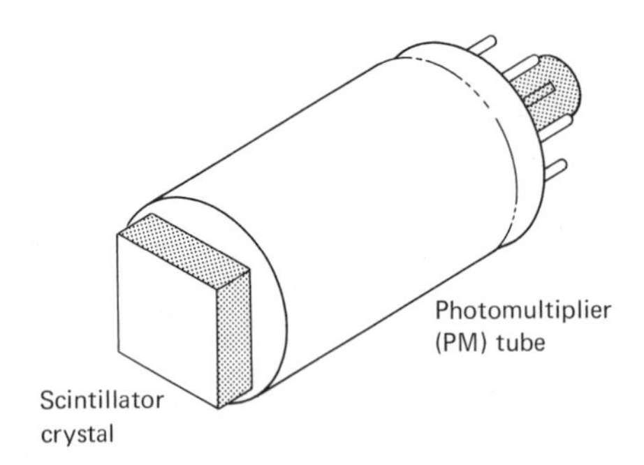

Scintillation counter

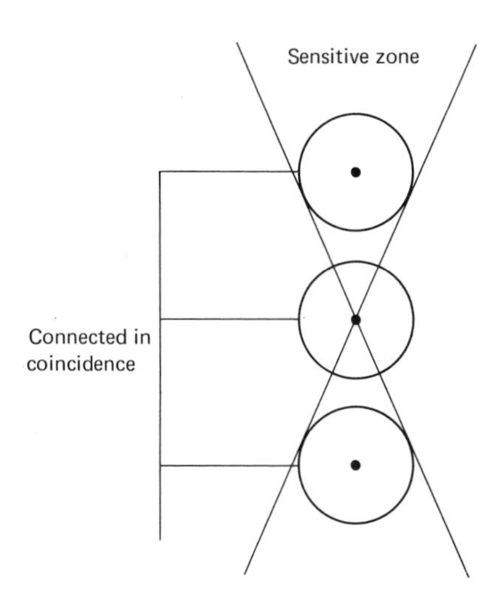

Geiger-Müller (GM) telescope

tors, intervening absorbers, and a thick scintillator can give a scientist all he needs to determine particle flux [number of particles per unit area per second], energy, direction of arrival and species. (36)

Scintillation counters can record the passage of a fast particle in less than one ten-millionth of a second. With the proper electronic circuitry, they can determine, with a precision of one-billionth of a second, whether or not two particles arrived at the same time.

The ions produced by a charged particle can also serve as seeds for the growth of much larger effects. In a photographic emulsion, for example, each silver atom freed from the silver iodide compound is the center on which a large grain or crystal of silver can grow during the photographic development process. The resulting trail of silver grains can be seen with the aid of a microscope.

In a cloud chamber (see diagram, page 108), a region of supersaturation is produced in a humid gas. This can be done

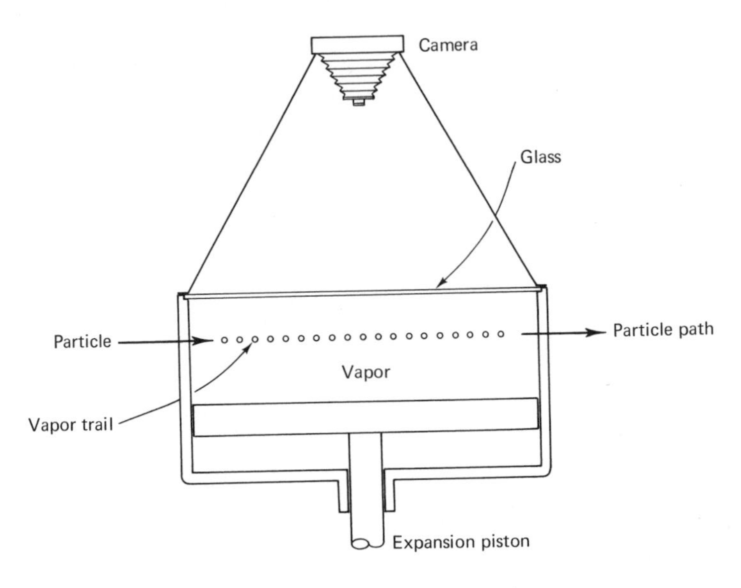

Cloud chamber

by a rapid expansion of the volume of the container to produce a rapid cooling of the gas, thus raising its humidity above 100%. Droplets form on the ion trails; each ion is the seed for a droplet, which grows to visible size in less than one thousandth of a second. The trails are recorded with a camera.

Operation of the bubble chamber is very similar to that of the cloud chamber. Liquid in a container is kept almost at its boiling point. A sudden reduction in pressure in the chamber sends the liquid over the boiling point for the lower pressure. Boiling begins, however, only on the ions left by particles passing just at that moment. A bright light flashed into the chamber at that time illuminates the bubble trails so that they can be seen and photographed. The main advantage of the bubble chamber for high-energy research is that a liquid is from a hundred to a thousand times as dense as the gas of a cloud chamber. Many more events can be seen in the liquid than in the same volume of gas.

Since 1961 many spark chambers have been developed for high-energy research. The structure of these can take many

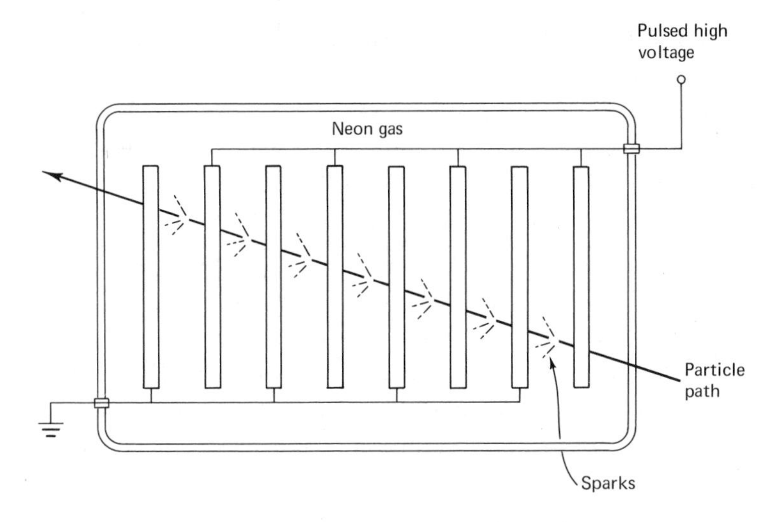

Spark chamber

forms, but usually numerous metal plates are spaced a few millimeters apart in a noble-gas atmosphere (see diagram). Upon command of other counters, the voltage of every other plate is raised by 10,000–30,000 volts. Any ion trail existing between the plates serves as the leader for a lightning stroke—a spark jumping from plate to plate. These are photographed by cameras looking between the plates. The whole process takes only one-millionth of a second, and the chief advantage of the spark chamber is this small response time. The apparatus can lie dormant while millions of particles go through, and then, when outside counters determine that the right combination has been produced, the spark chamber can capture the rare event. (See photograph section for an example of a spark-chamber picture).

Tracking the Particles

IN the photograph section of this book there are two pictures showing particle tracks as they were formed in various kinds of counting chambers (diagrams of such tracks appear on page

81). To interpret these photographs, we need to keep the following facts in mind.

The motion of charged particles can be directed by electrical or magnetic forces. Charged particles are repelled by electrical charges like their own and so move away from them; they are attracted by opposite charges and move toward them. A charged particle in motion is influenced by a magnetic field; that is, negatively charged particles will circle one way in a magnetic field and positively charged particles the opposite way. Charged particles can be accelerated as they pass through an attracting electrical field, acquiring energy proportional to the difference of potential [voltage] creating the field. This is what happens in an accelerator.

A charged particle (such as one from an accelerator) dissipates its energy as it passes through any material, including materials making up photographic emulsions and the gases or liquids in cloud chambers, bubble chambers, and spark chambers. The particle may pass close enough to an atom (of the medium it is traversing) to ionize it by knocking off one of the atom's electrons. This leaves the atom available to serve as a center for formation of a drop in a cloud chamber, as a boiling center in a bubble chamber, or as part of a pathway for an electric spark in a spark chamber. (Ionization, of course, affects only a tiny fraction of the atoms in the medium.)

Alternatively, the accelerated particle may either come very close to an atomic nucleus or hit it. If the energy of impact is greater than the forces holding the nucleus together, the target nucleus may break up into other particles—protons, neutrons, and various mesons—which may leave tracks, too. (Some of the energy also may emerge as electromagnetic radiation.)

The charged particle may be unstable, in which case it will decay into other particles, which then continue in motion through the chamber. Some of these may decay, too. Whatever happens will be recorded in the chamber. (Uncharged particles, such as neutrons, neutrinos, and lambda zeros, leave no trails; but when they strike something else or decay into charged components the new product will leave trails.)

To read a track-chamber photograph, then, we watch for ionization trails which suddenly change course or which branch into two or more trails. (Most of the parallel trails are caused by particles merely passing through, knocking off an occasional electron as they go by.) Each point of changed direction or branching indicates where a collision or a decay has taken place; the trails beyond that point are left by particles created in a collision (or by decay) or by particles altering their course because of it. Curved trails indicate the presence of a magnetic field in the track chamber. When an electron is removed from an atom, for example, the magnetic field winds the electron up in a tight little spiral.

Physicists determine the mass and energy of resulting particles by measuring the width, density, length, and direction of the trails, and from these they identify the particles.

The lefthand diagram on page 81 shows an example of pair production, or the conversion of energy to matter. A gamma ray (leaving no track) with energy greater than 1 million electron volts (entering the photograph from the left) passed close to the nucleus of an atom. It converted into an electron and a positron; since the gamma ray had no charge, its products also had to add up to zero—hence the creation of oppositely charged particles. Pair production is one of the most common processes by which a neutral high-energy photon produces charged particles.

As we study such track-chamber photographs, remember that we are interested in particles which make up the microstructure of matter; the trails are merely a means of studying them. (13)

7 ◊ *Cosmology and Astrophysics*

FROM the vantage point of 1974, the major scientific breakthroughs of this century have been relativity theory, quantum theory, and molecular biology. We considered the first briefly, the second more extensively, and discuss the third in some detail in part II. Another breakthrough may be imminent—in astrophysics and cosmology, the latter the study of the universe, the origin of the universe in particular. Space limitations preclude the in-depth treatment the subject warrants, but to omit it altogether would be almost immoral. The essential ideas are readily accessible, but rigor will have to be sacrificed, which means for example that Newtonian rather than relativistic physics will be employed.

It might be in order first to consider whether cosmology is a science. There is usually a sharp distinction in science between the relatively important "laws" and the relatively unimportant "initial conditions." For objects falling freely near the surface of the earth, for example, the law governing the height y of the object as a function of time t is given by

$$y = y_0 + v_0 t + \frac{1}{2} g t^2$$

where y_0 is the height at time $t = 0$ when we began to measure, v_0 is the initial velocity, and $g = 32$ feet per second per second is the acceleration due to the earth's gravitational attraction. The values of the initial conditions y_0 and v_0 may differ from one falling object to another, but the form of the law is always the same. In cosmology, however, there is only one object, the universe, and the initial conditions—creation itself—are no less important than the laws.

In science experiments are repeated. But how can we repeat the creation of the universe?

Let us now begin with the unremarkable remark that it is dark at night. A simple, natural description of the universe leads to the conclusion that it should be quite the contrary.

The Italian philosopher Giordano Bruno was burned at the stake in 1600 for his belief in an infinite, static universe made up of stars distributed uniformly throughout. His seemingly innocent belief has a startling consequence, which had nothing to do with his death, as it was not recognized until the nineteenth century.

Consider an infinite, static, homogenous volume containing many, many stars, n stars per unit volume on the average. (The smallest volume considered must contain many galaxies—clusters of stars, hundreds of billions in each cluster—if the universe is taken to be homogeneous.) The average star emits energy in the form of light, L units of light each second. The light radiated by a star spreads in all directions. Since the surface area of a sphere of radius r is $4\pi r^2$, the energy per second passing through any unit area of the sphere with the star at its center is $L/4\pi r^2$. If, out of the infinite volume, we focus our attention on a thin spherical shell with an inner radius r and thickness d—where d is very much less than r—the volume of the shell will be approximately the product of the inner surface area $4\pi r^2$ and the thickness d. The total number of stars in the shell will be n times the shell's volume, or $n \times 4\pi r^2 d$. Suppose an observer, located at the center of the shell, looks out in a cone that accepts a fraction f of all the light incident on him. f is small enough for all the light rays that reach the observer to be considered parallel to each other and perpendicular to a detector of unit area held by him. Our observer detects an amount of light per second equal to $(fn \times 4\pi r^2 d) \times (L/4\pi r^2)$. The first factor in parentheses represents the number of stars in the shell that are visible to our observer, the stars that are within the acceptance cone. The second factor is the light reaching the detector each second from a star in the acceptance cone at a distance anywhere from r to $r+d$ from the detector. On multiplying the two factors, we find that our observer detects an amount of light per second equal to $fnLd$. What is essential is that the distance r does not appear in the expression. For shells at a greater distance the contribution per star goes down, but the number of stars in a thickness d goes up in such a way that the two effects cancel each other out. This situation is

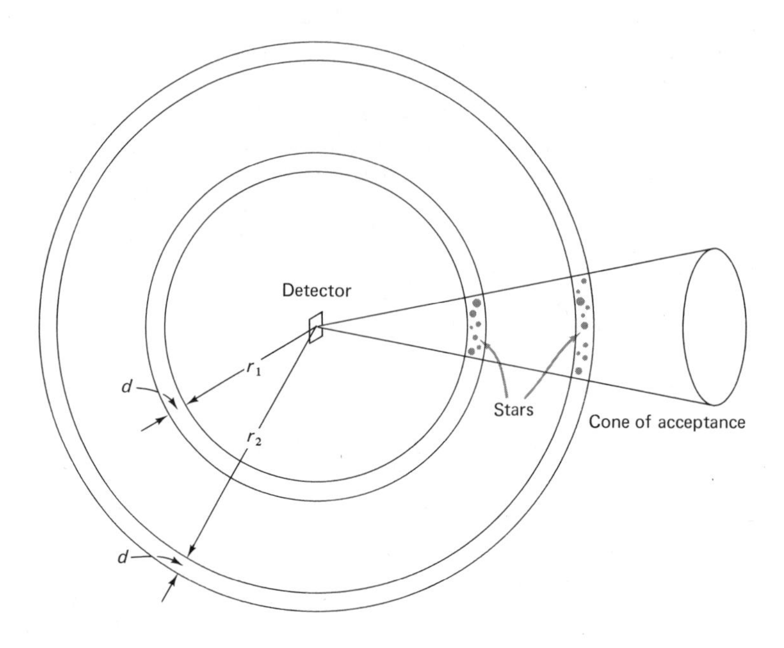

Light from stars reaching an observer

shown in the diagram, where the observer is at the center. Two shells are indicated, one at a distance r_1, the other at a distance r_2. If the second shell is twice as far away as the first ($r_2 = 2r_1$) the outer shell contains 2^2 or 4 times as many stars as the inner one—its volume is $4\pi r_2^2 d = 4\pi(2^2)r_1^2 d$—but each star in the outer shell that is within the acceptance cone contributes only one-fourth as much energy per second as a star in the inner shell in the acceptance cone. Thus each shell sends the same amount of light to the observer. And what is the final result? Since the number of shells that can be considered is infinite, and since each contributes the same amount of light, the light reaching the observer each second from the stars is infinite. Not only should the night not be dark, it should be infinitely bright.

However, we see that it is dark at night.

A weak point in the argument is the assumption that the stars are points of light in an infinite expanse. Stars are actually of considerable size, and light from a very distant star will almost certainly not reach our observer, for it will be blocked on the way by another star. This helps, but not nearly enough. A detailed analysis shows that our observer no longer receives an infinite amount of light—he receives only an intensity of light equal to that at the surface of a star! The surface of the sun, a run-of-the-mill star, is at a temperature of about 6000° C.

This is Olbers's paradox, named for the German astronomer Heinrich W. M. Olbers (1758–1840). (A subsidiary paradox is that it was probably named for the wrong man.) And this is the conclusion one is inexorably led to from Bruno's infinite, homogeneous, static, star-studded universe. Bruno would have been burned at a temperature very much higher than that at the stake had his conception of the universe been correct. It is not surprising that Olbers's paradox was shoved under the rug until the 1920s when it was pulled out, dusted off, and resolved—by the discovery that the universe is not static but is expanding.

A star at a distance r from us is racing away with a speed v proportional to r. This situation, called Hubble's law for the American astronomer Edwin P. Hubble (1889–1953) who first stated it, is written mathematically as

$$v = Hr$$

where H, the Hubble constant, has a value of about 10^{-10} year. (H may depend upon the time of measurement, and for this reason is sometimes called the Hubble parameter.)

Many of us are aware that the frequency, or pitch, of sound is altered when the source of the sound moves with respect to us. The whoooo of a train whistle is higher in pitch when the train approaches, and drops after the train passes. This is known as the Doppler effect. There is a similar Doppler effect for light waves, with color playing for light the role pitch plays for sound. The frequency of light from a star receding from us is less than the frequency of light from a stationary star. Its color is redder. Since the energy of light is proportional to its frequency—white hot is hotter, or more energetic, than red hot—the energy of light from a receding star is less than that from a stationary star, with the reduction greater the more distant the star. Detailed analysis shows that the temperature resulting from the light energy from these receding stars is neither infinite nor 6000° C, but far less. We can rest comfortably and coolly. Olbers's paradox is a paradox no more.

How are the velocities of distant galaxies measured? The light from stars in a distant galaxy is observed. Some of this light consists of sharply defined wavelengths corresponding to atomic processes occurring in the galaxy. Decrease in the frequency of the light from a receding source as a result of the

Doppler effect is associated with increase in wavelength, because for any wave the product of frequency and wavelength equals the velocity. The velocity of light in a vacuum is always constant, according to Einstein's theory of relativity. Suppose that instead of finding light with wavelengths of 4500, 5000, and 6000 wavelength units, which correspond to light from particular atoms believed to exist in the galaxy, we find light with wavelengths of 5400, 6000, and 7200 units. The ratio of each wavelength we find to the expected wavelength is 1.2. Doppler theory states that for velocities that are small compared to the velocity of light c the ratio of observed to expected wavelength is $1+v/c$. If $1+v/c=1.2$, then $v/c=0.2$; the velocity of recession is two-tenths the velocity of light.

Consider an extension of the idea of the Polish astronomer Nicholas Copernicus (1473–1543)—which should more justly be credited to Aristarchus of Samos (310–230 B.C.), but often an idea is named for the last man to argue for it—that the earth is not the center of our solar system: the cosmological principle, which states that all points and all directions in space are equivalent; no matter where you stand or where you look you will see on the average the same thing. (The region of space considered must be large enough to include many galaxies.) The cosmological principle is the penultimate in humility.

The steady state theory, introduced in 1948 by the Austrian-born astrophysicists Hermann Bondi and Thomas Gold, and separately by the British astrophysicist Fred Hoyle, is the extension of the cosmological principle that makes it the ultimate in humility. The theory states that what the observer sees not only is independent of where he looks from and where he looks, but of when he looks. There is nothing at all special about our location in space or about this era. This is called the perfect cosmological principle. As Bondi puts it, the cosmological principle says the universe has no geography, while the perfect cosmological principle says the universe has neither geography nor history.

This appears to contradict the fact that the universe is expanding. An observer in our galaxy at the small central circle should see the universe at two times, as shown in the diagram; dots represent galaxies and arrows represent the movement of the galaxies with respect to the observer. (Assume that he can see only up to a certain distance, denoted by the sphere.) At the later time, the galaxy farthest from him, which has the greatest velocity of recession, has disappeared from his sight. The number of galaxies per unit volume, the density, will decrease

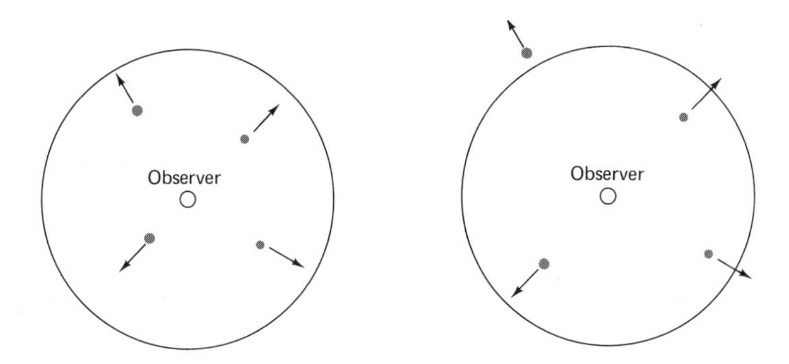

Galaxies (dots) seen by an observer at a certain time (*left*) and at a later time (*right*)

gradually with time. In the steady state theory, to keep the status quo, matter is assumed to be continuously created at such a rate as to just compensate for the diminution in density caused by expansion. The required rate of creation is so small that it cannot be detected experimentally. (The theory is also called the theory of continuous creation.)

The steady state theory offers some easy answers to questions that have long plagued man. "When was the universe created?" It has always existed. "When will it end?" Never. Billions of billions of years from now cosmologists may be arguing whether the steady state theory or the big bang theory is the correct one.

The perfect cosmological principle is the perfect solution for those who crave immortality. Unfortunately, it is probably wrong.

The "in" theory at the moment is the big bang. Its most recent major proponents were the Belgian astrophysicist Abbé Georges Lemaître (1894–1966) and Russian-American George Gamow (1904–1968). According to their theory, the universe started in a tremendous explosion involving an incredibly hot, dense gas of rudimentary matter. Much of the subsequent history of the universe was determined by what happened in the first second or two after creation. In attempting to reconcile the big bang with the cosmological principle there is a difficulty: if the universe began at one point at zero time, why are all points in space equivalent now?

To answer the question, consider a line as a model of the universe, with the big bang occurring at point 0 at time $t = 0$.

Particles will fly out from 0 with a range of velocities. Those with greater velocities will travel farther from 0 in a time T measured from the big bang, traveling a distance $x = vT$. If we let $H = 1/T$, we have $v = Hx$. (Note the resemblance to Hubble's law.) The equation applies not only for the one observer at 0 but for all observers. An observer astride any particle sees all other particles flying from him with velocities proportional to their distances from him. To understand this, consider two particles, one at $x_1 = v_1T$, the other at $x_2 = v_2T$, where distances x and velocities v are measured by an observer at 0. Let the distance and velocity of the second particle, as measured by an observer riding on the first particle, be X_2 and V_2. Then $X_2 = x_2 - x_1$, or, if we substitute the relationships above, $X_2 = (v_2 - v_1)T$. That is, the distance of the second particle (or of any other particle) from the first is proportional to its velocity with respect to the first. Thus, all observers, including the one at 0, see particles with velocities proportional to their distance away, and are, therefore, equivalent.

The argument is the same for a universe with more than one dimension but is somewhat more complicated. Take a balloon, paint some dots on it, and blow it up at a regular rate. Imagine yourself an observer stationed on one of the dots. The other dots are receding from you as the balloon expands. Analysis will show that the velocity of recession of a dot is proportional to its distance from your observation dot.

A measurement of the distances and velocities of galaxies determines T, which is none other than the age of the universe. Such measurements determine T to be about 13 billion years.

Astrophysicists, like other scientists, like to speculate on how things would be under different circumstances. Suppose certain forces that now play important roles did not exist.

If the strong interactions were turned off, matter would fly apart; if the electromagnetic interactions were turned off, chemical reactions would stop; and if gravity were turned off, we would float off the surface of the earth. (8)

What if the weak interactions were turned off?

The sun would stop shining and then the stars, one by one, would go out.

As we shall see it is just these weak interactions that help to produce the energy to keep the sun shining. The sun keeps its present size because the force of gravity, which tends to make it collapse, is balanced by the pressure produced by the heated particles in its interior. If these heat processes were turned off, then gravitation would cause the sun to shrink, and it then would heat up more due to the gravitational energy increase. Eventually it would burn itself out. This would take about 30 million years, but we would all have frozen solid, or would have been burned up in the original heating process long before!

Until the late nineteenth century, the gravitational collapse theory of solar radiation was believed to be the correct explanation of why the sun shone. The trouble began when the process of solar evolution was traced backward in time.

After Einstein's formula for the interconnection between mass and energy, $E = mc^2$, was revealed, it was widely conjectured that this must be the key to the sun's ability to give off so much radiation energy over such a comparatively long time. The problem was to devise some method for converting mass into energy that would work on the scale necessary to keep the sun shining.

In the 1930s the neutron and nuclear reactions, processes in which the nuclei are transformed into each other under suitable conditions, were discovered. In such reactions energy is ordinarily given off because the final nuclei are usually less massive than the initial nuclei. Because of the huge c^2 factor, a lot of energy is released. The problem of applying these ideas to the sun is twofold: (1) To find the right nuclear reactions that involve nuclei available in the sun. There is no point in invoking some reactions involving uranium, for example, since there is no uranium in the sun. (2) Defining the suitable conditions and making sure that the sun offers these conditions for any reaction that one has invented.

In a typical nuclear reaction one begins with two positively charged nuclei close to each other. (Positively charged since all the stable nuclei have protons in them.) The natural inclination of these nuclei is to repel each other since like charges repel.

However if they are pushed so close together that the strong, short-range, nuclear force or the even shorter ranged, weak force can take over, a nuclear reaction can occur.

On earth we accomplish this feat by bouncing one nucleus off another one at great energy in an accelerator, or by making the temperature of the nuclear amalgam hot enough so that in random collisions the nuclei bounce off each other frequently enough to be effective.

A good working temperature for the latter method is about 10 million degrees C. This is a rare temperature on earth, although it is produced artificially in atomic explosions, and perhaps in electron-proton plasmas that have been confined by magnetic fields and heated with electrical discharges. (See chapter 19.) However, it is a typical temperature for the interior of an average star like the sun. (Red giants are much cooler and white dwarfs are much hotter. Red giants are very young stars with low surface temperature and diameters many times that of the sun. White dwarfs are very old, bluish stars with high surface temperature and a mass close to that of the sun, but which can have a diameter as small as five times the diameter of the earth.)

As in any good cuisine the nuclear reactions that will cook depend very sensitively on the temperature of the star. There are two excellent reactions for the sun and similar stars. The one that dominates the resultant confection again depends in a crucial way on the temperature. The simplest such reaction was first suggested by the German scientist Carl Friedrich von Weizsäcker in 1937. It is a proton collision in which deuterium (heavy hydrogen) is made along with a positron

$$p + p \rightarrow D + e^{+} + \nu$$

It proceeds via the weak force and out comes a neutrino. The second reaction was proposed about the same time by the German-born American physicist Hans A. Bethe, and since it is really a series, or cycle of reactions, we give the series as

$$
\begin{aligned}
{}^{12}C + p &\rightarrow {}^{13}N + \gamma \\
{}^{13}N &\rightarrow {}^{13}C + e^{+} + \nu \\
{}^{13}C + p &\rightarrow {}^{14}N + \gamma \\
{}^{14}N + p &\rightarrow {}^{15}O + \gamma \\
{}^{15}O &\rightarrow {}^{15}N + e^{+} + \nu \\
{}^{15}N + p &\rightarrow {}^{12}C + {}^{4}He
\end{aligned}
$$

in which p is the proton, C is carbon, N is nitrogen, and O is oxygen. A remarkable feature of this reaction is that it begins and ends with carbon, and is known as the carbon cycle. No carbon is consumed and it acts here as a catalyst. In the cycle two neutrinos and three gamma rays are released.

In a given star both the Bethe and the von Weizsäcker reactions can take place simultaneously in principle. The theory shows that at low stellar temperatures von Weizsäcker dominates over Bethe and vice versa at high temperatures. (The crossover temperature between the two reactions is estimated to be about 13 million degrees.) Astrophysicists believe that the von Weizsäcker process is the dominant one in the sun. After deuterium is formed in the initial weak process

$$p + p \rightarrow D + e^{+} + \nu$$

we find some quite interesting results and an experimental prediction.

The newly formed D collides with another proton to produce a light isotope of helium

$$D + p \rightarrow {}^{3}He + \gamma$$

with the release of a photon. Now there are two possibilities. Two heliums can react according to the scheme

$${}^{3}He + {}^{3}He \rightarrow {}^{4}He + p + p$$

or, and this is the interesting case, beryllium can be formed via the process

$${}^{4}He + {}^{3}He \rightarrow {}^{7}Be + \gamma$$

This beryllium can now go into ordinary boron

$$^{7}Be + p \rightarrow {}^{8}B + \gamma$$

followed by

$$^{8}Be \rightarrow {}^{4}He + {}^{4}He + e^{+} + \nu$$

in which Be is beryllium and B is boron. The breakup of ^{8}B into two helium nuclei, a positron, and a neutrino is of special interest since this neutrino has a high energy, 10 MeV. This high energy enables the neutrino to trigger a ^{37}Cl to ^{37}Ar reaction.

The physicist Raymond Davis, Jr., has had an apparatus containing 100,000 gallons of perchloroethylene cleaning fluid nearly a mile underground in the Homestake gold mine at Lead, South Dakota. The astrophysical theory of neutrinos would suggest that Davis should have seen some two to seven events a day. But after a few years of observation, he hasn't seen any. It is still too early to say if this will require some profound change in our ideas about the sun, if there is some fluke in the experimental machinery, or if we have missed something in the weak interaction theory.

It will be of special interest to detect these neutrinos since they come directly from the interior of the sun, whereas sunlight comes from the surface where the temperature is relatively low: some thousands of degrees centigrade. A photon that is made deep inside the sun suffers innumerable collisions on its trip to the solar surface. The neutrino, since it interacts rarely, emerges from the depths just as it was made. (It has been estimated that it takes about a million years for a typical photon created in the sun's center to wander to the surface while a neutrino makes the trip in about 3 seconds.) (8)

Though no solar neutrinos—let alone neutrinos from other stars—have been detected, the work has been given the name neutrino astronomy. Astrophysicists are much concerned over the negative outcome of Davis's experiment.

From the nuclear reaction equations above you can see how different elements can be formed from primordial hydrogen. Part II describes how life can be formed from the different elements.

PART II

THE ATOM AND THE BODY

FOR years newspapers have carried accounts of arguments between those who favor testing nuclear weapons and those who maintain that more harm than good can come from such tests. Distinguished scientists can be found on both sides of the argument. Whom is one to believe?

The fact that scientists are so divided shows that not enough is known about the effects of radiation. And it would seem that, as long as not enough is known, for the sake of man it is better to err on the side of caution. But let us see what is known.

In part I, the atom and radiation were treated in isolation. Now we consider the atoms that unite to compose living things, and the interaction of radiation with these atoms.

The laws governing the molecules of living organisms are no different from the laws of chemistry, which are none other than the quantum mechanical laws of physics. There are no new laws in this part. The laws of molecular biology are the basic laws of Newton as generalized in quantum mechanics. This is not to diminish the significance of molecular biology, the field of some of the most important scientific advances of the last two decades.

Also, there is no evidence that the molecules of living things contain anything other than the ordinary constituents of inert matter.

Obviously, one cannot learn everything about molecular biology from a few chapters. But they should allow the reader to savor what has been brewing in the field and will stimulate his appetite for further reading. Recommended are *Biology and the Future of Man,* a report prepared by the National Academy of Sciences and edited by Philip Handler (Oxford University Press, 1970), and two books by the American biologist James D.

Watson, *Molecular Biology of the Gene* (W. A. Benjamin, 1965) and *The Double Helix* (Atheneum, 1968). The latter, an autobiographical account of Watson's work with Sir Francis Crick of Great Britain that led to their finding the structure of the DNA molecule, depicts science at its most crass—and, unfortunately, entertaining—level. It is charitable to concede that those operating in close proximity to the radiance of the Nobel prize sometimes become blinded by it. If you contemplate buying the book, then, make it the paperback edition, so that the financial rewards, at least, will be kept down. *Molecular Biology of the Gene* may be purchased in hardcover, for it may well become a classic.

If this part contains few new concepts, it makes up for it in terminology. We will meet sentences like "The r-RNA and the proteins to which it is firmly bound form the ribosomes, the RNA-rich microsomes that are attached to the endoplasmic reticulum." But we need not be frightened by such words. They are simply biological jargon, and need be no more difficult to absorb than the rules of cricket would be for a baseball fan.

Chapter 8 discusses the genetic effects of radiation without recourse, beyond a necessary minimum, to the biology of living organisms. Chapter 9 treats the details of the biology and chemistry of living tissue. Recently, some doubt has been cast on the inheritance scheme presented in this chapter. But since the resolution of the doubts may take some time, we will present the material as it was before the doubts were cast. Besides, a knowledge of what is presented here may well be necessary to understand what may replace it.

Chapter 10, on the mechanism of interaction of radiation with tissue, treats some concepts presented in chapter 8, on genetics, in greater detail. The reader will see that radiation isn't all bad. With proper control it helps more than it harms; it enables us not only to learn about living tissue but to cure disease.

8 ◊ The Genetic Effects of Radiation

The Machinery of Inheritance

THERE is nothing new under the sun, says the Bible. Nor is the sun itself new, we might add. As long as life has existed on earth, it has been exposed to radiation from the sun, so that life and radiation are old acquaintances and have learned to live together.

We are accustomed to looking upon sunlight as something good, useful, and desirable, and certainly we could not live long without it. The energy of sunlight warms the earth, produces the winds that tend to equalize earth's temperatures, evaporates the oceans and produces rain and fresh water. Most important of all, it supplies what is needed for green plants to convert carbon dioxide and water into food and oxygen, making it possible for all animal life (including ourselves) to live.

Yet sunlight has its dangers, too. Lizards avoid the direct rays of the noonday sun on the desert, and we ourselves take precautions against sunburn and sunstroke.

The same division into good and bad is to be found in connection with other forms of radiation—forms of which mankind has only recently become aware. Such radiations, produced by radioactivity in the soil and reaching us from outer space, have also been with us from the beginning of time. They are more energetic than sunlight, however, and can do more damage, and because our senses do not detect them, we have not learned to take precautions against them.

To be sure, energetic radiation is present in nature in only very small amounts and is not, therefore, much of a danger. Man, however, has the capacity of imitating nature. Long ago in dim prehistory, for instance, he learned to manufacture a kind of sunlight by setting wood and other fuels on fire. This involved a new kind of good and bad. A whole new technology became possible, on the one hand, and, on the other, the chance of death by burning was also possible. The good in this case far outweighs the evil.

In this century man has learned to produce energetic radiation in concentrations far surpassing those we usually encounter in nature. Again a new technology is resulting and again there is the possibility of death.

The balance in this second instance is less certainly in favor of the good over the evil. To shift the balance clearly in favor of the good, it is necessary for mankind to learn as much as possible about the new dangers in order that we might minimize them and most effectively guard against them.

To see the nature of the danger, let us begin by considering living tissue itself—the living tissue that must withstand the radiation and that can be damaged by it.

The average human adult consists of about 50 trillion cells—50 trillion microscopic, more or less self-contained, blobs of life. He begins life, however, as a single cell, the fertilized ovum.

After the fertilized ovum is formed, it divides and becomes two cells. Each daughter cell divides to produce a total of four cells, and each of those divides and so on.

There is a high degree of order and direction to those divisions. When a human fertilized ovum completes its divisions an adult human being is the inevitable result. The fertilized ovum of a giraffe will produce a giraffe, that of a fruit fly will produce a fruit fly, and so on. There are no mistakes, so it is quite clear that the fertilized ovum must carry "instructions" that guide its development in the appropriate direction.

These instructions are contained in the cell's chromosomes, tiny structures that appear most clearly (like stubby bits of tangled spaghetti) when the cell is in the actual process of division. Each species has some characteristic number of chromosomes in its cells, and these chromosomes can be considered in pairs. Human cells, for instance, contain twenty-three pairs of chromosomes—forty-six in all.

When a cell is undergoing division (mitosis), the number of chromosomes is temporarily doubled, as each chromosome brings about the formation of a replica of itself. (This process is called replication.) As the cell divides, the chromosomes are evenly shared by the new cells in such a way that if a particular chromosome goes into one daughter cell, its replica goes into the other. In the end, each cell has a complete set of pairs of chromosomes; and the set in each cell is identical with the set in the original cell before division.

In this way, the fundamental instructions that determine the characteristics of a cell are passed on to each new cell. Ideally, all the trillions of cells in a particular human being have identical sets of instructions.

Each cell is a tiny chemical factory in which several thousand different kinds of chemical changes are constantly taking place among the numerous sorts of molecules that move about in its fluid or that are pinned to its solid structures. These chemical changes are guided and controlled by the existence of as many thousands of different enzymes within the cell.

Enzymes possess large molecules built up of some twenty different, but chemically related, units called amino acids. A particular enzyme molecule may contain a single amino acid of one type, five of another, several dozen of still another and so on. All the units are strung together in some specific pattern in one long chain, or in a small number of closely connected chains.

Every different pattern of amino acids forms a molecule with its own set of properties, and there are an enormous number of patterns possible. In an enzyme molecule made up of five hundred amino acids, the number of possible patterns is unimaginable.

Every cell has the capacity of choosing among this unimaginable number of possible patterns and selecting those characteristic of itself. It therefore ends with a complement of specific enzymes that guide its own chemical changes and, consequently, its properties and its behavior. The instructions that enable a fertilized ovum to develop in the proper manner are essentially instructions for choosing a particular set of enzyme patterns out of all those possible.

The differences in the enzyme-guided behavior of the cells making up different species show themselves in differences in body structure. We cannot completely follow the long and intricate chain of cause-and-effect that leads from one set of enzymes to the long neck of a giraffe and from another set of enzymes to the large brain of a man, but we are sure that the chain is there. Even within a species, different individuals will have slight distinctions among their sets of enzymes and this accounts for the fact that no two human beings are exactly alike (leaving identical twins out of consideration).

Each chromosome can be considered as being composed of small sections called genes, usually pictured as being strung along the length of the chromosome. Each gene is considered to be responsible for the formation of a chain of amino acids in a fixed pattern. The formation is guided by the details of the gene's own structure (which are the instructions earlier referred to). This gene structure, which can be translated into an enzyme's structure, is now called the genetic code.

If a particular enzyme (or group of enzymes) is, for any reason, formed imperfectly or not at all, this may show up as some visible abnormality of the body—an inability to see color, for instance, or the possession of two joints in each finger rather than three. It is much easier to observe physical differences than some delicate change in the enzyme pattern of the cells. Genes are therefore usually referred to by the body change they bring about, and one can, for instance, speak of a gene for color blindness.

A gene may exist in two or more varieties, each producing a slightly different enzyme, a situation that is reflected, in turn, in slight changes in body characteristics. Thus, there are genes governing eye color, one of which is sufficiently important to be considered a gene for blue eyes and another a gene for brown eyes. One or the other, but not both, will be found in a specific place on a specific chromosome.

The two chromosomes of a particular pair govern identical sets of characteristics. Both, for instance, will have a place for genes governing eye color. If we consider only the most important of the varieties involved, those on each chromosome of the pair may be identical; both may be for blue eyes or both may be for brown eyes. In that case, the individual is homozygous for that characteristic and may be referred to as a homozygote. The chromosomes of the pair may carry different varieties: a gene for blue eyes on one chromosome and one for brown eyes on the other. The individual is then heterozygous for that characteristic and may be referred to as a heterozygote. Naturally, particular individuals may be homozygous for some types of characteristics and heterozygous for others.

When an individual is heterozygous for a particular characteristic, it frequently happens that he shows the effect associated with only one of the gene varieties. If he possesses both a gene for brown eyes and one for blue eyes, his eyes are just as brown as though he had carried two genes for brown eyes. The gene for brown eyes is dominant in this case while the gene for blue eyes is recessive.

How does the fertilized ovum obtain its particular set of chromosomes in the first place?

Each adult possesses gonads in which sex cells are formed. In the male, sperm cells are formed in the testes; in the female, egg cells are formed in the ovaries.

In the formation of the sperm cells and egg cells there is a key step—meiosis—a cell division in which the chromosomes group into pairs and are then apportioned between the daughter cells, one of each pair to each cell. Such a division, unaccompanied by replication, means that in place of the usual twenty-three pairs of chromosomes in each other cell, each sex cell has twenty-three individual chromosomes, a half set, so to speak.

In the process of fertilization, a sperm cell from the father enters and merges with an egg cell from the mother. The fertilized ovum that results now has a full set of twenty-three pairs of chromosomes, but of each pair, one comes from the father and one from the mother.

In this way, each newborn child is a true individual, with its characteristics based on a random reshuffling of chromosomes. In forming the sex cells, the chromosome pairs can separate in either fashion (*a* into cell 1 and *b* into cell 2, or vice versa). If each of twenty-three pairs does this randomly, nearly 10 million different combinations of chromosomes are possible in the sex cells of a single individual.

Furthermore, one can't predict which chromosome combination in the sperm cell will end up in combination with which in the egg cell, so that by this reasoning, a single married couple could produce children with any of 100 trillion possible chromosome combinations.

It is this that begins to explain the endless variety among living beings, even within a particular species.

It only begins to explain it, because there are other sources of difference, too. A chromosome is capable of exchanging pieces with its pair, producing chromosomes with a brand-new pattern of gene varieties. Before such a crossover, one chromosome may have carried a gene for blue eyes and one for wavy hair, while the other chromosome may have carried a gene for brown eyes and one for straight hair. After the crossover, one would carry genes for blue eyes and straight hair, the other for brown eyes and wavy hair.

Mutations

SHIFTS in chromosome combinations, with or without cross-overs, can produce unique organisms with characteristics not quite like any organism that appeared in the past nor likely to appear in the reasonable future. They may even produce novelties in individual characteristics since genes can affect one another, and a gene surrounded by unusual neighbors can produce unexpected effects.

Matters can go further still, however, in the direction of novelty. It is possible for chromosomes to undergo more serious changes, either structural or chemical, so that entirely new characteristics are produced that might not otherwise exist. Such changes are called mutations.

We must be careful how we use this term. A child may possess some characteristics not present in either parent through the mere shuffling of chromosomes and not through mutation.

Suppose, for instance, that a man is heterozygous to eye color, carrying one gene for brown eyes and one for blue eyes. His eyes would, of course, be brown since the gene for brown eyes is dominant over that for blue. Half the sperm cells he produces would carry a single gene for brown eyes in its half set of chromosomes. The other half would carry a single gene for blue eyes. If his wife were similarly heterozygous (and therefore also had brown eyes), half her egg cells would carry the gene for brown eyes and half the gene for blue.

It might follow in this marriage, then, that a sperm carrying the gene for blue eyes might fertilize an egg carrying the gene for blue eyes. The child would then be homozygous, with two genes for blue eyes, and he would definitely be blue-eyed. In this way, two brown-eyed parents might have a blue-eyed child and this would not be a mutation. If the parents' ancestry were traced further back, blue-eyed individuals would undoubtedly be found on both sides of the family tree.

If, however, there is no record of, say, anything but normal color vision in a child's ancestry, and he is born color-blind, that could be assumed to be the result of a mutation. Such a mutation could then be passed on by the normal modes of inheritance and a certain proportion of the child's eventual descendants would be color-blind.

A mutation may be associated with changes in chromosome structure sufficiently drastic to be visible under the microscope. Such chromosome mutations can arise in several ways. Chromosomes may undergo replication without the cell itself dividing. In that way, cells can develop with two, three, or four times the normal complement of chromosomes, and organisms made up of cells displaying such polyploidy can be markedly different from the norm. This situation is found chiefly among plants and among some groups of invertebrates. It does not usually occur in mammals, and when it does it leads to quick death.

Less extreme changes take place, too, as when a particular chromosome breaks and fails to reunite, or when several break and then reunite incorrectly. Under such conditions, the mechanism by which chromosomes are distributed among the daughter cells is not likely to work correctly. Sex cells may then be produced with a piece of chromosome (or a whole one) missing, or with an extra piece (or whole chromosome) present.

In 1959, such a situation was found to exist in the case of persons suffering from a long-known disease called Down's syndrome. (This is more commonly known as Mongolism or Mongolian idiocy though it has nothing to do with the Mongolian people.) Each person so afflicted has forty-seven chromosomes in place of the normal forty-six. It turned out that the twenty-first pair of chromosomes (using a convention whereby the chromosome pairs are numbered in order of decreasing size) consists of three individuals rather than two. The existence of this chromosome abnormality clearly demonstrated what had previously been strongly suspected—that Down's syndrome originates as a mutation and is inborn.

Most mutations, however, are not associated with any noticeable change in chromosome structure. There are, instead, more subtle changes in the chemical structure of the genes that make up the chromosome. Then we have gene mutations.

The process by which a gene produces its own replica is complicated and, while it rarely goes wrong, it does misfire on occasion. Then, too, even when a gene molecule is replicated perfectly, it may undergo change afterward through the action upon it of some chemical or other environmental influence. In either case, a new variety of a particular gene is produced and, if present in a sex cell, it may be passed on to descendants through an indefinite number of generations.

Of course, chromosome or gene mutations may take place in ordinary cells rather than in sex cells. Such changes in ordinary cells are somatic mutations. When mutated body cells divide, new cells with changed characteristics are produced. These changes may be trivial, or they may be serious. It is often suggested, for instance, that cancer may result from a somatic mutation in which certain cells lose the capacity to regulate their growth properly. Since somatic mutations do not involve the sex cells, they are confined to the individual and are not passed on to the offspring.

Mutations that take place in the ordinary course of nature, without man's interference, are spontaneous mutations. Most of these arise out of the very nature of the complicated mechanism of gene replication. Copies of genes are formed out of a large number of small units that must be lined up in just the right pattern to form one particular gene and no other.

Ideally, matters are so arranged within the cell that the necessary changes giving rise to the desired pattern are just those that have a maximum probability. Other changes are less likely to happen but are not absolutely excluded. Sometimes through the accidental jostling of molecules a wrong turn may be taken, and the result is a spontaneous mutation.

We might consider a mutation to be either good or bad in the sense that any change that helps a creature live more easily and comfortably is good and that the reverse is bad.

It seems reasonable that random changes in the gene pattern are almost sure to be bad. Consider that any creature, including man, is the product of millions of years of evolution. In every generation those individuals with a gene pattern that fit them better for their environment won out over those with less effective patterns—won out in the race for food, for mates, and for safety. The more fit had more offspring and crowded out the less fit.

By now, then, the set of genes with which we are normally equipped is the end product of long ages of such natural selection. A random change cannot be expected to improve it any more than random changes would improve any very complex, intricate, and delicate structure.

Yet over the aeons, creatures have indeed changed, largely through the effects of mutation. If mutations are almost always for the worse, how can one explain that evolution seems to progress toward the better and that out of a primitive form as simple as an amoeba, for instance, there eventually emerged man?

In the first place, environment is not fixed. Climate changes, conditions change, the food supply may change, the nature of living enemies may change. A gene pattern that is very useful under one set of conditions may be less useful under another.

Suppose, for instance, that man had lived in tropical areas for thousands of years and had developed a heavily pigmented skin as a protection against sunburn. Any child who, through a mutation, found himself incapable of forming much pigment, would be at a severe disadvantage in the outdoor activities engaged in by his tribe. He would not do well and such a mutated gene would never establish itself for long.

If a number of these men migrated to northern Europe, however, children with dark skin would absorb insufficient sunlight during the long winter when the sun was low in the sky, and visible for brief periods only. Dark-skinned children would, under such conditions, tend to suffer from rickets.

Mutant children with pale skin would absorb more of what weak sunlight there was and would suffer less. There would be little danger of sunburn so there would be no penalty counteracting this new advantage of pale skins. It would be the dark-skinned people who would tend to die out. In the end, you would have dark skins in Africa and pale skins in Scandinavia, and both would be fit.

In the same way, any child born into a primitive hunting society who found himself with a mutated gene that brought about nearsightedness would be at a distinct disadvantage. In a modern technological society, however, nearsighted individuals, doing more poorly at outdoor games, are often driven into quieter activities that involve reading, thinking, and studying. This may lead to a career as a scientist, scholar, or professional man, categories that are valuable in such a society and are encouraged. Nearsightedness would therefore spread more generally through civilized societies than through primitive ones.

Then, too, a gene may be advantageous when it occurs in low numbers and disadvantageous when it occurs in high numbers. Suppose there were a gene among humans that so affected the personality as to make it difficult for a human being to endure crowded conditions. Such individuals would make good explorers, farmers, and herdsmen, but poor city dwellers. Even in our modern urbanized society, such a gene in moderate concentration would be good, since we still need our outdoorsmen. In high concentration, it would be bad, for then the existence of areas of high population density (on which our society now seems to depend) might become impossible.

In any species, then, each gene exists in a number of varieties upon which an absolute good or bad cannot be unequivocally stamped. These varieties make up the gene pool, and it is this gene pool that makes evolution possible.

A species with an invariable set of genes could not change to suit altered conditions. Even a slight shift in the nature of the environment might suffice to wipe it out.

The possession of a gene pool lends flexibility, however. As conditions change, one combination of varieties might gain over another and this, in turn, might produce changes in body characteristics that would then further alter the relative goodness or badness of certain gene patterns.

Thus, over the past million years, for example, the human brain has, through mutations and appropriate shifts in emphasis within the gene pool, increased notably in size.

Some gene mutations produce characteristics so undesirable that it is difficult to imagine any reasonable change in environmental conditions that would make them beneficial. There are mutations that lead to the nondevelopment of hands and feet, to the production of blood that will not clot, to serious malformations of essential organs, and so on. Such mutations are unqualifiedly bad.

The badness may be so severe that a fertilized ovum may be incapable of development; or, if it develops, the fetus miscarries or the child is stillborn; or, if the child is born alive, it dies before it matures so that it can never have children of its own. Any mutation that brings about death before the gene producing it can be passed on to another generation is a lethal mutation.

A gene governing a lethal characteristic may be dominant. It will then kill even though the corresponding gene on the other chromosome of the pair is normal. Under such conditions, the lethal gene is removed in the same generation in which it is formed.

The lethal gene may, on the other hand, be recessive. Its effect is then not evident if the gene it is paired with is normal. The normal gene carries on for both.

When this is the case, the lethal gene will remain in existence and will, every once in a while, make itself evident. If two people, each serving as a carrier for such a gene, have children, a sperm cell carrying a lethal may fertilize an egg cell carrying the same type of lethal, with sad results.

Every species, including man, includes individuals who

carry undesirable genes. These undesirable genes may be passed along for generations, even if dominant, before natural selection culls them out. The more seriously undesirable they are, the more quickly they are removed, but even outright lethal genes will be included among the chromosomes from generation to generation provided they are recessive. These deleterious genes make up the genetic load.

The only way to avoid a genetic load is to have no mutations and therefore no gene pool. The gene pool is necessary for the flexibility that will allow a species to survive and evolve over the aeons and the genetic load is the price that must be paid for that. Generally, the capacity for a species to reproduce itself is sufficiently high to make up, quite easily, the numbers lost through the combination of deleterious genes.

The size of a genetic load depends on two factors: the rate at which a deleterious gene is produced through mutation, and the rate at which it is removed by natural selection. When the rate of removal equals the rate of production, a condition of genetic equilibrium is reached and the level of occurrence of that gene then remains stable over the generations.

Even though deleterious genes are removed relatively rapidly, if dominant, and lethal genes are removed in the same generation in which they are formed, a new crop of deleterious genes will appear by mutation with every succeeding generation. The equilibrium level for such dominant deleterious genes is relatively low, however.

Deleterious genes that are recessive are removed much more slowly. Those persons with two such genes, who alone show the bad effects, are like the visible portion of an iceberg and represent only a small part of the whole. The heterozygotes, or carriers, who possess a single gene of this sort, and who live out normal lives, keep that gene in being. If people in a particular population marry randomly and if one out of a million is born homozygous for a certain deleterious recessive gene (and dies of it), one out of five hundred is heterozygous for that same gene, shows no ill effects, and is capable of passing it on.

It may be that the heterozygote is not quite normal but does show some ill effects—not enough to incommode him seriously, perhaps, but enough to lower his chances slightly for mating and bearing children. In that case, the equilibrium level for that gene will be lower than it would otherwise be.

It may also be that the heterozygote experiences an actual advantage over the normal individual under some conditions. There is a recessive gene, for instance, that produces a serious disease called sickle-cell anemia. People possessing two such genes usually die young. A heterozygote possessing only one of these genes is not seriously affected and has red blood cells that are, apparently, less appetizing to malaria parasites. The heterozygote therefore experiences a positive advantage if he lives in a region where the incidence of certain kinds of malaria is high. The equilibrium level of the sickle-cell anemia gene can, in other words, be higher in malerial regions than elsewhere.

Here is one subject area in which additional research is urgently needed. It may be that the usefulness of a single deleterious gene is greater than we may suspect in many cases, and that there are greater advantages to heterozygousness than we know. This may be the basis of what is sometimes called hybrid vigor. In a world in which human beings are more mobile than they have ever been in history and in which intercultural marriages are increasingly common, information on this point is particularly important.

It is easier to observe the removal of genes through death or through failure to reproduce than to observe their production through mutation. It is particularly difficult to study their production in human beings, since men have comparatively long lifetimes and few children, and since their mating habits cannot well be controlled.

For this reason, geneticists have experimented with species much simpler than man—smaller organisms that are short-lived, produce many offspring, and that can be penned up and allowed to mate only under fixed conditions. Such creatures may have fewer chromosomes than man does and the sites of mutation are more easily pinned down.

An important assumption made in such experiments is that the machinery of inheritance and mutation is essentially the same in all creatures and that therefore knowledge gained from very simple species (even from bacteria) is applicable to man. There is overwhelming evidence to indicate that this is true in general, although there are specific instances where it is not completely true and scientists must tread softly while drawing conclusions.

The animals most commonly used in studies of genetics and mutations are certain species of fruit flies, called *Drosophila*. The American geneticist Hermann J. Muller (1890–1967) devised techniques whereby he could study the occurrence of lethal mutations anywhere along one of the four pairs of chromosomes possessed by *Drosophila*.

A lethal gene, he found, might well be produced somewhere along the length of a particular chromosome once out of every two hundred times that chromosome underwent replication. This means that out of every two hundred sex cells produced by *Drosophila*, one would contain a lethal gene somewhere along the length of that chromosome.

That particular chromosome, however, contained at least five hundred genes capable of undergoing a lethal mutation. If each of those genes is equally likely to undergo such a mutation, then the chance that any one particular gene is lethal is 1 out of 200×500, or 1 out of 100,000.

This is a typical mutation rate for a gene in higher organisms generally, as far as geneticists can tell (though the rates are lower among bacteria and viruses). Naturally, a chance for mutation takes place every time a new individual is born. Fruit flies have many more offspring per year than human beings, since their generations are shorter and they produce more young at a time. For that reason, though the mutation rate may be the same in fruit flies as in man, many more actual mutations are produced per unit time in fruit flies than in men.

This does not mean that the situation may be ignored in the case of man. Suppose the rate for production of a particular deleterious gene in man is 1 out of 100,000. It is estimated that a human being has at least ten thousand different genes, and therefore the chance that at least one of the genes in a sex cell is deleterious is 10,000 out of 100,000 or 1 out of 10.

Furthermore, it is estimated that the number of gene mutations that are weakly deleterious are four times as numerous as those that are strongly deleterious or lethal. The chances that at least one gene in a sex cell is at least weakly deleterious then would be $4 + 1$ out of 10, or 1 out of 2.

Naturally, these deleterious genes are not necessarily spread out evenly among human beings with one to a sex cell. Some sex cells will be carrying more than one, thus increasing the number that may be expected to carry none at all. Even so, it is supposed that very nearly half the sex cells produced by humanity carry at least one deleterious gene.

Even though only half the sex cells are free of deleterious genes, it is still possible to produce a satisfactory new generation of men. Yet one can see that the genetic load is quite heavy and that anything that would tend to increase it would certainly be undesirable, and perhaps even dangerous.

We tend to increase the genetic load by reducing the rate at which deleterious genes are removed, that is, by taking care of the sick and retarded, and by trying to prevent discomfort and death at all levels.

There is, however, no humane alternative to this. What's more, it is, by and large, only those with slightly deleterious genes who are preserved genetically. It is those persons with nearsightedness, with diabetes, and so on, who, with the aid of glasses, insulin, or other props, can go on to live normal lives and have children in the usual numbers. Those with strongly deleterious genes either die despite all that can be done for them even today or, at the least, do not have a chance to have many children.

The danger of an increase in the genetic load rests more

heavily, then, at the other end—at measures that (usually inadvertently or unintentionally) increase the rate of production of mutant genes. It is to this matter we will now turn.

Our modern technological civilization exposes mankind to two general types of genetic dangers unknown earlier: synthetic chemicals (or unprecedentedly high concentrations of natural ones) absent in earlier eras, and intensities of energetic radiation equally unknown or unprecedented.

Chemicals can interfere with the process of replication by offering alternate pathways with which the cellular machinery is not prepared to cope. In general, however, it is only those cells in direct contact with the chemicals that are so affected, such as the skin, the intestinal linings, the lungs, and the liver (which is active in altering and getting rid of foreign chemicals). These may undergo somatic mutations, and an increased incidence of cancer in those tissues is among the drastic results of exposure to certain chemicals.

Such chemicals are not, however, likely to come in contact with the gonads where the sex cells are produced. While individual persons may be threatened by the manner in which the environment is being permeated with novel chemicals, the next generation is not affected in advance.

Radiation is another matter. In its broadest sense, radiation is any phenomenon spreading out from some source in all directions. Physically, such radiation may consist of waves or of particles. Of the wave forms the two best known are sound and electromagnetic radiations.

Sound carries very low concentrations of energy. This energy is absorbed by living tissue and converted into heat. Heat in itself can increase the mutation rate but the effect is a small one. The body has effective machinery for keeping its temperature constant and the gonads are not likely to suffer unduly from exposure to heat.

Electromagnetic radiation comes in a wide range of energies, with visible light about in the middle of the range (see diagram on page 77). Electromagnetic radiations less energetic than light (such as infrared waves and microwaves) are converted into heat when absorbed by living tissue. The heat thus formed is sufficient to cause atoms and molecules to vibrate more rapidly, but this added vibration is not usually sufficient to pull molecules apart and therefore does not bring about chemical changes.

Light will bring about some chemical changes. It will break up silver compounds and produce tiny black grains of metallic silver (the chemical basis of photography). Living tissue, however, is largely unaffected—the retina of the eye being one obvious exception.

Ultraviolet light, which is more energetic than visible light, correspondingly can bring about chemical changes more easily. It will redden the skin and stimulate the production of pigment. It will even interfere with replication to some extent. At least there is evidence that persistent exposure to sunlight brings about a heightened tendency to skin cancer. Ultraviolet light is not very penetrating, however, and its effects are confined to the skin.

Electromagnetic radiations more energetic than ultraviolet light, such as X rays and gamma rays, carry sufficient concentrations of energy to bring about changes not only in molecules but in the very structure of the atoms making up those molecules. An X ray or gamma ray, crashing into an atom, will jar electrons loose. What is left of the atom will carry a positive electric charge. An atom fragment carrying an electric charge is an ion. X rays and gamma rays are therefore examples of ionizing radiation. Radiations may consist of particles, too, and if these carry sufficient energy they are also ionizing.

Ionizing radiation is capable of imparting so much energy to molecules as to cause them to vibrate themselves apart, producing not only ions but also high-energy uncharged molecular fragments called free radicals.

The direct effect of ionizing radiation on chromosomes can be serious (see photograph section). Enough chemical bonds may be disrupted so that a chromosome struck by a high-energy wave or particle may break into fragments. Even if the chromosome manages to remain intact, an individual gene along its length may be badly damaged and a mutation may be produced.

If only direct hits mattered, radiation effects would be less dangerous than they are, since such direct hits are comparatively few. However, near-misses may also be deadly. A streaking bit of radiation may strike a water molecule near a gene and may break up the molecule to form a free radical. The free radical will be sufficiently energetic to bring about a chemical reaction with almost any molecule it strikes. If it happens to strike the neighboring gene before it has disposed of that energy, it will produce the mutation as surely as the original radiation might have.

Furthermore, ionizing radiations (particularly of the electromagnetic variety) tend to be penetrating, so that the interior of the body is as exposed as is the surface. The gonads cannot hide from X rays, gamma rays, or cosmic particles.

All these radiations can bring about somatic mutations—all can cause cancer, for instance.

What is worse, all of them increase the rate of genetic mutations so that their presence threatens generations unborn as well as the individuals actually exposed.

The delicate structure of the genes and chromosomes is particularly vulnerable to the impact of high-energy radiation. Chromosomes can be broken by such radiation and this is the main cause of actual cell death. A cell that is not killed outright by radiation may nevertheless be so damaged as to be unable to undergo replication and mitosis.

If a cell is of a type that will not, in the course of nature, undergo division, the destruction of the mitosis machinery is not in itself fatal to the organism. A creature like the fruit fly *Drosophila,* which, in its adult stage, has very few cell divisions going on among the ordinary cells of its body, can survive radiation doses a hundred times as great as would suffice to kill a man.

In a human being, however—even in an adult who is no longer experiencing overall growth—there are many tissues whose cells must undergo division throughout life. Hair and fingernails grow constantly, as a result of cell division at their roots. The outer layers of skin are steadily lost through abrasion and are replaced through constant cell division in the deeper layers. The same is true of the lining of the mouth, throat, stomach, and intestines. Too, blood cells are continually breaking up and must be replaced in vast numbers.

If radiation kills the mechanism of division in only some of these cells, it is possible that those that remain reasonably intact can divide and eventually replace or do the work of those that can no longer divide. In that case, the symptoms of radiation sickness are relatively mild in the first place and eventually disappear.

Past a certain critical point, when too many cells are made incapable of division, this is no longer possible. The symptoms, which show up in the growing tissues particularly (as in the loss of hair and the lowering of the blood cell count), grow steadily more severe and death follows.

Where radiation is insufficient to render a cell incapable of division, it may still induce mutations, and it is in this fashion that skin cancer, leukemia, and other disorders may be brought about.

Mutations can be brought about in the sex cells, too, of course, and when this happens it is succeeding generations that are affected and not merely the exposed individual. Indeed, where the sex cells are concerned, the relatively mild effect of mutation is more serious than the drastic one of nondivision. A fertilized ovum that cannot divide eventually dies and does no harm; one that can divide but is altered may give rise to an individual with one of the usual kinds of major or minor physical defects.

The effect of high-energy radiation on the genetic mechanism was first demonstrated experimentally in 1927 by Muller. Using *Drosophila* he showed that after large doses of X rays, flies experienced many more lethal mutations per chromosome than did similar flies not exposed to radiation (see photograph section).

Later experiments, by Muller and by others, showed that the number of mutations was directly proportional to the quantity of radiation absorbed. Doubling the quantity of radiation absorbed doubled the number of mutations, tripling the one tripled the other, and so on. This means that if the number of mutations is plotted against the amount of radiation absorbed, a straight line can be drawn.

It is generally believed that the straight line continues all the way down without deviation to very low radiation absorptions. This means there is no "threshold" for the mutational effect of radiation. (An apparent threshold has been found in mice; see page 151.) No matter how small a dosage of radiation the gonads receive, this will be reflected in a proportionately increased likelihood of mutated sex cells with effects that will show up in succeeding generations.

In this respect, the genetic effect of radiation is quite different from the somatic effect. A small dose of radiation may affect growing tissues and prevent a small proportion of the cells of those tissues from dividing. The remaining, unaffected cells take up the slack, however, and if the proportion of affected cells is small enough, symptoms are not visible and never become visible. There is thus a threshold effect: the radiation absorbed must be more than a certain amount before any somatic symptoms are manifest.

Matters are quite different where the genetic effect is concerned. If a sex cell is damaged and if that sex cell is one of the pair that goes into the production of a fertilized ovum, a damaged organism results. There is no margin for correction. There is no unaffected cell that can take over the work of the damaged sex cell once fertilization has taken place.

Suppose only one sex cell out of a million is damaged. If so, a damaged sex cell will, on the average, take part in one out of every million fertilizations. And when it is used, it will not matter that there are 999,999 perfectly good sex cells that might have been used—it was the damaged cell that *was* used. That is why there is no threshold in the genetic effect of radiation and why there is no safe amount of radiation insofar as genetic effects are concerned. However small the quantity of radiation absorbed, mankind must be prepared to pay the price in a corresponding increase of the genetic load.

If the straight line obtained by plotting mutation rate against radiation dose is followed down to a radiation dose of zero, it is found that the line strikes the vertical axis slightly above the origin. The mutation rate is more than zero even when the radiation dose is zero. The reason for this is that it is the dose of man-made radiation that is being considered. Even when man-made radiation is completely absent there still remains the natural background radiation.

It is possible in this manner to determine that background radiation accounts for considerably less than 1% of the sponta-

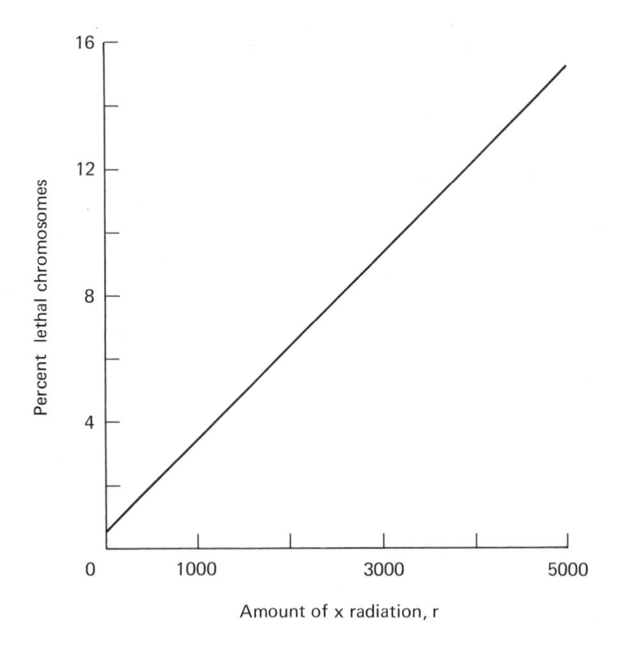

Mutation rate as a function of radiation dose

neous mutations that take place. The other mutations must arise out of chemical misadventures, out of the random heat-jiggling of molecules, and so on. These, it can be presumed, will remain constant when the radiation dose is increased.

This is a hopeful aspect of the situation for it means that, if the background radiation is doubled or tripled for mankind as a whole, only that small portion of the spontaneous mutation rate that is due to the background radiation will be doubled or tripled.

Let us suppose, for instance, that fully 1% of the spontaneous mutations occurring in mankind is due to background radiation. In that case, the tripling of the background radiation produced in the United States by man-made causes would triple that 1%. In place of 99 nonradiational mutations plus 1 radiational, we would have 99 plus 3. The total number of mutations would increase from 100 to 102—an increase of 2%, not an increase of 200% that one would expect if all spontaneous mutations were caused by background radiation.

Another difference between the genetic and somatic effects of radiation rests in the response to changes in the rate at which radiation is absorbed. It makes a considerable difference to the body whether a large dose of radiation is absorbed over the space of a few minutes or a few years.

When a large dose is absorbed over a short interval of time, so many of the growing tissues lose the capacity for cell division that death may follow. If the same dose is delivered over years, only a small bit of radiation is absorbed on any given day and only small proportions of growing cells lose the capacity for division at any one time. The unaffected cells will continually make up for this and will replace the affected ones. The body is, so to speak, continually repairing the radiation damage and no serious symptoms will develop.

Then, too, if a moderate dose is delivered, the body may show visible symptoms of radiation sickness but can recover. It will then be capable of withstanding another moderate dose, and so on.

The situation is quite different with respect to the genetic effects, at least as far as experiments with *Drosophila* and bacteria seem to show. Even the smallest doses will produce a few mutations in the chromosomes of those cells in the gonads that eventually develop into sex cells. The affected gonad cells will continue to produce sex cells with those mutations for the rest of the life of the organism. Every tiny bit of radiation adds to the number of mutated sex cells being constantly produced. There is no recovery, because the sex cells, after formation, do not work in cooperation, and affected cells are not replaced by those that are unaffected.

This means (judging by the experiments on lower creatures) that what counts, where genetic damage is in question, is not the rate at which radiation is absorbed but the total sum of radiation. Every exposure an organism experiences, however small, adds its bit of damage.

Accepting this hard view, it would seem important to make every effort to minimize radiation exposure for the population generally.

Since most of the man-made increase in background radiation is the result of the use of X rays in medical diagnosis and therapy, many geneticists are looking at this with suspicion and concern. No one suggests that their use be abandoned, for certainly such techniques are important in the saving of life and the mitigation of suffering. Still, X rays ought not to be used lightly, or routinely as a matter of course.

It might seem that X rays applied to the jaw or the chest would not affect the gonads, and this might be so if all the X rays could indeed be confined to the portion of the body at which they are aimed. Unfortunately, X rays do not uniformly travel a straight line in passing through matter. They are scattered to a certain extent; if a stream of X rays passes through the body anywhere, or even through objects near the body, some X rays will be scattered through the gonads.

It is for this reason that some geneticists suggest that the history of exposure to X rays be kept carefully for each person.

A decision on a new exposure would then be determined not only by the current situation but by the individual's past history.

Such considerations were also an important part of the driving force behind the movement to end atmospheric testing of nuclear bombs. While the total addition to the background radiation resulting from such tests is small, the prospect of continued accumulation is unpleasant.

Although genetic findings on such comparatively simple creatures as fruit flies and bacteria seem to apply generally to all forms of life, it seems unsafe to rely on these findings completely in anything as important as possible genetic damage to man through radiation. During the 1950s and 1960s, therefore, there have been important studies on mice, particularly by W. L. Russell at Oak Ridge National Laboratory, Oak Ridge, Tennessee.

While not as short-lived or as fecund as fruit flies, mice can nevertheless produce enough young over a reasonable period of time to yield statistically useful results.

Mice are far closer to man in the scheme of life than is any other creature that has been studied genetically on a large scale, and their reactions (one might cautiously assume) are likely to be closer to those that would be found in man.

Almost at once, when the studies began, it turned out that mice were more susceptible to genetic damage than fruit flies were. The induced mutation rate per gene seems to be about fifteen times that found in *Drosophila* for comparable X-ray doses. The only safe course for mankind then is to err, if it must, strongly on the side of conservatism. Once we have decided what might be safe on the basis of *Drosophila* studies, we ought then to tighten precautions several notches by remembering that we are very likely more vulnerable than fruit flies are.

Counteracting the depressing nature of this finding was that of a later, quite unexpected discovery. It was well established that in fruit flies and other simple organisms, it was the total dosage of absorbed radiation that counted and that whether this was delivered quickly or slowly did not matter.

This proved not to be so in the case of mice. In male mice, a radiation dose delivered slowly produced only from one-quarter to one-third as many mutations as did the same total dose delivered rapidly.

In the male, cells in the gonads are constantly dividing to produce sex cells. The latter are produced by the billions. It might be, then, that at low radiation dose rates, a few of the gonad cells are damaged but that the undamaged ones produce a flood of sperm cells, drowning out the few produced by the damaged gonad cells. The same radiation dose delivered in a short time might, however, damage so many of the gonad cells as to make the damaged sex cells much more difficult to flood out.

A second possible explanation is that there is present within the cells themselves some process that tends to repair damage to the genes and to counteract mutations. It might be a slow-working, laborious process that could keep up with the damage inflicted at low dosage rates but not at high ones. High dosage rates might even damage the repair mechanism itself. That, too, would account for the fewer mutations at low dosage rates than at high ones.

To check which of the two possible explanations was nearer the truth, Russell performed similar tests on female mice. In the female mouse (or the female human being, for that matter) the egg cells have completed almost all their divisions before the female is born. There are only so many cells in the female gonads that can give rise to egg cells, and each one gives rise to only a single egg cell. There is no possibility of damaged egg cells being drowned out by floods of undamaged ones because there are no floods.

Yet it was found that in the female mouse the mutation rate also dropped when the radiation dose rate was decreased. In fact, it dropped even more drastically than was the case in the male mouse.

Apparently, then, there must be actual repair within the

cell. There must be some chemical mechanism inside the cell capable of counteracting radiation damage to some extent. In the female mouse, the mutation rate drops very low as the radiation dose rate drops, so that it would seem that almost all mutations might be repaired, given enough time. In the male, the mutation rate drops only so far and no farther, so that some mutations (about one-third is the best estimate so far) cannot be repaired.

If this is also true in the human being (and it is at least reasonably likely that it is), then the greater vulnerability of our genes as compared with those of fruit flies is at least partially made up for by our greater ability to repair the damage.

This opens a door for the future, too. The workings of the gene-repair mechanism ought (it is to be hoped) eventually to be puzzled out. When it is, methods may be discovered for reinforcing that mechanism, speeding it, and increasing its effectiveness. We may then find ourselves no longer completely helpless in the face of genetic damage.

On the other hand, it is only fair to point out that the foregoing appraisal may be an overoptimistic view. Russell's experiments involved just seven genes and it is possible that these are not representative of the thousands that exist altogether. Much research remains to be carried out.

If, then, we cannot help hoping that natural devices for counteracting radiation damage may be developed in the future, we must, for the present, remain rigidly cautious.

As long as man-made radiation exists, there will be some absorption of it by human beings. The advantages of its use in our modern society are such that we must be prepared to pay some price. This is not a matter of callousness. We have come to depend a great deal for comfort, and even for extended life, upon the achievements of our technology, and any serious crippling of that technology will cost us lives. An attempt must be made to balance the values of radiation against its dangers; we must balance lives against lives. This involves hard judgments.

A 10% increase in mutation rate, whatever it might mean

in personal suffering and public expense, is not likely to threaten the human race with extinction, or even with serious degeneration.

The human race as a whole may be thought of as somewhat analogous to a population of dividing cells in a growing tissue. Those affected by genetic damage drop out and the slack is taken up by those not affected.

If the number of those affected is increased, there would come a crucial point, or threshold, where the slack could no longer be taken up. The genetic load might increase to the point where the species as a whole would degenerate and fade toward extinction—a sort of racial radiation sickness.

We are not near this threshold now, however, and can, therefore, as a species, absorb a moderate increase in mutation rate without danger of extinction.

On the other hand, it is not correct to argue, as some do, that an increase in mutation rate might be actually beneficial. The argument runs that a higher mutation rate might broaden the gene pool and make it more flexible, thus speeding up the course of evolution and hastening the advent of supermen—brainier, stronger, healthier than we ourselves are.

The truth seems to be that the gene pool, as it exists now, supplies us with all the variability we need for the effective working of the evolutionary mechanism. That mechanism is functioning with such efficiency that broadening the gene pool cannot very well add to it, and if the hope of increased evolutionary efficiency were the only reason to tolerate man-made radiation, it would be insufficient.

The situation is rather analogous to that of a man who owns a good house that is heavily mortgaged. If he were offered a second house with a similar mortgage, he would have to refuse. To be sure, he would have twice the number of houses, but he would not need a second house since he has all the comfort he can reasonably use in his first house—and he would not be able to afford a second mortgage.

What humanity must do, if additional radiation damage is

absolutely necessary, is to take on as little of that added damage as possible, and not pretend that any direct benefits will be involved. Any pretense of that sort may well lure us into assuming still greater damage—damage we may not be able to afford under any circumstances and for any reason. (12)

9 ◊ The Biology and Chemistry of Life

How does radiation produce mutations? What are the specific chemical and biological effects of radiation on living tissue? To answer these questions we must first learn about the chemistry and biology of living tissue. A good deal of our recent knowledge of biology has come from radiation from radioactive isotopes.

Cell Theory: DNA Is the Secret of Life

THE German anatomist Theodor Schwann (1810–83) wrote: "We have seen that all organisms are composed of essentially like parts, namely cells; that these cells are formed and grow in accordance with essentially the same laws; hence that these processes must everywhere result from the operation of the same forces."

The nature of life has excited the interest of human beings from the earliest times. Although it is still not known what life is, the characteristics that set living things apart from lifeless matter are well known. One feature common to all living things, from one-celled creatures to complex animals like man, is that they are all composed of microscopic units known as cells.

The cell is the smallest portion of any organism that exhibits the properties we associate with living material. (An organism is a complete living plant or animal.) In spite of the immense variety of sizes, shapes, and structures of living things, they all have this in common: they are composed of cells, and living cells contain similar components that operate in similar ways. One might say that life is a single process and that all living things operate on a single plan.

The cell theory, based on the concept that higher organisms consist of smaller units called cells, was formulated in 1838 by two German biologists, Theodor Schwann and Mathias-Jacob Schleiden (1804–81), a botanist. The theory had far-reaching effect upon the study of biological phenomena. It suggested that living things had a common basis of organization. Appreciation of its full significance, however, had to await more precise knowledge of the structure and activities of cells.

Some organisms—for instance, amoebae—consist of a single cell each and are therefore called unicellular organisms. Higher animals are multicellular, containing aggregations of cells grouped into tissues and organs. A man, for instance, consists of millions of different kinds of cells performing a variety of different functions. Cells of higher animals differ vastly from one another in size, shape, and function; they are specialized cells.

There is a remarkable similarity, moreover, in the molecular composition and metabolism of all living things. (Metabolism is the sum of the life-sustaining activities in a living organism, including nutrition, production of energy, and synthesis of new living material.) This similarity has been taken to mean that life could have originated only once in the past and had a specific chemical composition on which its metabolic processes depended. This structure and metabolism were handed down to subsequent living things by reproduction, and all variations thereafter resulted from occasional mutation, or changes in the nature of the heredity-transmitting units. One of the most extraordinary of all the attributes of life is its ordered complexity, both in function and structure.

The basic structure of a cell is shown in the diagram on page 158. Each cell consists of a dense inner structure called the nucleus [not to be confused with the nucleus of the atom],

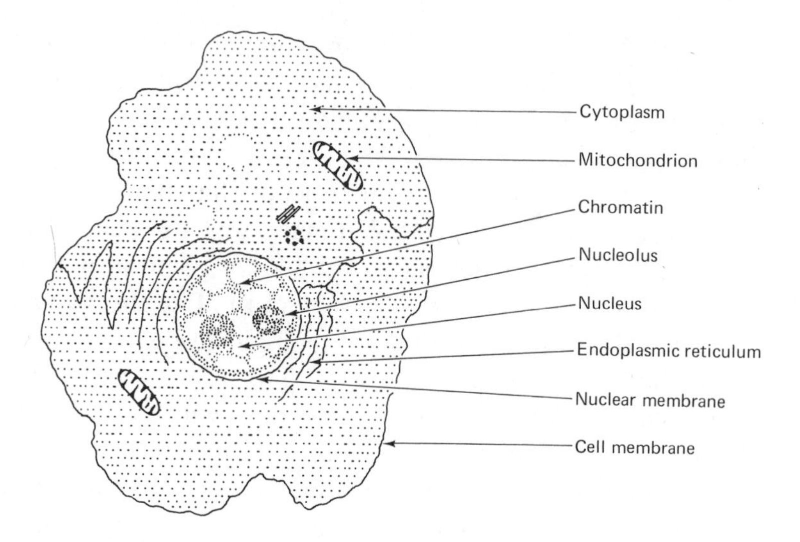

Generalized diagram of a cell, showing the organelles, or "little organs," of its internal structure

which is surrounded by a less dense mass of cytoplasm. The nucleus is separated from the cytoplasm by a double envelope, called the nuclear membrane, which is peppered with perforations. The cytoplasm contains a network of membranes, which form the boundaries of countless canals and vesicles (or pouches), and is laden with small bodies called ribosomes. This membranous network is called the endoplasmic reticulum and is distinct from the mitochondria, which are membranous organelles (little organs) structurally independent of other components of the cytoplasm. The outer coat of the cell is called the cell membrane, or plasma membrane, and forms the cell boundary.

The nucleus, which in many cells is the largest and most central body, is of special importance. It contains the chromosomes, the carriers of the cell's heredity-controlling system. These contain granules of a material called chromatin, which is rich in a nucleic acid, DNA (deoxyribonucleic acid). The chromosomes usually are not readily seen in the nucleus except when the cell, along with its nucleus, is dividing. When the nucleus is not dividing, a spherical body, the nucleolus, can be seen. (In some nuclei there may be more than one.) When the nucleus is dividing, the nucleolus disappears.

Not all cells possess all these structures. For instance, the red cells of the blood do not have a nucleus, and in other cells the endoplasmic reticulum is at a minimum. The generalized diagram is valid for a great majority of the cells of higher organisms.

The cell structures are visible with an electron microscope. They contain the chemical components of the cell. The chief classes of these constituents are the carbohydrates (sugars), the lipids (fats), the proteins, and the nucleic acids. However, a cell also contains water (about 70% of the cell weight is due to water) and several other organic and inorganic compounds, such as vitamins and minerals.

Carbohydrates serve mostly as foodstuff within the cell. They can be stored in several related forms. Further, they may serve a number of functions outside the cell, especially as structural units. In this way structure and function are correlated.

Lipids in the cell occur in a great variety of types, including alcohols, fats, and steroids. They are found in all fractions of the cell. Their most important functions seem to be to form membranes and to give these membranes specific permeability. They are also important as stores of chemical energy, mostly in the form of neutral fats.

The proteins occur in many cell structures and are of many kinds: Enzymes, the catalysts for the cell's metabolic processes, are proteins, for instance. The nucleic acids are DNA and RNA (ribonucleic acid), which function together to manufacture the cell's proteins. These two types of materials are interrelated in their function and both are essential. Although our insight into the mutual dependence of these two materials has greatly increased in recent years and although we know the relation between them is a fundamental factor in such events as reproduction, mutation, and differentiation (or specialization) of cells, our understanding of their interplay is far from complete. Real understanding of the relation between them would give us insight into the essence of growth—both normal and abnormal—or, indeed, one could almost say, into the complexity of life itself.

Practically all the DNA of most cells is concentrated in the nucleus. RNA, on the other hand, is distributed throughout the cell. Some RNA is present in the nucleus, but most of it is associated with minute particles in the cytoplasm known as microsomes, some of which are especially rich in RNA and are accordingly named ribosomes. These are much smaller particles than the mitochondria.

One of the most remarkable characteristics of cells is their ability to grow and divide. New cells come from preexisting cells. When a cell reaches a certain stage in its life, it divides into two parts. These parts, after another period of growth, can in turn divide. In this way plants and animals grow to their normal size and injured tissues are repaired. Cell division occurs when some of the contents of the cell have been doubled by replication, or copying (to be discussed later). The division of a cell results in two roughly equal new parts, the daughter cells. The process of cell division is known as mitosis and is illustrated in the diagram on the facing page.

Mitosis is a continuous process; the following stages of the process are designated only for convenience. During interphase the cell is busy metabolizing, synthesizing new cellular materials, and preparing for self-duplication by synthesizing new chromosomes. In prophase the chromosomes, each now composed of two identical strands called chromatids, shorten by coiling, and the nucleolus and nuclear membrane disappear. During metaphase the chromosomes line up in one plane near the cell equator. At anaphase the sister chromatids of each chromosome separate, and each part moves toward the ends, or poles, of the cell. During telophase the chromosomes uncoil and return to invisibility; a new nucleus, nucleolus, and nuclear membrane are reconstituted at each end, and division of the cell body occurs between the new nuclei, forming the two new cells. Each daughter cell thereby receives a full set of chro-

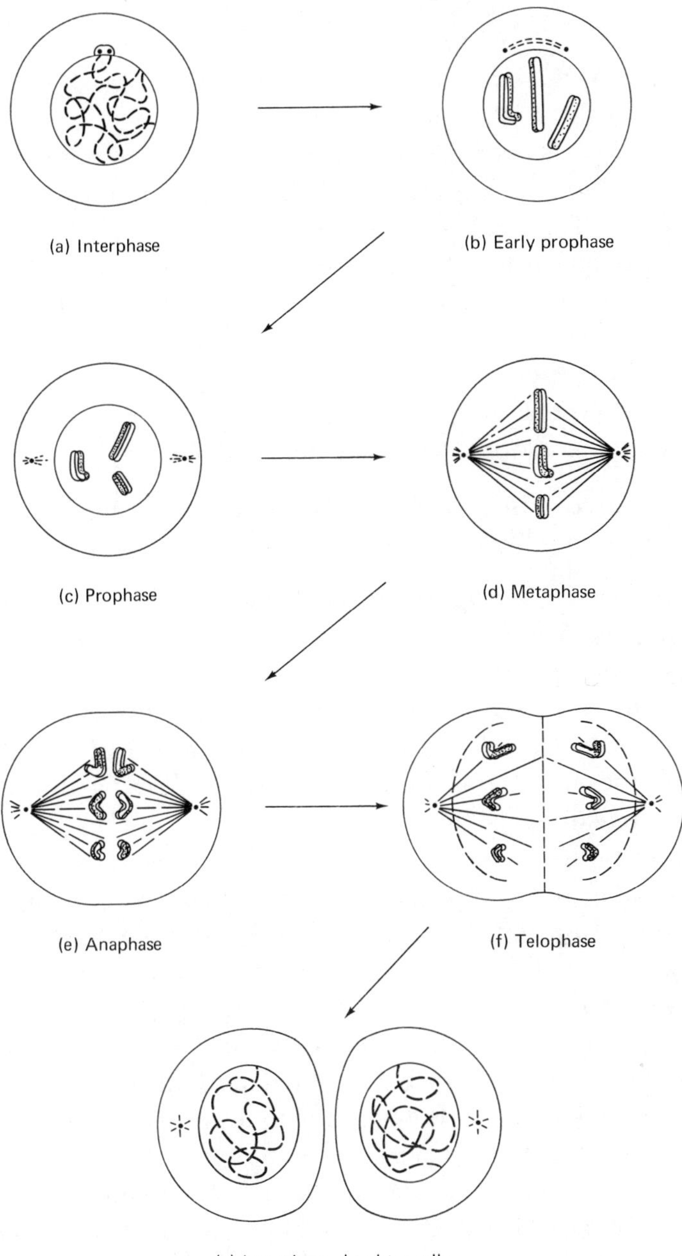

Stages of the mitotic cycle in a hypothetical cell with three chromosomes. (In the metaphase, a spindle apparatus appears, the fibers of which assist the daughter chromosomes to move to opposite poles.)

mosomes, and, since the genes are in the chromosomes, each daughter cell has the same genetic complement.

All life processes use up energy and therefore require fuel. The mitochondria have a central role in the reactions by which the energy of sugars is supplied for cellular activity. The importance of this vital activity is obvious. Here, however, we are concerned with the processes, involving nucleic acids and proteins, that can be described as making up the gene-action system. The gene-action system is the series of biochemical events that regulate and direct all life processes by transcription of the genetic information contained in molecules of DNA. (29)

Radioactive isotopes have been a powerful means enabling biologists to analyze living matter.

Radioactive isotopes occur as minor constituents in many natural materials, from which they can be concentrated by fractionation procedures. In a very limited number of cases, more significant amounts of a radioactive isotope, for example, radium or radioactive lead, can be found in nature. Most radioactive isotopes in use today, however, are prepared artificially by nuclear reactions.

The great usefulness of radioactive isotopes, as we shall see later, is that they can be detected and identified by proper instruments. Biochemists have long recognized the desirability of tagging or labeling a molecule to permit tracing or keeping track of the label and consequently of the molecule as it moves through a reaction or process. Since the radiations emitted by radioactive isotopes can be detected and measured, we can readily follow a molecule tagged with a radioactive atom.

The earliest biochemical studies employing radioactive isotopes go back to 1924, when the Hungarian chemist Georg von Hevesy (1885–1966) used natural radioactive lead to investigate a biological process. It was only after World War II, however, when artificially made radioactive isotopes were readily available, that the technique of using isotopic tracers became popular.

In our investigations of life processes, we are especially interested in three radioactive isotopes: ^{3}H, the hydrogen atom of mass 3; ^{14}C, the atom of carbon with atomic weight 14; and ^{32}P, the atom of phosphorus with atomic weight 32. These radioactive isotopes are important because the corresponding stable isotopes of hydrogen, carbon, and phosphorus are present in practically all cellular components that are important in maintaining life. With the three radioactive isotopes, therefore, we can tag or label the molecules that participate in life processes.

Hydrogen 3 is a weak beta emitter; that is, it emits beta particles with a very low energy (0.018 MeV) and therefore with a very short range. (The range is the distance a particle travels before stopping.) Carbon 14 is also a weak beta emitter (0.154 MeV), although the beta particles emitted by ^{14}C have a higher energy and therefore a longer range than those emitted by ^{3}H. The beta particles emitted by phosphorus 32 are quite energetic (1.69 MeV) and have a longer range.

To biologists, then, the essential feature in the use of radioactive isotopes is the possibility of preparing labeled samples of any organic molecule involved in biological processes. With labeled samples it is possible to distinguish the behavior and keep track of the course of molecules involved in a particular biological function.

In this capacity the isotope may be likened to a dynamic and revolutionary type of "atomic microscope," which can actually be incorporated into a living process or a specific cell. Just as a real microscope permits examination of the structural details of cells, isotopes permit examination of the chemical activities of molecules, atoms, and ions as they react within cells.

DNA Synthesis: The Autobiography of Cells

THE many characteristic features of each living species, its complex architecture, its particular behavior patterns, the ingenious modifications of structure and function that enable it to compete and survive—all these must pass, figuratively speaking, through the eye of an ultramicroscopic needle before they are brought together as a new, individual organism. The thread that passes through the eye of this needle is a strand of the filamentous molecule, deoxyribonucleic acid (DNA). Let us now outline the research that led to these conclusions.

One of the fundamental laws of modern biology—which states that the DNA content of somatic cells is constant for any given species—was first set forth in a research report of 1948. This finding means that in any given species, such as a mouse or a man, all cells except the germinal cells contain the same amount of DNA. Germinal cells, that is, the sperm cells of the male semen and the female egg, contain exactly half the amount of DNA of the somatic cells. This must be the case, since DNA is the hereditary material, and each individual's heredity is shaped half by his father and half by his mother. One ten-trillionth of an ounce of DNA from a father and one ten-trillionth of an ounce of DNA from the mother together contain all the specifications to produce a new human being.

A large amount of DNA must be manufactured by an individual organism as it develops from a fertilized egg (one single cell) to an adult containing several million cells. For instance, a mouse cell contains about 7×10^{-12} gram of DNA. A whole mouse contains in its body approximately 25 milligrams of DNA, and all this DNA was synthesized by the cells as the mouse grew to adulthood. Since the amount of DNA per cell remains constant and since each cell divides into two cells, it is apparent that each new cell receives the amount of DNA characteristic of that species.

Once we realize that a cell that is making new DNA (as most cells do) must divide to keep the amount of DNA per cell constant, it follows that a cell that is making DNA is one that is soon destined to divide. If we can now mark newly made DNA with a radioactive isotope, we can actually mark and thus identify cells that are preparing to divide. The task can be divided into two parts: (1) to label the newly made DNA and (2) to detect the newly made, labeled DNA.

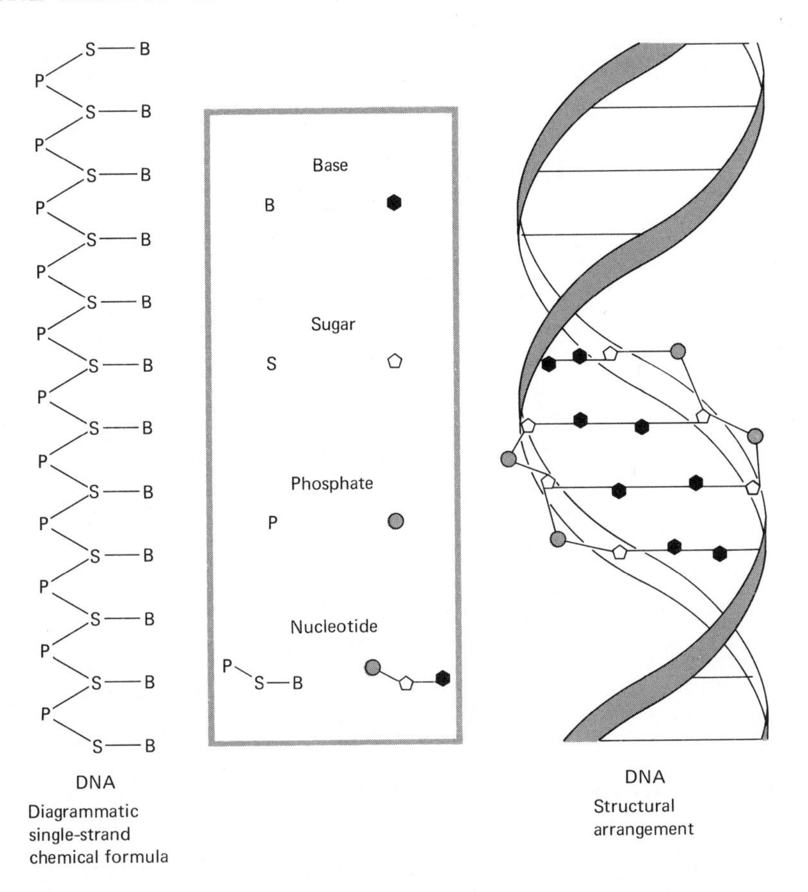

Diagrammatic structure of the DNA molecule as proposed by J. D. Watson and F. H. C. Crick

The essential structure of the large DNA molecule is shown in the diagram. According to the model of J. D. Watson and F. H. C. Crick, the molecule consists of two strands of smaller molecules twisted around each other to form a double helix. Each strand consists of a sequence of the smaller molecules linked linearly to each other. These smaller molecules are called nucleotides, and each consists of three still smaller molecules, a sugar (deoxyribose), phosphoric acid, and a nitrogen base. Each nucleotide and its nearest neighbor are linked together (between the sugar of one and the phosphoric acid of the neighbor). This leaves the nitrogen base free to attach itself,

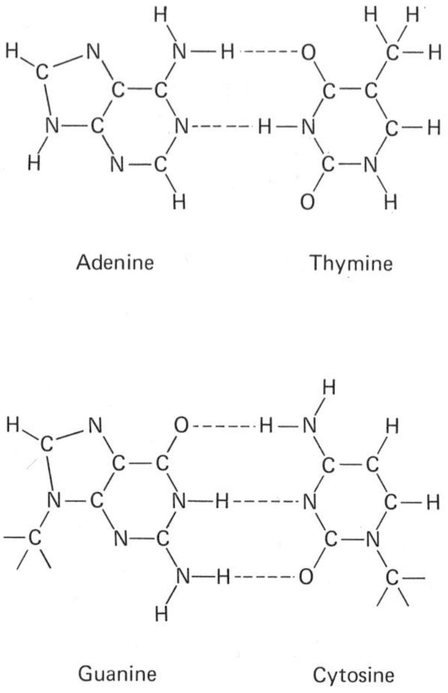

Adenine Thymine

Guanine Cytosine

The pairing of the nucleotide bases that make up DNA. C, H, N, and O denote carbon, hydrogen, nitrogen, and oxygen atoms. The broken lines represent the hydrogen bonds that join the nucleotides.

through hydrogen bonding, to another nitrogen base in the opposite strand of the helix.

In the DNA of higher organisms, there are only four types of nitrogen bases: adenine, guanine, thymine, and cytosine. Adenine in either strand of the helix pairs only with thymine in the opposite strand, and vice versa, and guanine pairs only with cytosine, and vice versa, so that each strand is complementary in structure to the other strand. The full structure resembles a long twisted ladder, with the sugar and phosphate molecules of the nucleotides forming the uprights and the linked nitrogen bases forming the rungs. Each upright strand is essentially a mirror image of the other, although the two ends of any one rung are dissimilar.

When DNA is replicated, or copied, as the organism grows, the two nucleotide strands separate from each other by disjoining the rungs at the point where the bases meet, and each strand then makes a new and similarly complementary strand by extracting the appropriate materials from its surroundings. The result is two double-stranded DNA molecules, each of which is identical to the parent molecule and contains the same genetic material. When the cell divides, each of the two daughter cells gets one of the new double strands; each new cell thus always has the same amount of DNA and the same genetic material as the parent cell.

(All that has been said so far about DNA replication depends upon an assumption that the DNA molecule is in some way untwisted to allow separation of two helical strands, but there is no compelling reason to believe that such an untwisting does indeed take place, nor do we know, if the untwisting does take place, how it is accomplished.)

Of the four bases in DNA, three are also found in the other nucleic acid, RNA; but the fourth, thymine, is found only in DNA. Therefore, if thymine could be labeled and introduced into a number of cells, including a cell in which DNA is being formed, we would specifically label the newly synthesized DNA, since neither the old DNA nor the RNA would make use of the thymine. We could in this way mark cells preparing to divide. (Actually, thymine itself is not taken up in mammalian cells, but its nucleoside is. A nucleoside is the base plus the sugar, or, in other words, the nucleotide minus the phosphoric acid.) The nucleoside of thymine is called thymidine, and we say that thymidine is a specific component of DNA and can be used, both in laboratory studies and in living organisms, for labeling DNA.

Thymidine labeled with radioactive compounds is available as ^{14}C-thymidine (thymidine with a stable carbon atom replaced by a radioactive carbon atom) and as ^{3}H-thymidine (thymidine in which a stable hydrogen atom has been replaced by ^{3}H, tritium). Thus, when cells actively making DNA are exposed to radioactive thymidine, they incorporate it, and the DNA becomes radioactive.

We have thus found a way to complete the first part of the task, the labeling of new DNA. We still must find out how to

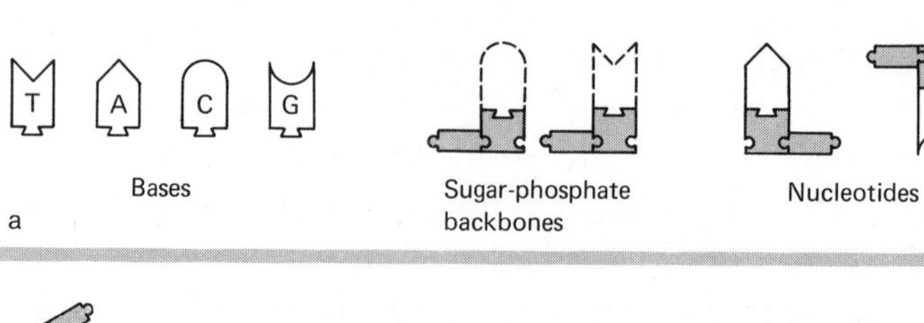
T
A
C
G
Bases
Sugar-phosphate backbones
Nucleotides
a

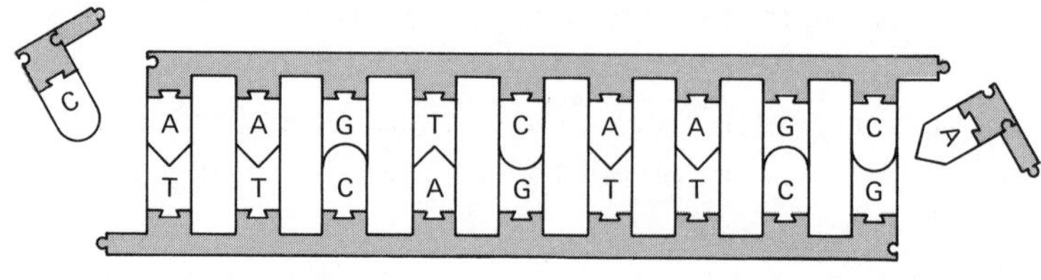
C
A A G T C A A G C
T T C A G T T C G
A
b

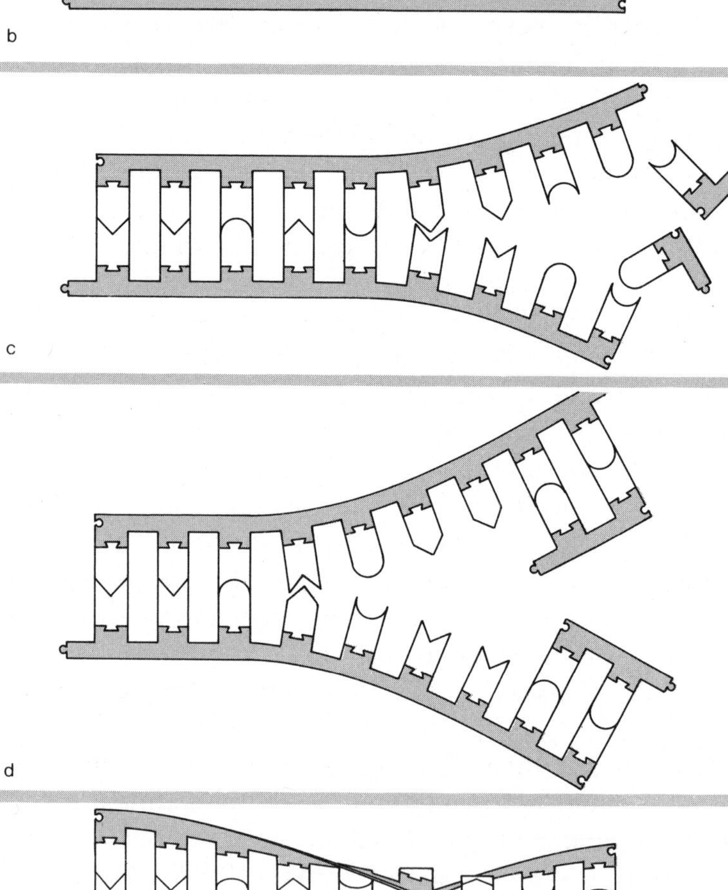
c
d

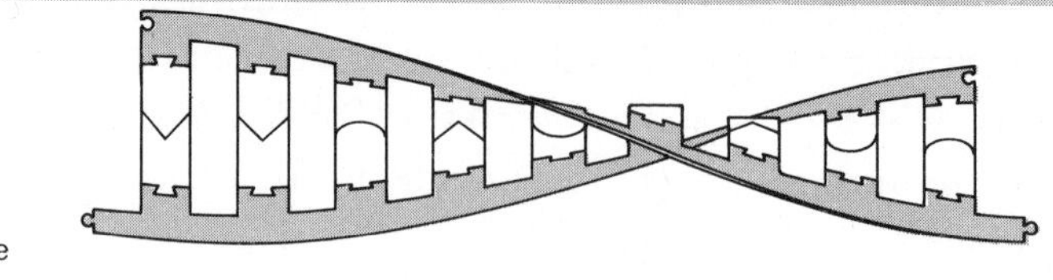
e

How the DNA molecule replicates: (*a*) The constituent submolecules. (*b*) Assembly of subunits in complete DNA molecule. A, G, C, and T denote the four bases adenine, guanine, cytosine, and thymine. (The molecule is actually twisted as in *e* but is shown flat for easier viewing of the subunits.) (*c*) "Unzipping" of the double nucleotide strand. (*d*) The forming of a new strand by each individual strand. (*e*) One of the resulting two DNA molecules in twisted double-strand configuration.

distinguish labeled DNA among the many components of the cell. We might do it with a system based on measuring the amount of radioactivity incorporated into the DNA of cells exposed to radioactive thymidine, as an approximation of the frequency of cell division in the group of cells. However, a better method for studying cells synthesizing DNA, and thus preparing to divide, is the use of high-resolution autoradiography.

Autoradiography is based on the same principle as photography. Just as photons of light impinging on a photographic emulsion produce an image, so do beta particles (or alpha particles) emitted by decomposing radioactive atoms. A photographic emulsion is a suspension of crystals of a silver halide (usually silver bromide) embedded in gelatin. When crystals of silver bromide are struck by beta particles, the silver atoms are ionized and form a latent image, so called because it is invisible to our eyes. After the emulsion is developed and fixed, each little aggregate of reduced silver atoms becomes a visible black speck on the emulsion. The distribution and combination of the specks make up the photographic image (see diagram).

The distinction of having made the first autoradiograph belongs to Antoine Henri Becquerel; and to another Frenchman, Jean Alexandre Eugène Lacassagne (1843–1924), goes

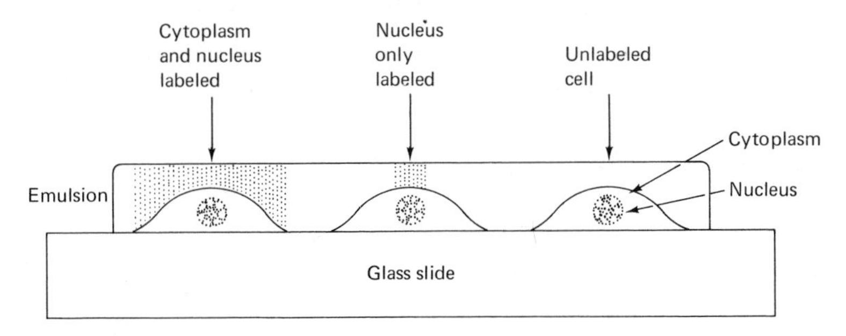

Schematic diagram of an autoradiograph

the credit for having introduced this technique into biological studies.

Much of our knowledge of the cell cycle stems from the use of high-resolution autoradiography.

Autoradiography enables us to find out which cells are dividing in a cell population and how many of them do so.

RNA Synthesis: How to Translate One Language into Another

WE have mentioned previously that there are two main types of nucleic acids: DNA, the genetic material itself, and RNA, the molecule that translates the genetic message from DNA into terms the cell can use as instructions for making protein. Cells differ from each other on the basis of kinds of proteins they contain, and, since differences among cells determine differences among organisms, it follows that differences in the composition of DNA serve to explain the variety in living organisms populating the world. However, if differences between two organisms can be explained by differences in the chemical composition of their respective DNA's, how can we explain differences between cells of the same organism? How can we explain that cells of the human pancreas secrete insulin, whereas other cells in man produce no insulin? Or how can we explain that certain cells make bone and others make fat? If indeed all cells in the same organism contain the same amount and kind of DNA (since all DNA in an organism derives from the duplication of the DNA of the fertilized egg cell and its descendants), it would seem, at first glance, that DNA is not the molecule responsible for differences among the cells. The clarification of this apparent contradiction is found in the remarkable properties of the other nucleic acid, the translator molecule, RNA.

In the first place, there are at least three different kinds of RNA. The largest quantity is a special kind called ribosomal RNA, or r-RNA. It is found in close conjunction with proteins and makes up the structural frame upon which the protein-synthesizing machinery is built. The r-RNA and the proteins to which it is firmly bound form the ribosomes, the RNA-rich microsomes that are attached to the endoplasmic reticulum. Proteins are synthesized on ribosomes. We shall see later what determines the differences among proteins and how these differences are dictated directly by RNA and indirectly by DNA.

Besides r-RNA, there is a kind of RNA called soluble RNA, or transfer RNA, or s-RNA. It combines with r-RNA to complete the sequence of events that synthesizes the proteins. A bond between r-RNA and s-RNA is established by a third RNA molecule called messenger RNA, or template RNA, or m-RNA. This m-RNA molecule is truly the messenger that carries the genetic message from DNA to the protein-synthesizing apparatus.

The DNA→RNA→protein sequence has been compared to the activities of a newspaper staff. DNA is the editor; m-RNA molecules are copyboys who carry the editorials to the typesetters, the r-RNA and s-RNA, who then take the "letters" of nucleic acid and set them into slots in accordance with the editor's directions. There are also workers who melt down outworn letters and still other workers who make new letters for further use; these are the enzymes, special kinds of proteins. If we wish to continue the analogy, we may say that each kind of cell in the organism has a different subeditor, who writes that cell's own editorial. Actually we might say that all cells have the same board of editors in common, but only one editor functions in any given type of cell. In biological terms this means that only a portion of all the cellular DNA is active in each cell.

The active DNA is the DNA that makes m-RNA that will carry instructions to the protein-synthesizing machinery of that type of cell. Cells of the same organism therefore differ from each other on the basis of the segment of DNA that is active in making m-RNA.

RNA synthesis is investigated with radioactive tracers in the same way as DNA synthesis. If we can mark, with a radioactive atom, a small molecule that is incorporated into newly formed RNA, we can then trace the course of the labeled RNA molecule with a radiation-detection device. DNA had one advantage in this regard—the fact that one compound, thymidine, was a precursor of DNA, a specific material that could be incorporated only into DNA. We do not know similar specific precursors of RNA. But we know several precursors that are predominantly incorporated into RNA. All these precursors can be labeled with either ^{3}H or ^{14}C, and their incorporation into RNA can be measured.

As in DNA synthesis, we can use autoradiography to follow the incorporation of precursors into RNA. The advantage of high-resolution autoradiography in DNA studies is the possibility of identifying particular cells that are synthesizing nucleic acid. This advantage is apparently lost in the case of RNA. The reason is that, at any given time, only a few cells are making DNA, whereas practically all cells are synthesizing RNA constantly. The only exceptions are cells in the midpoint of mitosis. At the beginning (prophase) and at the end of cell division (telophase), RNA is synthesized. If we want a quantitative measurement of RNA synthesis, other methods are considerably more precise. But autoradiography can still give us valuable information.

If we look at cells soon after they have been exposed to an RNA precursor, we find that the radioactivity detectable by autoradiography is only in the nuclei of the cells. No radioactivity can be detected in the cytoplasm, although we know that the cytoplasm of living cells contains large amounts of r-RNA and s-RNA. One or two hours later, however, radioactive RNA appears in the cytoplasm as well as in the nucleus. What autoradiography is telling us is that RNA is made in the nucleus and then is slowly transferred to the cytoplasm.

Autoradiography cannot tell us whether the RNA that has been newly synthesized in the nuclei of cells is m-RNA, s-RNA, or r-RNA. The methods necessary to make this distinction are based on the chemical fractionation of the tissue, isolation of RNA, determination of its amount by quantitative analysis, and determination of the amount of radioactivity by physical methods [which are not examined here].

Protein Synthesis: The Molecules That Make the Difference

PROTEINS occupy a central position in the structure and functioning of living matter and are intimately connected with all the metabolic reactions that maintain life. Some proteins serve as structural elements of the body, for instance, hair, wool, and the scleroproteins of bone and collagen, the latter an important constituent of connective tissue. Other proteins are enzymes, which are extremely important since they regulate all metabolic reactions. Most of the proteins in the tissues of actively functioning organs, such as the liver and the kidney, are enzymes. Other proteins participate in muscular contraction, and still others are hormones or oxygen carriers. Special proteins called histones are associated with gene function, and the antibodies that an organism produces to defend itself from bacteria are also proteins.

The differences in proteins, especially in enzymes, account for differences among cells. It is now appropriate to ask what makes one protein different from another. We know that the structure of a protein depends upon several factors, such as the molecular weight. But the main differences among proteins depend upon the sequence, or order, of the amino acids that are linked together in the protein molecules.

Amino acids are the fundamental structural units of proteins. There are twenty amino acids found frequently in mammalian proteins, and these molecules may be linked to one another to form a chain called a polypeptide chain. The structure of a protein then depends on (1) the quantity of each amino acid present: (2) the sequence of amino acids in the polypeptide chain; (3) the length of the polypeptide chain, that is, the molecular weight; and (4) the folding and the side (nonlinear) arrangement of the polypeptide chain molecules, that is, the secondary and tertiary structures.

How can we investigate protein synthesis by using radioactive isotopes? Since proteins are made up of amino acids, the logical conclusion, after what we have learned about DNA synthesis and RNA synthesis, is that the best way would be to mark an amino acid and follow its incorporation into a molecule of protein. We could label a mixture of several amino acids, but, for the sake of clarity, we will describe the incorporation of a single labeled amino acid.

Suppose we have the amino acid leucine labeled with ^{14}C and we inject a solution containing it into an experimental animal. Since leucine is incorporated into proteins, if we isolate the proteins and determine both the amount of proteins and the amount of radioactivity, we can measure fairly accurately the rate of protein synthesis. Autoradiography, by the way, is of little help in studying most protein synthesis because all cells are always synthesizing proteins and so are all labeled after a single exposure to a radioactive amino acid. With RNA precursors autoradiography at least told us where RNA was being made, but with amino acids we do not even get this information because proteins are synthesized both in the nucleus and in the cytoplasm.

Under these circumstances radiochemical methods are better for studying protein synthesis. Proteins are isolated from the residue left after a nucleic-acid extraction process, and the amount of protein is determined by a simple colorimetric analysis based on comparison of the color of the solution with a standard color. The amount of radioactivity (remember that we are now using a precursor labeled with ^{14}C) can be determined with a counter.

Cell Cycle and Gene Action: Life Is the Secret of DNA

For a biologist interested in the mechanism of cell proliferation, the most important event in the life of a cell was, until very recently, cell division. As we mentioned, when a cell divides into two daughter cells, it undergoes a process called mitosis; mitosis itself is subdivided into four stages called prophase, metaphase, anaphase, and telophase. Mitosis in most cells takes less than one hour. Between one mitosis and the next, there can be an interval, from a few hours to several days in length, during which a cell is said to be in interphase. The entire period between the midpoints of two successive mitoses is called the cell cycle.

Until a few years ago, we knew very little about interphase. In fact, in one classic book on histology (the study of tissues), while a description of mitosis required almost twelve pages, interphase was dismissed in less than six lines! The reason for this lack of interest was, of course, the fact that no adequate methods were available for studying metabolic activities of cells in interphase. The methods of high-resolution autoradiography and radiochemical analysis of synchronized cell populations have become available only in the past few years.

We now know that metabolic activities during interphase are of primary importance in understanding the mechanism of cell division. It is, in fact, the orderly sequence of metabolic events occurring in interphase that leads from one mitosis to the next.

Try to imagine the cell cycle in the diagram as a race track, with the individual cells as cars that race around it. You are sit-

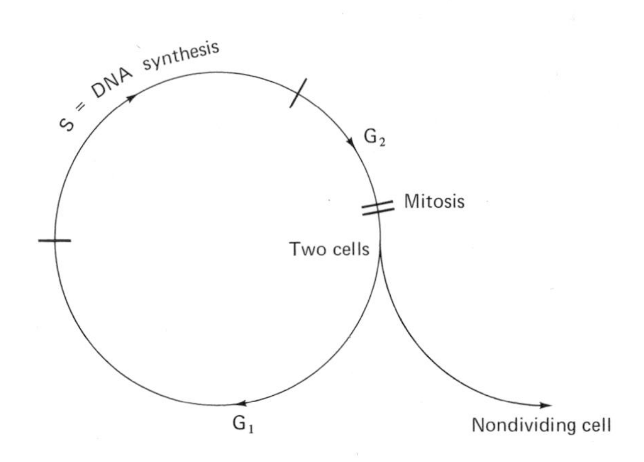

The cell cycle

ting at the finish wire, which is mitosis (we chose mitosis because it is easy to recognize when the cell is observed with the aid of a microscope). At a certain time during the race, all the cars in a portion of the track, say a 200-yard sector of the backstretch, are sprayed with a blue dye as they race by. These cars are now marked, just as cells synthesizing DNA are marked if briefly exposed to tritiated thymidine, the common radioactive precursor of DNA. As soon as these cars have been sprayed, you observe all the cars as they pass the finish line in front of you. At first, you will see cars that were nearest the wire and were not sprayed; then the dye-marked cars will pass; and finally more unmarked cars, those that had passed the finish line but had not reached the spray area when the marking was done, will come by. If you replace the words spray, cars, and wire with the words radioactivity, cells, and mitosis, you have described the cell cycle and the flow of cells in the cycle.

Now, if all cars were going at the same speed, you could calculate with great accuracy the time taken for any one car to go around the track, or from the finish line to the backstretch, or through the spray sector, and so on. However, since cars move at different speeds, you can only obtain an average time for all sprayed cars. Similarly, since individual cells behave differently, you can only obtain averages of the times these cells spend in the various portions of the cell cycle.

These cell-cycle portions are four in number, according to nomenclature originated by A. Howard and S. R. Pelc, two English investigators who first described the cycle: (1) mitosis; (2) G_1, which is the period between mitosis and DNA synthesis; (3) S phase, which is the period during which DNA is replicated; and (4) G_2, which is the period between DNA synthesis and the next mitosis. Only cells in the S phase (DNA synthesis) are marked when exposed to a radioactive precursor of DNA.

Because it has several important implications in biology and medicine, it is important to remember that DNA synthesis occurs only during the short, well-defined S period of the cell cycle. Other synthetic processes go on throughout the cycle.

For instance, all cells can be labeled by a brief exposure to a radioactive amino acid, a precursor of proteins; this means that protein synthesis occurs throughout the entire cell cycle, including mitosis. When we use a radioactive RNA precursor, all cells except those in anaphase and metaphase are labeled; this means that RNA synthesis occurs throughout the entire cycle except during anaphase and metaphase. But a radioactive tag on a DNA precursor reveals that only during the S phase is there DNA synthesis.

It is also important to remember that a cell that has synthesized DNA is a cell that, with a few exceptions, will divide in the very near future. Thus, for an understanding of the mechanisms that control cellular proliferation, it is important to investigate the factors that control DNA synthesis. Our recent knowledge of the cell cycle has therefore led to a shift in the focus of investigation from mitosis to DNA synthesis.

Another point to remember is that not all cells keep going through the cell cycle indefinitely. As shown in the diagram on page 175, when a cell divides, the daughter cells have two alternatives, either to go through another cycle or to leave it altogether. Cells that leave the cycle are called differentiated cells and will eventually die without any further division. Many cells in an adult organism also have lost the capacity to make DNA and therefore the capacity to divide. These cells often have other specialized functions in the body; examples are nerve cells and muscle cells.

The synthesis of other macromolecules (giant molecules, like DNA) connected with the gene-action system is another field of active investigation. We have described how we can investigate the synthesis of proteins and RNA with radioactive isotopes, and we have given some information on the gene-action system, which is shown in the diagram on page 178 and discussed again shortly.

The genetic material of a cell is DNA. The DNA molecule is in the form of a double-stranded helix that is supported by a protein backbone. Genes are often described as simply seg-

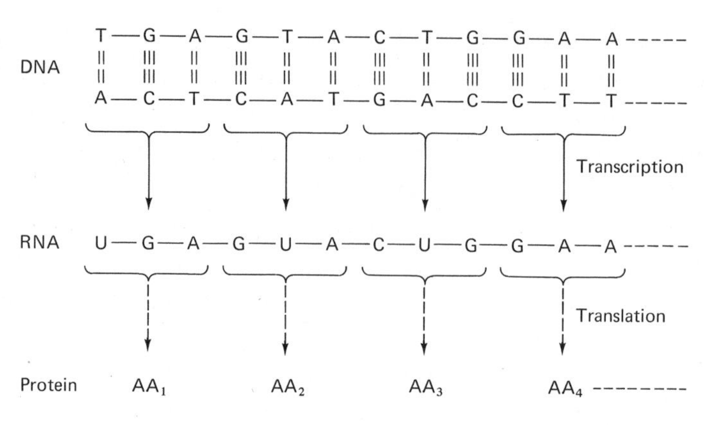

The gene-action system

ments of DNA. They differ from each other only in the order in which the four nucleotide bases that make up DNA are arranged. (Look again at the diagrams on pages 166 and 168.) Since a single gene is usually made up of several hundred bases, it is easy to imagine the infinite variety of genes that could exist by simply changing the order of the four bases several hundred times.

Not all genes in the cells of a living organism are active. In fact, most of them are inactive, or, as geneticists say, repressed. What represses genes to make them inactive is not known, but many investigators believe the activity, or lack of it, is regulated by proteins called histones. If a gene is repressed, nothing happens; it remains inactive, presumably until something removes the repressing factor. But an active gene sets in motion a train of events that results in activation of one of the processes of life: The gene's DNA directs the manufacture of RNA, which in turn brings about the synthesis of a specific protein to carry out a specific metabolic process. In other words, all the activities of the cell are dictated by active genes (the DNA molecules) through the mediation of RNA and are executed by proteins.

Here is what happens, as nearly as scientists can reconstruct it: the DNA of a particular active gene manufactures a molecule of m-RNA by the same kind of replication that it uses for making more DNA. In m-RNA the sequence of bases is the same as in the parent DNA segment; for this reason, m-RNA is also called DNA-like RNA. As the diagram of the gene action system shows, a cytosine molecule in m-RNA corresponds to a cytosine molecule in DNA, a guanine to a guanine, and so on, except that the m-RNA has uracil in all the places where thymine occurs in DNA. The order of the nucleotides in the m-RNA is the same as that in the DNA, so the m-RNA carries the genetic code of the gene that made it. This process, all of which occurs in the cell nucleus, is one of copying, or transcription, rather than translation, since the same codewords (the nucleic-acid bases) are reproduced.

The new m-RNA molecule then travels from the nucleus to the cytoplasm and attaches itself to an unoccupied ribosome (see diagram, page 180). Here it fits to a molecule of r-RNA and blends its shape geometrically, or spatially, with the shape of the r-RNA in lock-and-key or jigsaw-puzzle fashion. The combined new RNA molecule is now capable of manufacturing a specific protein.

At this point an s-RNA molecule arrives, bringing with it one amino acid molecule, which then combines with other amino acids in the specific order dictated by the RNA to form a specific protein. After the amino acids have been formed into the protein molecule, they detach themselves from the s-RNA molecule. The s-RNA molecule has two recognition sites by which it matches up to its neighbors: One recognizes, or fits, the amino acid, and the other recognizes a corresponding triplet of bases on m-RNA. There is thus a particular s-RNA molecule for each amino acid and a particular triplet of bases on the m-RNA molecule for each triplet of bases that is specific to the s-RNA molecule.

In this process the machinery has translated the nucleic-acid code into the protein code; that is, it has translated a sequence of the bases into a sequence of amino acids. This process is therefore called translation of the genetic message. Once the protein has been synthesized, it will become active in performing some of the cell's metabolic activities.

The gene-action system actually is somewhat more elabo-

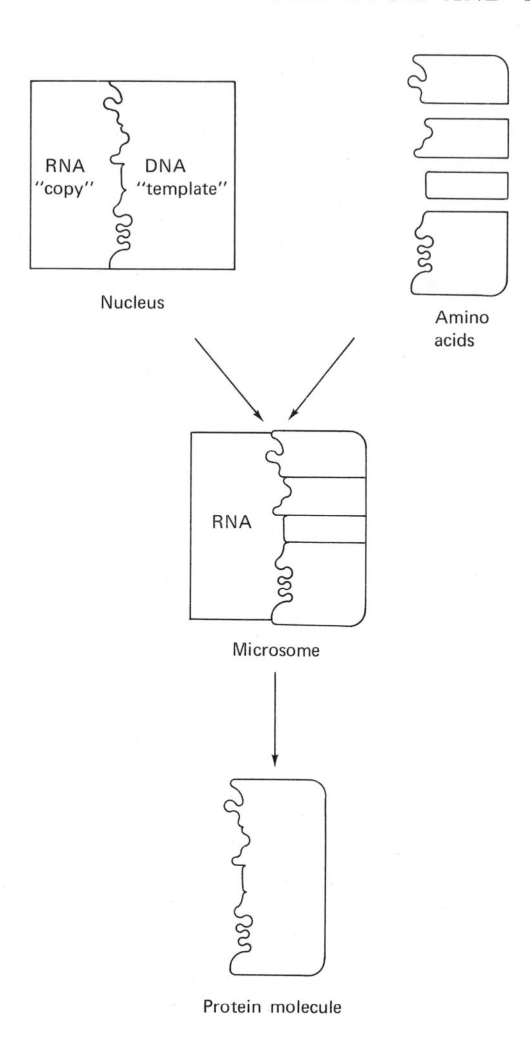

Protein synthesis in a ribosome (microsome), and its control by DNA in the nucleus, using RNA as an intermediary

rate than this. There are feedback mechanisms, genes that control the activity of other genes, either directly or through the production of specific proteins, and so on. However, the scheme just outlined gives a fair, if simplified, idea of how the genetic message is carried to the entire cell and how it is translated into actual life processes.

Our knowledge of the cell cycle and of the gene-action system has been useful in determining how organisms grow and how cancer cells behave. It has been determined that certain normal adult cells divide more frequently than some cancer cells and that the growth of cancers depends not so much on the speed of cellular proliferation as on the number of cells actually dividing.

Knowledge of the cell cycle has also brought new insight to the control of cell division, as in studies related to the therapy of cancer. The most important problem now is, not the control of cell division, but the control of the synthesis of DNA.

Our information on the gene-action system provides broad new opportunity for the investigation of many life processes. Hormone action, processes by which the body develops immunity to disease, and even cell division itself are apparently regulated through the gene-action system. This, in turn, offers possibilities for investigations meant to control these processes.

It is difficult to chart the future course of modern molecular biology, but it is not difficult to predict that the next few years will bring to biology the same kind of sweeping advances that revolutionized physics a few decades ago. The DNA molecule has been called the atom of life. When man has mastered the genetic code, he'll hold a vast power in his hands—power over the nature of coming generations. (29)

10 ⸙ *Your Body and Radiation*

HAVING considered the biology of cells—which radiation was an aid in learning about—and the genetic effects of radiation on cells, the next subject is the somatic (bodily) effects, both good and bad, of radiation on cells.

Unfortunately (or fortunately) man makes a rather touchy experimental animal. So, we'll be forced to take most of our information from experience with cells, large and small cooperative mammals, occasional bacteria and other plants, and whatever human incidents or accidents we can find.

Radiation, unlike Caesar's Gaul, can be divided into only two parts, ionizing and exciting. Ionizing radiation is the sort that is of chief interest to us here. This is simply because the ionization (or the splitting into charged fragments) of any of the body's atoms, which were not meant to be ionized, creates havoc in the body's well-ordered economy.

Charged particles will be attracted to other charged particles (those that make up the matter they approach) of the opposite sign or repelled by those of the same sign. This electrical force is so strong that charged particles penetrate matter (including tissue) very poorly. Their energy is spent so quickly pushing and pulling the protons and electrons they meet near the surface of what they strike that it's all used up before they get far.

The neutral particles, chiefly neutrons and photons, have no charge to use in this way. Their ability to ionize comes about by direct collision. Much like cue balls on a pool table, they either knock another pool ball hard enough to ionize it by sheer impact, or pretty much miss it completely. Since the particles in molecules are relatively about as far apart as stars in the universe, neutral cue balls miss lots more often than they hit. So, they get much farther than their charged brothers before losing all their energy.

Each particle has its own favorite targets. Charged particles mostly pick on the equally charged electrons orbiting outside the atomic nuclei. This is largely because the electrons occupy and enclose so much more space than the nucleus. Photons also do most of their dirty work on the lighter electrons. [A photon would have to be very energetic to affect the heavy nucleus.]

Neutrons have about the same mass as hydrogen atoms, so that hydrogen nuclei (protons) are their favorite target. Thus living tissue, rich as it is in H_2O, gets badly punished by neutrons. Other kinds of nuclei get some punishment, but the heavier they are the less they get. [A very heavy nucleus will suffer only a very small recoil when hit by a low energy neutron.] Since neutrons are uncharged they tend to brush through electron clouds with little permanent damage, much like baseballs through a thick fog.

As these particles lose energy they come to different ends. Electrons and protons simply slow down by progressive bounces until they can undergo chemical reaction with the molecules around them. A proton, for example, eventually becomes either an atom of hydrogen, H, or the common ion of water, H_3O, the hydronium ion. (The simple hydrogen ion H^+ reacts immediately with H_2O to form positively charged H_3O^+.)

A positron slows down about as much as an electron does, but instead of simply coming to rest, it ends up by being captured by an electron. The pair then "commits suicide," usually becoming two photons, in the process known as annihilation. Photons themselves are eventually absorbed completely and usually produce heat.

Neutrons are eventually captured by nuclei to produce new and heavier nuclei that are often unstable. These unstable nuclei are radioactive and they spit out photons and/or charged particles until new but stable nuclei finally result.

Particles of radiation, like bullets, have little effect unless they possess kinetic energy.

The energy per particle may be small, but with such large numbers of particles, the total energy may be quite large. Then, too, chemical molecules are held together by energies of only a few eV so that the collision of even a 1-keV particle can cause considerable chemical havoc (keV means kilovolt): 32.5 eV will cause the ionization of almost any molecule in living tissue.

Molecules, especially biological ones, are most often held together by electron pair (covalent) bonds. The diagram below shows the most common of biological molecules, water, being split by ionizing radiation. If the bond is split on one side (A) or the other (B), two ions (charged chemical fragments) result. Such ions are usually quite reactive chemically. Now the splitting of the molecule is, itself, disturbing to the biochemistry of the organism. But the further reactions of the ions damage other molecules nearby as well, adding a sort of insult to injury!

The bond can also be split down the middle (C). In this case the products are not charged and, therefore, not ions. Instead they are called free radicals, because they contain at least one unpaired electron. Because of the tendency of these free electrons to pair themselves and form covalent bonds, free radicals are much more reactive than ions and the net unpleasantness in tissue is even greater. It is believed that most of the biochemical damage of radiation comes via these free radicals.

Once these radicals are formed there may be enough of them around to react with one another (D). Reforming the original molecule (Da) leaves everything pretty much as before. Formation of hydrogen (Db), while not pleasant, results in little net damage; a few loose H_2 molecules can be tolerated by the body. However (Dc) leads to formation of hydrogen peroxide, H_2O_2, which is distinctly unhealthy. In fact, chemical poisoning by hydrogen peroxide resembles radiation illness in many respects.

At least as bad is the combination of a free radical with oxygen (E). The resulting HO_2 radical seems to have even more undesirable habits than hydrogen peroxide.

From this one can see the possibility of two kinds of molecular damage. In the first, called the direct effect, a biologically important molecule is struck directly by an incoming particle and split into biologically useless fragments. Probably the most important molecules in the living cell are the DNA molecules of the cell nucleus. These carry the master blueprints needed by the cell to reproduce itself properly. Direct destruction of a DNA molecule results in a cell that can live but not divide. On

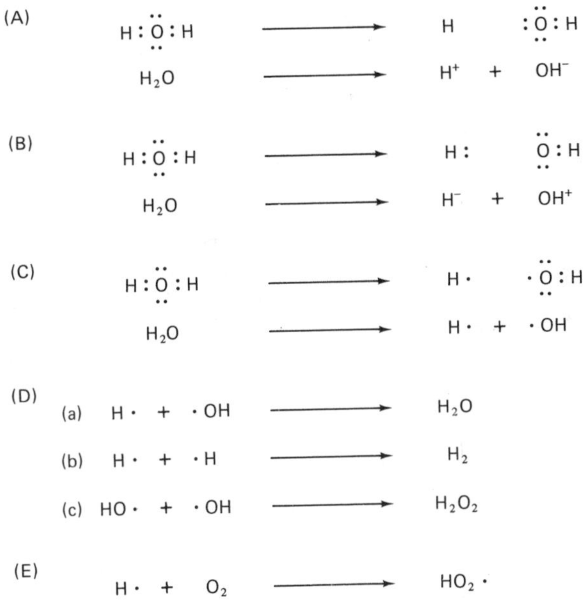

(A)

$$H:\underset{\cdot\cdot}{\ddot{O}}:H \longrightarrow H \qquad :\underset{\cdot\cdot}{\ddot{O}}:H$$

$$H_2O \longrightarrow H^+ + OH^-$$

(B)

$$H:\underset{\cdot\cdot}{\ddot{O}}:H \longrightarrow H: \qquad \underset{\cdot\cdot}{\ddot{O}}:H$$

$$H_2O \longrightarrow H^- + OH^+$$

(C)

$$H:\underset{\cdot\cdot}{\ddot{O}}:H \longrightarrow H\cdot \qquad \cdot\underset{\cdot\cdot}{\ddot{O}}:H$$

$$H_2O \longrightarrow H\cdot + \cdot OH$$

(D)

(a) $H\cdot + \cdot OH \longrightarrow H_2O$

(b) $H\cdot + \cdot H \longrightarrow H_2$

(c) $HO\cdot + \cdot OH \longrightarrow H_2O_2$

(E)

$$H\cdot + O_2 \longrightarrow HO_2\cdot$$

Chemical effects of ionizing radiation. The dots in the upper sections of A, B, and C represent electrons in the outer shell of the atom. Throughout, $^+$ and $^-$ superscripts denote positive and negative ions, respectively. A dot associated with a neutral atom or molecule (see lower part of C; D and E) indicates that there is at least one unpaired electron.

dying of old age it leaves no daughters behind to carry on its work. Such progressive cell death without replacement soon leads to the malfunction and eventual death of the irradiated tissue. If this dying tissue is essential to the organism, and cannot be replaced in time, the entire organism will degenerate and die prematurely.

An indirect effect occurs if a less critical molecule, usually water, is split into reactive ions or radicals. If these reactive fragments then drift over to react with such critical molecules as the DNA of the nucleus, damage will be much the same as if they had been struck directly. In direct action no time lag exists between collision and destruction since the particles themselves travel at nearly the speed of light. With indirect action the diffusion of ions and radicals may be sufficiently slow that chemical protective agents may be placed in their path, and thus sacrificed to protect the most critical molecules.

As a charged particle speeds along through tissue it collides with parts of atoms every once in a while, much as a bullet shot into a thin forest collides with leaves and branches. At each collision about 100 eV of energy is lost and about ten free radicals and ions are left behind in a little clump called a spur.

Electrons (and positrons), being rather light, ricochet fairly easily, and their spur tracks are a bit erratic. The spur track of a 1-MeV electron might look like this:

This is the case whether these particles impinge on the tissue as primary beta rays or have been secondarily produced by photons (from a primary beam of X or gamma rays) colliding with electrons.

Protons and alpha particles, being much heavier, ricochet less easily. They resemble cannon balls more than bullets, and leave short, dense spur tracks. The track of a 1-MeV proton looks like this:

p ···············——

This is the case, again, whether a primary proton is involved or a secondary one produced by a primary fast neutron.

Two features of this behavior are significant. (1) For a given energy loss (1 MeV in this case), an electron travels much farther than a proton so that the spacing between electron spurs is greater; and (2) as either particle reaches the end of its track, spur spacing decreases.

In general, spur spacing decreases with increasing particle mass, increasing particle charge, and decreasing particle energy.

In radiobiology, particles can be compared on the basis of their average linear energy transfer (LET). This is just the average amount of energy lost per unit of particle spur-track length. The average amount is specified, to even out the effect of a particle that is slowing down near the end of its path and to allow for the fact that secondary particles from photon or fast-neutron beams are not all of the same energy. Some average LET values are given in the table on page 188, expressed in keV of energy lost per micron [10^{-6} meter] of spur-track length in tissue. As in the case of spur spacing, LET increases with increasing mass, increasing charge, and decreasing energy.

For a first approximation, photons and neutrons may be thought of as producing electrons or protons, respectively, of about half the photon or neutron energy. The average LET of a 10-MeV fast-neutron beam would be about the same as that of a 5-MeV proton, or 8 keV per micron.

For charged particles the relationship between average LET and penetration is fairly obvious. The higher the LET (the more keV of energy lost per micron of travel) the sooner its energy is used up and the shorter its range (maximum penetration). Of course, LET itself drops with increasing energy, and a high-energy particle will have a greater range than would be predicted from simple proportion to a low-energy particle.

PARTICLE	MASS / MASS OF PROTON	CHARGE	ENERGY (KEV)	AVERAGE LET (KEV/ MICRON)	TISSUE PENETRATION (MICRONS)
Electron	0.00055	−1	1	12.3	0.01
			10	2.3	1
			100	0.42	180
			1000	0.25	5000
Proton	1	+1	100	90	3
			2000	16	80
			5000	8	350
			10,000	4	1400
			200,000	0.7	300,000
Deuteron	2	+1	10,000	6	700
			200,000	1.0	190,000
Alpha	4	+2	100	260	1
			5000	95	35
			200,000	5	20,000

Of course, in reverse order, as a charged particle loses energy in tissue its LET increases, whereupon it loses energy even faster, whereupon its LET increases still more, and so on. This is the explanation for the rapid decrease in spur spacing toward the end of the tracks shown in the diagrams on pages 186 and 187. It also partly explains why we use average LET figures for comparing radiation biologically, but measured range for computing penetration.

The increase of ionization at the end of a charged-particle track has been made use of in the treatment of deep-seated tumors. Initially low-LET (by reason of high energy) charged particles will depost a relatively small fraction of their energy near the skin and a larger fraction in the tumor. By the time they reach the tumor, they have lost enough energy so that their LET has increased. (39)

Neutral particles penetrate deeper than charged particles of the same energy, and the more energy they have, the deeper they penetrate.

Like drugs, radiation can either heal or poison, depending on the amount given. Thus, amounts of radiation delivered are spoken of as doses and measurement of these amounts is called dosimetry.

The original international unit of radiation, the roentgen (abbreviated r), depends on the ionization of air and, while useful for X and gamma rays, has proved inadequate for the measurement of forms of radiation (such as neutrons) that ionize tissue better than they do air.

The most widely used unit today is the rad (*r*adiation *a*bsorbed *d*ose). The rad is defined as that quantity of radiation that delivers 100 ergs of energy to 1 gram of substance—in this case tissue. In practice the two units seldom differ by more than a few percent when we refer to tissue, so that we will consider them equivalent from here on.

The rem (*r*oentgen *e*quivalent, *m*an) is a biological, rather than a physical, unit of radiation damage. It represents that quantity of radiation that is equivalent—in biological damage of a specified sort—to 1 rad of 250 KVP X rays (kilovolt peak, a mixture of photons from 250 KeV down to about 50 KeV. When not otherwise noted, doses will refer to this relatively standard radiation).

The ratio of rem to rad is called the relative biological effectiveness (RBE). While the rem and RBE are related to the average LET for a given type of radiation, the numerical relation is not entirely clear. Hence, both rem and RBE comparisons are subject to greater errors than rad comparisons.

A teaspoonful of castor oil will have very different effects on a 25-gram mouse and a 70,000-gram man. What's important in dosimetry of any sort is not so much the total dose to the whole system as the dose per gram. That's why a physician prescribes different doses of medicine for different-sized people. The rad has already been defined as energy per gram to take this into account. Thus when we say that 1000 rads leads to the death of nearly any mammal we mean the delivery of $1000 \times 100 = 100{,}000$ ergs of radiation to each gram of the mammal's tissue.

Now 100,000 ergs of energy is not much by our usual standards. It would only raise the temperature of a gram of tissue about 0.0025°C—much less than the rise in body temperature brought about by a couple of deep breaths. Where, then, does radiation get its power to damage?

The answer lies in its ability to change critical molecules, either by direct ionization or by indirect chemical reaction. Heat or mechanical energy is spread out more or less evenly among tissue molecules; no one molecule receives enough energy to injure it. But each particle of radiation packs enough wallop to smash to bits the chemical molecules it encounters.

Damage from radiation tends to be largely cumulative so that, in general, a few thousand rads will be lethal whether given over seconds or years. The maximum lifetime dose that can be considered "safe" has been set at about 250 rads.

It is not only the number of rads that is important, however, but also where they were absorbed. For example, one might tolerate a dose of several thousand rads to a limited part of the body in an emergency or for the cure of a serious disorder, where a whole-body dose of this size would certainly be lethal. This amounts to tolerating a badly burned hand to save the body, where the same burn over the whole body would be fatal.

Shielding against charged particles is relatively easy, since they interact so well with matter. Thin sheets of metal, paper, or even the air itself often provide nearly complete shielding.

The uncharged X-ray and gamma photons, and neutrons, tell us a different story. Lead is about as opaque to gamma rays as Kleenex is to light. Because complete absorption is unlikely, shielding materials are compared in terms of their half-value thickness for a given radiation. This is the thickness that will reduce the radiation intensity to one-half its unshielded value. The same sort of law applies as for half-lives; two thicknesses yield ¼, three ⅛, four $^1/_{16}$, five $^1/_{32}$ the dose, etc. The half-value thickness of lead for cobalt 60 gamma rays is about 1 centimeter, for example, while the half-value thicknesses of earth and water are about 5 centimeters. Earth and water often are used if space permits, because they are inexpensive.

When these matters were first studied, the possibility existed that the chief effect of radiation would be the disruption of the organization or communication system of the body, which would leave the cells relatively uninjured. Considerable evidence has accumulated, however, to indicate that the cell, and particularly its nucleus, is the primary site of radiation damage. Disorganization of the body follows only after a sufficient number of cells have been so injured that they can no longer carry out their normal functions.

The average cell lifetime in a body is much less than that of the body itself, just as individual lifetimes in a human society are much less than that of the society itself. Accordingly one of the most important functions of a cell is its own reproduction. As a cell dies it is replaced by the daughters of its sister cells.

The chief function of the cell's nucleus appears to be that of overall control of the cell. It acts as an executive whose records and blueprints are contained in the biochemical coding of its DNA molecules, which are organized into chromosomes. The very ability of the cell to repair itself depends on its ability to read these blueprints. Their destruction or damage leaves the cell without proper reference material on which to base needed biochemical decisions.

If damage to the coding on a DNA molecule is slight and occurs during that part of a cell's lifetime when it contains two sets of DNA, the damage can apparently be repaired properly and the cell returns to normal. If the damage is more severe and leaves the cell with one or more destroyed blueprints, it may be able to maintain metabolic activity but loses its previous identity. This cellular amnesia may render the cell unfit to continue in its environment, and it or its daughters may eventually die away.

However, the amnesic cell may be quite vigorous in its environment and yet have forgotten its previous identity so completely as to have passed out of the growth control of the body. All normal cells are controlled in their growth by the needs and demands of the body as a whole. But this susceptibility to control apparently requires a specific normal complement of DNA molecules in the nucleus. An uncontrolled cell becomes a cancer cell; it may die away, it may simply remain—growing slowly and doing little damage (a benign tumor), or it may invade and destroy its host (a malignant cancer). If damage is still more severe the cell will lose its ability to divide properly and will die, usually after a period of confused growth of giant cells. If damage is more severe than this, not only DNA but other cell components are damaged beyond repair. Cellular activity slows markedly, the cell becomes visibly abnormal, and it dies quickly. (For illustrations of abnormal cell development, see photograph section.)

Radiobiological damage is considered to occur, then, in this light. If a critical DNA molecule in a cell nucleus is damaged by radiation, the cell becomes deranged. If enough cells become deranged, the tissue, then the organ, and finally the body become disordered to a degree that depends on the severity of the damage to the DNA molecules and the number and relative importance of the cells deranged. Thus radiation damage ranges from the unimportant death of a few replaceable cells (much like a mild sunburn) through the induction of tumors and cancers, to premature aging, acute illness (radiation sickness), or nearly immediate death.

A few human cells, notably those of the nerves (neurones) and the red cells of the blood (erythrocytes), are incapable of division. The latter cannot divide because they have no nucleus. But, as they die in the normal course of events (their average life span is about 120 days), new ones are quickly supplied by the hematopoietic (blood-forming) systems of the spleen and bone marrow. Nerve cells, on the other hand, have nuclei and are able to regenerate a lost fibre (axone) or other cell part. But the nuclei themselves simply seem to be incapable of division; thus when once destroyed, they can never be supplied.

Two aspects of cell death from such nucleoplasmic poisons as radiation should be stressed. The first is that the low redundancy (the low degree of repetition of identical molecules) in the nucleus makes a little damage there much more serious than the same amount of damage in the much more redundant cytoplasm. The other aspect is the delay between nucleoplasmic damage and obvious derangement of the rest of the cell. Damage to the nucleus isn't obvious until the cell tries to divide.

But, by the same token, much more radiation is needed to kill the cell outright than to damage the nucleus badly enough to ensure the cell's death at its next division. This is simply because the most important and easily damaged molecules of the cytoplasm, probably its enzymes (the protein catalysts responsible for the cell metabolism) occur in relative profusion. Nearly all the many enzyme molecules of a particular kind must be destroyed before the cell dies for lack of them.

Radiation, as a poison, has some effects that are similar to those of the lead salts. It can be delivered all over the body at once; it can be delivered to certain parts of the body; its effects are cumulative; and some body tissues are more affected than others.

Also, like insoluble lead salts, radiation effects can be abscopal (literally, "away from the place you're watching"). This means that the results of local damage can be felt elsewhere, usually in a different form. A bad burn on the leg, for another example, can give one a bad headache as well as a feeling of being miserable all over.

In three respects radiation, as a poison, is nearly unique. (Certain chemical poisons are said to be radiomimetic, because they mimic the actions of radiation.) The first lies in its nearly specific effect on the critical molecules of the nucleoplasm. Few chemicals can behave in this way, because they must diffuse slowly through the cytoplasm before they can reach the nucleus and are usually captured on their way through.

Second, the damage is done at high speeds. Unlike chemical poisons radiation cannot be removed before its full damage is complete.

Finally, radiation can destroy the body's normal immune response. A body's best defense against outside invaders is its ability to rapidly develop specific biochemical defenses, called antibodies, that act against the invaders. Many antibodies continue in production for life once they begin, and make one immune to further attack by that particular invader (mumps, for example). Others last only as long as the invasion is active. Radiation can destroy the body's ability to produce or maintain these antibodies and, in this way, leaves one prey to serious infection by any invader that comes along.

Nausea, vomiting, hemorrhage, diarrhea, loss of weight, and severe anemia are symptoms that, taken together with the possibility of exposure to radiation, constitute the syndrome of acute radiation illness.

The severity of the symptoms of poisoning will depend on the dose, of course. But it will also depend very much on the individual as well. Therefore, we can discuss only what will happen to an individual on the average. In addition, most of our information on radiation effects has been obtained from mice and other laboratory animals. This information must be extended to humans by the laws of scaling, that is, according to the increase in scale, or relative size, of human beings compared to the experimental organism. The sort of individual response one finds in a population of animals to different doses of radiation (or of nearly any other poison for that matter) is shown in the following graph. Very few animals are affected until the dose is somewhat over 500 rads, while 50% are affected by 1000 rads.

Although the curve given is that for the death of mice exposed to ^{60}Co gamma rays, similar curves of the same sort apply equally well to other dose-effect studies. Thus, with a change of dose scale, it would fit for the percent of humans cured of infection by increasing doses of penicillin or the number of little boys developing tummy aches on eating increasing numbers of green apples.

Since the largest percentage of individuals will be affected by the smallest change in dose at the 50% level we ordinarily use this point for reference. The curve is steepest at 50% so that

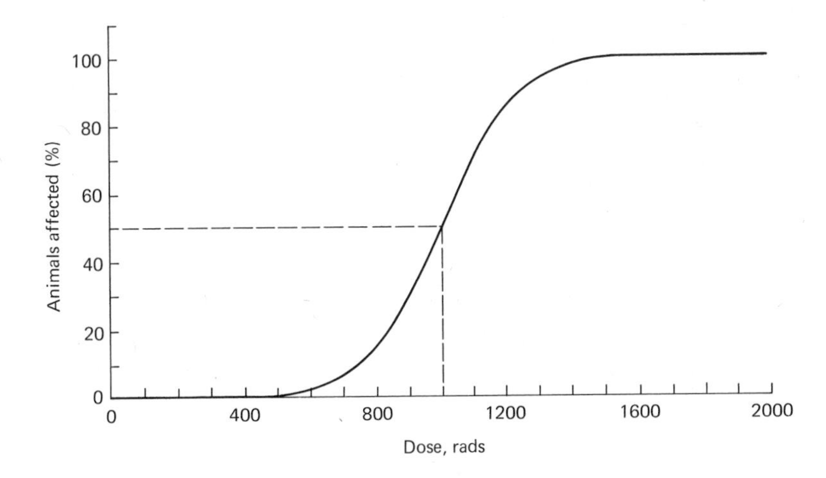

A dose-effect distribution. This type of curve is called "sigmoid" from the Greek for the letter S, because its shape is that of a rather lazy S.

one can observe there the greatest percentage increase in subjects affected for a small increase in dose. Thus a D-50 value is that dose at which 50% of the population studied showed the effect described.

Just how nasty a poison can be depends, of course, on its natural nastiness (LET) and on the total dose (rads). If living creatures weren't (living, that is), that would be all there was to it. But cells and people alike are able to recover from unpleasantness, the degree of recovery depending not only on the unpleasantness itself but also on the time between successive doses. There is also, for most poisons, a threshold dose below which no effect is observed.

Radiation can occur steadily or in intermittent chunks. Furthermore, it can be delivered to the whole body or to parts of it. Each of these factors will have its own effect.

If the radiation occurs steadily, the situation is similar to that of Napoleon, who, it seems, was being steadily poisoned with arsenic and mercury most of his life. (He may have taken them as tonics. See chapter 24.) When poisons come in slowly enough (when the dose rate is low), the body can recover from the damage about as rapidly as it occurs. This seems to be the case with radiation from the natural background. We can apparently recover from 10 rads delivered at the rate of 0.15 rad per year without anything like the damage we might suffer if given 10 rads all at once. Our natural recovery is seemingly treading water quite nicely at 0.15 rad per year.

In much the same way we can usually handle several spaced-out doses much better than a single dose of the same total amount. Then, too, recovery depends on the kind of radiation. The body recovers from low-LET radiation much more rapidly and completely than from high-LET radiation. The latter appears to damage seriously the systems necessary for recovery.

From this one would expect damage from low-LET radiation to increase with increasing dose rate, and this is indeed the case. The rapid administration of low-LET radiation can jam the recovery mechanism badly where slow administration would not. But if the recovery mechanism itself is badly hurt by even small doses of a more complex poison (high-LET radiation), an increased dose rate can add little to what has already been done.

Finally, recovery can be greatly affected by the site of administration. Much more radiation is usually needed to injure an appendage or isolated organ than is needed to cause death if given over the whole body. Still, if the dose to a region of the body is high enough, the effect produced locally can eventually bring about the death of the body.

A very interesting aspect of chronic low-LET irradiation is that low levels at low dose rates can have the effect of lengthening life. This is quite commonly observed with laboratory animals, and is sometimes called the 102% effect because the lifespan is increased about 2%. Even more startling is the effect on the flour beetle, *Tribolium confusum* (so called because he never seems to know where he's going. Of course, in flour, it doesn't matter much). His life expectancy is increased about 30% by a dose of 3000 rads delivered at rates up to 10 rads per minute. The explanation probably lies in reduction of infection.

Up to now we have tacitly assumed a sort of normal human body for our doses, that of a healthy young adult. One would expect the ill and the aged to tolerate radiation less well, and this is, in fact, true. Less obvious is the effect on youth. Since the young are growing, their cells are dividing rapidly. But cells are most susceptible to radiation when dividing, so we might expect the young to be more sensitive than adults. This, too, has been confirmed by experiment and by observations of exposed humans. The unborn are most sensitive (see photograph section). After birth sensitivity declines as maturity is reached, then rises again as old age is reached.

Increased body temperature and a higher metabolic rate also seem to increase radiosensitivity over the increase to be expected from increased cell division. The exact nature of these effects, however, is not known definitely and certainly is not well understood. One clear effect has been observed, though: hibernation halts radiation damage almost completely. An animal given a lethal dose, then permitted to hibernate immediately after exposure will die—not within a certain number of days after exposure but within that number of days after he awakens! This is true even if he hibernates for many months. The reason is probably a combination of lower body temperature, lowered metabolism, and virtually nonexistent cell division during hibernation. (39)

Recent experiments indicate that radiation may interfere with the functioning of an animal's nervous system. Monkeys were trained to pull an overhead handle at a signal—either a flashing light or a tone. Scientists measured the time it took the monkeys to respond after the signal was given. Then they exposed the animals to high doses of mixed gamma and neutron radiation. When the signal was given again, the animals did not respond correctly, and appeared confused. Why? The scientists believe that high-level radiation interferes with the way the brain receives information from the sense organs.

It also appears that low levels of radiation may affect functioning of the nervous system. Sleeping animals, such as rats and cats, are immediately aroused when subjected to low radiation. Scientists have also measured brain waves of animals exposed to low-level radiation and found electrical changes similar to those produced by certain drugs. (1)

Can unaided human senses detect radiation? A strong enough dose rate can be felt; the effect is probably that of direct stimulation of the very sensitive touch receptors in the fingertips. An intense beam of charged particles ionizes the air, producing a glow much like that in a neon tube which can be seen.

Fast-moving electrons cause nearly any material to give off light. This is called Cerenkov light and is the blue glow often seen in color photographs of nuclear reactor cores.

But a number of people have reported instances of flashes and bright spots before their eyes during irradiation of the head. Whether this is due to Cerenkov light in the eyeball or to a primary action on the visual cells is unknown.

It has been shown quite clearly that rats can detect X rays by smell. This ability may be due to a primary effect on the olfactory cells. However, rats also can smell lower concentrations of ozone than humans, and ozone is produced whenever oxygen is ionized by radiation.

No examples of tasting or hearing radiation are known as yet. But certain hypersensitive mice, which go into convulsions on hearing the ringing of a doorbell, seem to do so much more readily if first exposed to 100 millirads or so of radiation.

In theory, at least, there is no reason to imagine that radiation cannot stimulate the nervous system, if only by indirect action. One rad will produce about 10^{13} highly reactive ions and radicals in 5 grams of tissue. This is much less than the amount of "perfume" that a female cockroach need pass on to a 5-gram male to stimulate an active interest on his part! (39)

Besides harmful effects of radiation on our bodies and helpful uses of radiation as an analytical tool in the study of life processes, radiation can be used in medical diagnosis and therapy.

Medical diagnosis would be set back half a century without the X-ray picture, whether traditional X rays are used or photons from radionuclides. Today, too, the ability of tracer radionuclides to follow the chemistry of normal processes is used routinely in the diagnosis and study of dozens of disorders. In one year recently, well over half-a-million atomic cocktails were given for the diagnosis of serious disorders—a tribute to the beneficial use of radiation.

X rays long provided almost the only alternative to surgery for treatment of cancer and were often chosen for a fair number of other disorders. Today radionuclides are helping this effort along, partly by providing higher energy photons for greater penetration and lower LET at the skin, and partly for their ability to localize in certain tissues. (39)

A radioisotope may be used either as a source of radiation energy or as a tracer: an identifying and readily detectable marker material. The location of this material can be determined with a suitable instrument even though an unweighably small amount of it is present in a mixture with other materials. In general, tracers are used for analysis and diagnosis, and radiant-energy emitters are used for treatment (therapy).

Radioisotopes offer two advantages. First, they can be used in extremely small amounts. As little as one-billionth of a gram can be measured with suitable apparatus. Secondly, they can be directed to various definitely known parts of the body. For example, radioactive sodium iodide behaves in the body just like normal sodium iodide found in the iodized salt used in many homes. The iodine concentrates in the thyroid gland where it is converted to the hormone thyroxin. Other radioactive, or tagged, atoms can be routed to bone marrow, red blood cells, the liver, the kidneys, or made to remain in the blood stream.

Of the three types of radiation, alpha particles are of such low penetrating power that they cannot be used for measurement from outside the body. Beta particles have a moderate penetrating power, therefore they produce useful therapeutic results in the vicinity of their release. Gamma rays are highly energetic, and they can be readily detected by counters used outside the body.

In one way or another, the key to the usefulness of radioisotopes lies in the energy of the radiation. When radiation is used for treatment, the energy absorbed by the body is used either to destroy tissue, particularly cancer, or to suppress some function of the body. Expressed in terms of the usual work or heat units, ergs or calories, the amount of energy associated with a radiation dose is small. The significance lies in the fact that this energy is released in such a way as to produce important changes in the molecular composition of individual cells within the body.

When a radioisotope is used as a tracer, the energy of the radiation triggers the counting device. This differentiates the substance being traced from other materials naturally present.

Once ordinary salt gets into the blood stream, for example, it has no characteristic by which anyone can decide what its source was, or which sodium atoms were added to the blood and which were already present, except when some of the atoms are tagged by being made radioactive. Then the radioactive atoms are readily identified and their quantity can be measured with a counting device.

A radioactive tracer, it is apparent, corresponds in chemical nature and behavior to the thing it traces. It is a true part of it, and the body treats the tagged and untagged material in the same way. A molecule of hemoglobin carrying a radioactive iron atom is still hemoglobin, and the body processes affect it just as they do an untagged hemoglobin molecule.

It should be evident that tracers used in diagnosis—to identify disease or improper body function—are present in such small quantities that they are relatively harmless. Their effects are analogous to those from the radiation that every one of us continually receives from natural sources within and without the body. Therapeutic doses—those given for medical treatment—by contrast, are given to patients with a disease that is in need of control, that is, the physician desires to destroy selective cells or tissues that are abnormal. In these cases, therefore, the skill and experience of the attending physician must be applied to limit the effects to the desired benefits, without damage to healthy organs.

Normal blood is about 1 percent sodium chloride or ordinary salt. This fact makes possible the use of ^{24}Na in some measurements of the blood and other fluids. The diagram illustrates this technique. A sample of ^{24}NaCl solution is injected into a vein in an arm or leg. The time the radioisotope arrives at another part of the body is detected with a shielded radiation counter. The elapsed time is a good indication of the presence or absence of constrictions or obstructions in the circulatory system.

The passage of blood through the heart may also be measured with the aid of sodium 24. Since this isotope emits gamma rays, measurement is done using counters on the out-

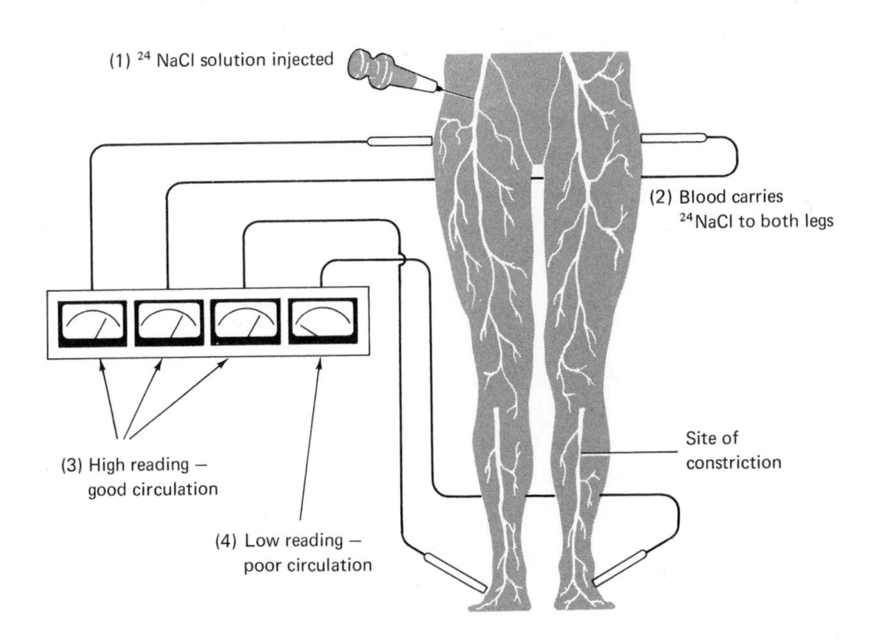

Radioisotope test for circulation of blood

side of the body, placed at appropriate locations above the different sections of the heart. (31)

The fact that cancer cells usually are rapidly dividing cells, and thus more sensitive than other cells, has long been the basis of cancer treatment by radiation. Another treatment utilizes the capture of slow-moving neutrons in otherwise stable nuclides to change the latter into radionuclides. Thus if a fissionable nuclide can be made to localize within a cancer, low-energy neutrons of a few eV can be beamed into the tumor to cause the release of many MeV of radiation energy. (39)

11 ◈ Animals in Atomic Research

More than 2500 years ago, the doctors of ancient India used black Bengali ants to close intestinal wounds. An Indian surgeon would arrange several of these large insects along the opening of the patient's wound. The ants would immediately bite, clamping the edges of the wound together in the bulldog grip of their fierce jaws. Then the doctor would remove the bodies of the ants. The heads were left clinging to the closed wound like a row of black buttons.

Harsh treatment? Not 25 centuries ago, when the only fibers available for stitching wounds quickly decomposed within the body.

Like the doctors of India, men through the centuries have depended on animals for treating the sick and injured. At the same time, these creatures have been—and still are—living books from which man reads the story of life.

Research with animals has furnished scientists with a vast storehouse of knowledge. In fact studies of animals led the way to most great achievements in medicine and biology.

A little over 2000 years later, animal experiments led William Harvey (1578–1657) to one of the most important biomedical discoveries of all time. Harvey, an Englishman, unraveled the true nature of the circulatory system—the twisting network of vessels and organs that channels blood through the body.

Harvey's discovery in 1628 followed 14 years of research with animals. He studied at least fifteen different species, from insects to dogs, and found that the blood pulses away from the heart through one set of vessels (arteries) and back to the heart through another set (veins).

These experiments revealed that the circulatory system is the route over which food, oxygen, and water travel to every cell in the body. It is also the network through which waste is removed, and a route by which disease can penetrate the body's innermost strongholds.

One of the most feared diseases of Harvey's day was smallpox. Eventually an English country doctor, Edward Jenner (1749–1823), conquered smallpox in 1796 with the help of a cow. Jenner noticed that cowpox, a cattle disease that sometimes made farmers and dairy workers mildly ill, was a cousin of smallpox. He obtained some infected matter from an inflammation on a milkmaid stricken with cow pox and injected it into other persons. Thereafter the patients Jenner had so inoculated were immune to smallpox. Jenner named his technique vaccination after *vaccinia,* the Latin word for cowpox (based, in turn, on *vacca,* Latin word for cow).

What occurred in the bodies of Jenner's patients that enabled them to ward off smallpox? The answer came a century later after two European scientists, working independently, conducted extensive studies on anthrax, a disease that animals sometimes transmit to man, usually with fatal results.

One of these men was Robert Koch (1843–1910) a German physician. It is still less than 100 years since Koch, working with anthrax bacteria, first proved that microorganisms—germs—cause disease in higher animals.

Searching for the cause of anthrax, then a serious problem for Europe's farmers, Koch placed bits of tissue from animals sick with anthrax into serum from healthy rabbits. Koch transferred the bacterial culture through several successive samples of rabbit serum. Then he inoculated a healthy mouse with part of the last sample. The mouse contracted anthrax, just as had mice infected directly with blood of animals killed by the disease.

This experiment seemed to prove that the bacteria carry anthrax, yet skeptics still doubted. Maybe something else in the serum produced the disease, they said. Doubts were silenced, however, when the great French scientist Louis Pasteur (1822–95) confirmed Koch's findings. From infected serum, he prepared a culture of almost pure anthrax bacteria. Even a tiny drop of the culture killed experimental guinea pigs and rabbits.

Armed with his germ theory of disease, Pasteur went on to show that vaccination could prevent many other diseases. The germ theory of disease explains how vaccination works: When a disease agent enters the body, the body calls up proteins called antibodies. These are the body's defense against foreign substances such as invading germs. Particular antibodies are produced by the body to match specific diseases. If you vaccinate someone with a mild form of a disease, his body produces antibodies effective against that disease. These proteins stave off possible future attacks of the full-strength illness, and the vaccinated person is then immune to the disease.

Pasteur soon developed a successful vaccine for anthrax, by cooking anthrax bacteria at a high temperature until the germs were weakened. Vaccinated cows, sheep, and goats stayed healthy, but unprotected animals that were exposed to anthrax quickly died.

Anthrax conquered, Pasteur turned to rabies, another disease transmitted from animals to man. Rabies is caused by an incredibly small virus—so small that it takes a very high-powered electron microscope to see it. Pasteur did not have a modern microscope, therefore he could not see the rabies virus. But he suspected that whatever caused rabies was so tiny he could not isolate it in a test tube. The only way to obtain a sample for a rabies vaccine, he conjectured, was to grow the tiny agent in the tissue of a living animal. Pasteur did this by infecting rabbits with rabies. Then he made a vaccine of weakened samples of the infected rabbit tissue and injected it into dogs. It worked. The dogs were able to ward off rabies. The first human test of the new rabies vaccine came on July 6, 1885, when Pasteur saved the life of a young boy who had been severely bitten by a rabid dog.

A half century after Pasteur's work, experiments with dogs led to the discovery of a treatment for diabetes, a disease that afflicts millions of people. Today, doctors know that a diabetic's body lacks insulin. Insulin is a hormone, a chemical messenger that helps cells take glucose from the blood and convert it to needed energy. Without glucose, the body looks elsewhere for fuel to burn. It turns cannibal and feeds on its own fats and proteins. The diabetic patient slowly wastes away.

Insulin, produced by the pancreas gland, was discovered in the early 1920s by two Canadian researchers, Dr. Frederick G. Banting (1891–1941) and medical student Charles H. Best. This important advance resulted from experiments with twenty dogs, conducted in a stuffy Toronto laboratory. Millions of diabetics who otherwise would have died are alive today because of insulin injections.

Many other discoveries in modern biology and medicine stem directly from animal studies. These include discovery of the endocrine glands, techniques of modern heart surgery, and proof that injuries and burns can be treated with blood transfusions.

All told, research animals in this country include about 30 million mice, 12 million rats, a million guinea pigs, a half million hamsters, 150,000 cats, and hundreds of thousands of rabbits and dogs.

At the same time, millions of more unusual animals—from amoebas to alligators, from beetles to salmon—also aid scientific research. Much of this research includes studies of the effects of nuclear radiation on life.

Experiments with animals have shown that the effects of exposure to ionizing radiation depend largely on the part of the body that receives the dose. A dog, for example, might survive a strong dose of radiation to a leg, but become ill or die if its stomach were exposed to the same dose. (Similarly, a cutting wound in a man's arm is usually less serious than an equally severe wound into his abdomen.) Early studies with animals also demonstrated that the tissues most easily damaged by radiation include the bone marrow and spleen, which manufacture ingredients of the blood.

Interestingly, effects of radiation also vary with the kind of animal. Usually, simple organisms withstand larger amounts of radiation than complex organisms. This is because the complicated cellular organization of a complex organism is more easily disturbed than the fundamental cellular network of simple animals. Scientists find, for example, that a snail can live through a dose of up to 20,000 rads. Ants have withstood doses of radiation up to 200,000 rads. But anything more than 450 rads is usually fatal to a man.

Effects of radiation on living organisms are classified in two categories—acute and chronic. Acute effects are usually noticeable almost at once. Chronic effects are the result of small but continuing doses of radiation. They may not be detected for many years.

One of the most interesting relationships between normal life processes and radiation is that between radiation and aging. This relationship is a puzzler. Does a small, nonfatal amount of radiation shorten or lengthen an animal's normal life-span? Experiments have indicated that either effect may result.

Certainly, exposure to radiation may cut short an animal's life by leaving it vulnerable to infection. Moreover, experiments indicate that radiation does seem to cause genuine signs of aging in some animals. Female beagles exposed to doses of 100 rads (chronic dosage) or 300 rads (acute dosage) were found to be aging rapidly only a few years later. The signs of old age were wrinkled skin, worn teeth, graying coats, and heart disease.

Investigations into the effects of gamma rays and X rays on both rats and dogs show that, while the animals may survive a single acute dose, their life-span is shortened in proportion to the strength of the dose received. Other studies, however, provide seemingly contradictory results. Normally the tails of rats and mice stiffen with increasing age. Scientists subjected the tails of rats and mice to high-level radiation, wondering if a dose of this size would result in such a proven sign of aging. It did not.

Some other examples: Rats exposed for a lifetime to 0.8 rad per day lived for 600 days. Control rats, not irradiated, lived only 460 days. Mice, rats, and guinea pigs exposed to 1 rad a week for life lived longer than their expected life-span; and, as was mentioned earlier, flour beetles exposed to small doses of radiation also lived an exceptionally long time.

Much more research must be carried out before scientists can explain fully the relationship between radiation and aging, however. The aging process itself is not completely understood. Radiation studies with animals are providing clues. (1)

Engebi Island, on Eniwetok's northeast reef, is the home of a wholly self-contained colony of Pacific rats living in a network of burrows in the shallow coral sands. After 1948 Engebi was exposed repeatedly to atomic detonations, and in 1952 the whole island was swept clean of growth and overwashed by waves from the thermonuclear explosion of Operation Ivy. On each of these occasions, exposure of the rat colony to radiation was intense. The island environment was so altered that radiobiologists believed it impossible that any of the rats had survived.

Contrary to all expectations, however, the original colony had not been eliminated. Biologists visiting Engebi in 1953 and 1954 found the rats apparently flourishing. New generations of rats were being born and were subsisting on grasses and other plants in an environment still slightly radioactive. In 1955 analysis of the bones of rats revealed the presence of strontium 89 and strontium 90 in amounts approaching what was assumed to be the maximum amount that would not cause bodily harm. The rats' muscle tissues contained radioactive cesium 137. But no physical malformations were found in the rats. All animals appeared in sound physical condition, despite these body burdens of radioactivity. By 1964 the rat population had so increased that it apparently had reached equilibrium with available food supplies.

Questions relating to the reestablishment of the colony are intriguing. Why are new generations of these warm-blooded animals continuing to thrive after the colony was exposed to devastating nuclear effects? Is there a different dose-effect relation for these rats than for other animals? Even if it is assumed, as it must be, that some members of the colony survived the original nuclear heat and radioactivity because they were shielded by concrete bunkers or other man-made structures, how is it that there have been no observable effects among rats existing for years in an area that continually exposed them to radiation? (4)

Radioactivity in a waste disposal pool has revealed a fascinating science mystery. The pool is in a sunny meadow near the Oak Ridge National Laboratory in Tennessee. Yellow-legged mud-dauber wasps use radioactive mud from the slick sides of the pool to build their clumplike nests. The wasps cannot carry their hot cargo more than a few hundred feet, so their nests are all close to the pool. The level of radiation right at poolside is high enough to endanger any human who stayed there very long, but does not seem to harm the wasps.

Strangely, however, a close relative of the yellow-legged mud dauber—the pipe-organ mud-dauber wasp—will not touch mud that is radioactive. Nevertheless, nests of both types are found side by side on the outside of buildings near the lake. The nests of the yellow-legged wasps give off beta and gamma radiation at a relatively high rate, without apparent effects on the wasps. The pipe-organ-wasp nests are almost never radioactive.

Both species have now been bred in the laboratory at Oak Ridge. The reactions of each to radiation are being tested. Yellow-legged mud daubers will use either radioactive or normal mud to build nests, scientists find. The pipe-organ mud daubers use only normal mud. Moreover, hungry pipe-organ wasps will not even eat honey that has been placed in the path of radiation from a cobalt source, but yellow-legged wasps will.

Why do the pipe-organ wasps shy away from radiation? Do they have some sort of built-in radiation detector? Why doesn't radiation bother the yellow-legged mud daubers? The mystery is still unsolved.

Early experiments with guinea pigs demonstrated that the body can repair itself even after receiving a sizable dose of radiation. The experimental animals were irradiated, but their legs were shielded from exposure. The blood-forming tissues were put out of commission all through the animals' bodies except in their legs. Tissue in the legs took over the job of producing vital blood ingredients until damaged tissue elsewhere recovered.

Studies of the one-celled amoeba, one of the simplest animals, have provided a hint of the way irradiated cells repair themselves. Amoebas reproduce by cell division. Irradiated amoebas are slow to divide, but if cytoplasm from a nonirradiated amoeba is injected into an irradiated amoeba, the latter divides at normal speed. Thus cytoplasm may hold the secret of cellular repair.

Researchers have long sought a successful method of transplanting tissues and organs from healthy persons to sick or injured people. The main obstacle has been the antibody reaction in the body of the person who is to receive the transplant, or graft. Antibodies attack grafts from other individuals just as they combat invading disease germs.

Strong doses of radiation, however, temporarily destroy antibody reaction. Using X rays, doctors have successfully grafted tissue and organs from one experimental animal to another. The first successful case of skin grafting following a strong dose of X rays was performed on a mouse.

Experiments with rodents show that the animals' bodies are most receptive to grafts one or two days after exposure to radiation. Why then isn't this technique used on man?

The reason is that the dose required to stop the antibody

reaction in animals is greater than the fatal dose. Moreover, suppressing the antibodies leaves the animal vulnerable to infection. The method, therefore, is not yet safe for use on humans. Radiation may still hold the key to successful transplantation, but before anyone can be certain much additional research will be needed.

Radiation is already widely used in treating human cancer. Yet, new forms of radiation treatment for cancer are still being tested on animals. For instance, radiation released from a substance inside the brain of a mouse has been used to destroy tumors in that organ.

How does radiation find the target in a tumor deep inside an animal's brain? Certain chemical compounds can be injected in the animal. These tend to gather in or near the brain tumors. When atoms in some of these are struck by a beam of neutrons, they capture the neutrons and become radioactive. Then they give off radiation close to the site of the tumor and nowhere else.

Scientists are also experimenting with radioisotopes that can be sent through blood vessels to deliver internal radiation against cancers lurking in organs such as the lungs. Some cancers have been successfully treated this way in rabbits, using radioactive yttrium 90. (1)

When studying the possible hazards of the exposure of skin to radiation, the animal that scientists select—because it has a skin similar to man's—is a kind of miniature swine (see photograph section). These swine have about the same amount of hair on their skin as humans, and their skin is just about as thick.

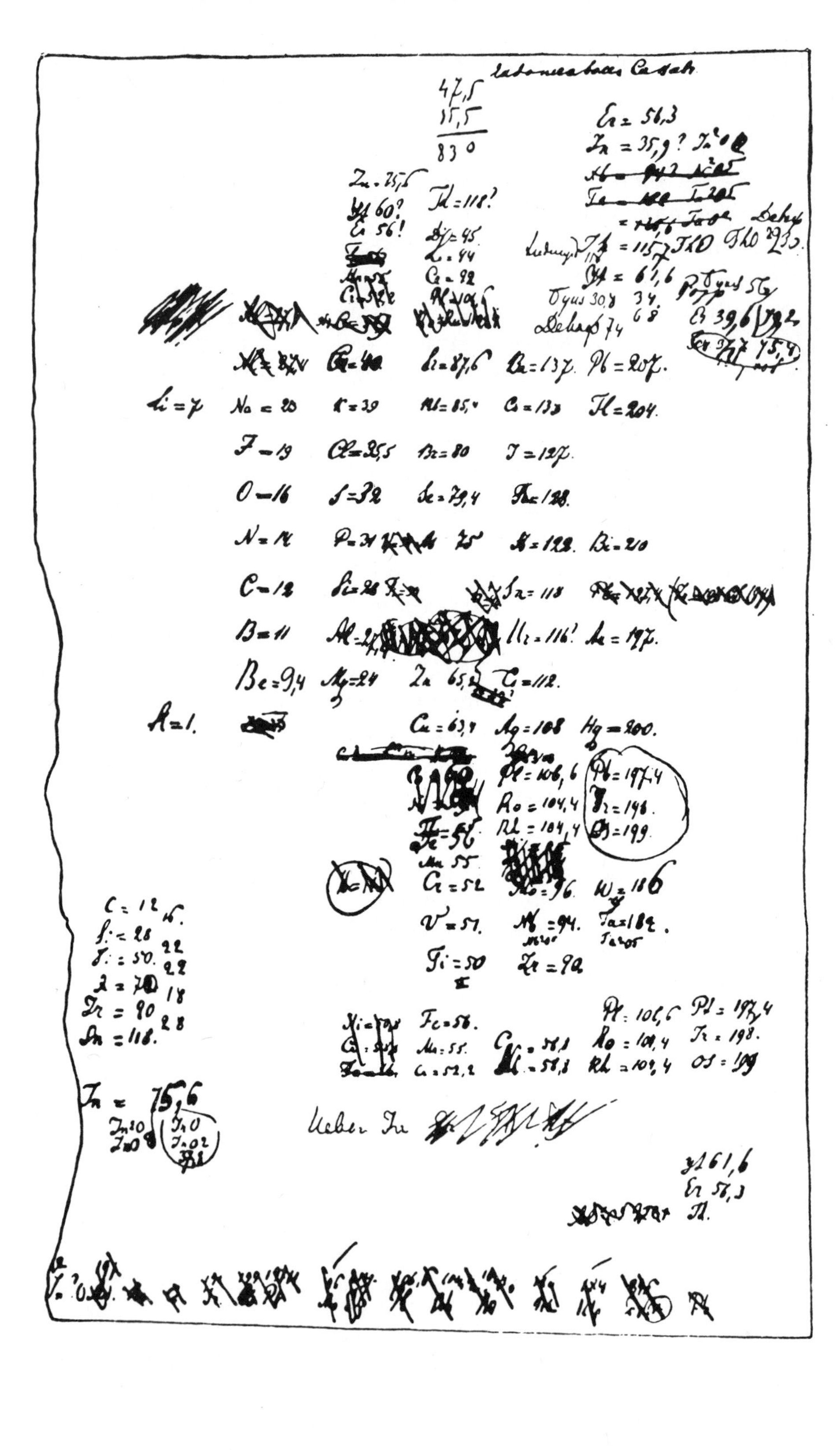

Ca = 40
Sr = 87,6
Ba = 137
Pb = 207
Li = 7
Na = 23
K = 39
Rb = 85,4
Cs = 133
Tl = 204
F = 19
Cl = 35,5
Br = 80
J = 127
O = 16
S = 32
Se = 79,4
Te = 128
N = 14
P = 31
Sb = 122
Bi = 210
C = 12
Si = 28
Sn = 118
B = 11
Al = 27,4
Ur = 116
Au = 197
Be = 9,4
Mg = 24
Zn = 65,2
Cd = 112
H = 1
Cu = 63,4
Ag = 108
Hg = 200
Pd = 106,6
Pt = 197,4
Ro = 104,4
Ir = 198
Rh = 104,4
Os = 199
Fe = 56
Mn = 55
Cr = 52
W = 186
V = 51
Nb = 94
Ta = 182
Ti = 50
Zr = 90
Er = 56,3
Yt = 61,6

ОПЫТЪ СИСТЕМЫ ЭЛЕМЕНТОВЪ,

ОСНОВАННОЙ НА ИХЪ АТОМНОМЪ ВѢСѢ И ХИМИЧЕСКОМЪ СХОДСТВѢ.

			Ti = 50	Zr = 90	? = 180.
			V = 51	Nb = 94	Ta = 182.
			Cr = 52	Mo = 96	W = 186.
			Mn = 55	Rh = 104.4	Pt = 197,4.
			Fe = 56	Ru = 104.4	Ir = 198
			Ni = Co = 59	Pl = 106.6	Os = 199.
H = 1			Cu = 63.4	Ag = 108	Hg = 200
	Be = 9.4	Mg = 24	Zn = 65.2	Cd = 112	
	B = 11	Al = 27,4	? = 68	Ur = 116	Au = 197?
	C = 12	Si = 28	? = 70	Sn = 118	
	N = 14	P = 31	As = 75	Sb = 122	Bi = 210?
	O = 16	S = 32	Se = 79,4	Te = 128?	
	F = 19	Cl = 35.5	Br = 80	I = 127	
Li = 7	Na = 23	K = 39	Rb = 85.4	Cs = 133	Tl = 204
		Ca = 40	Sr = 87,6	Ba = 137	Pb = 207.
		? = 45	Ce = 92		
		?Er = 56	La = 94		
		?Yt = 60	Di = 95		
		?In = 75,6	Th = 118?		

Д. Менделѣевъ

OPPOSITE. Mendeleev's rough draft of the periodic table of the elements. ABOVE. An early version of Mendeleev's periodic table (1869). The heading means, "tentative system of the elements," and the subheading, "based on atomic weights and chemical similarities." (O. N. Pisarzhevsky, *Dmitry Ivanovich Mendeleyev.* Moscow: Foreign Languages Publishing House, 1954)

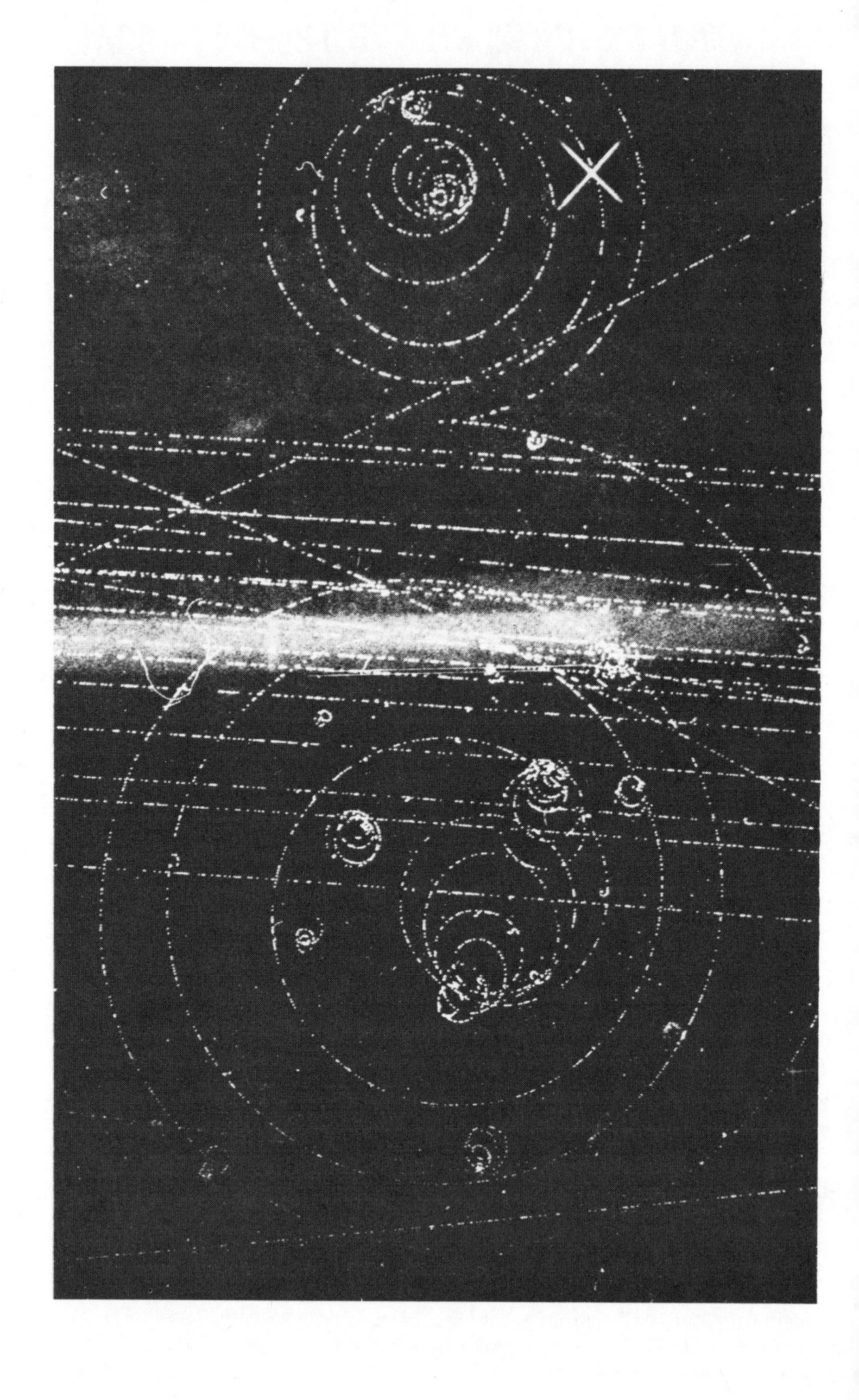

OPPOSITE. Electron-positron pair production (bubble-chamber photograph). A neutral particle, leaving no track, has decayed into the electron-positron pair, which have left spiral tracks. ABOVE. Proton-antiproton annihilation (bubble-chamber photograph). See diagrams, page 81. (Brookhaven National Laboratory)

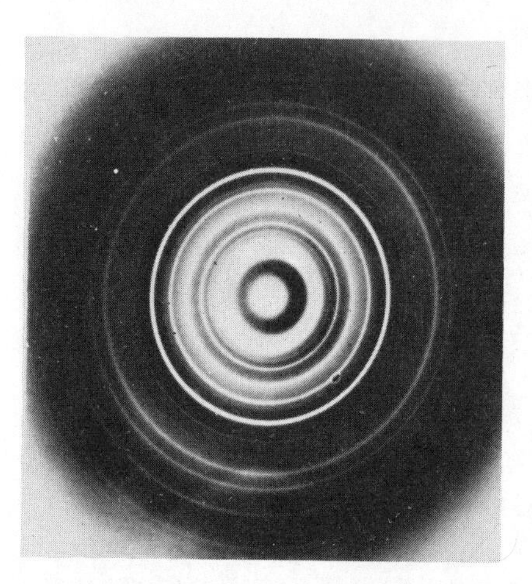

Diffraction pattern produced by electrons shot through a thin foil (Estate Dr. Lester Germer, and Bell Laboratories)

BELOW. Spark-chamber photograph of a neutrino collision. This is a 10-ton aluminum spark chamber. The long straight spark track is that of a mu-meson created by an incident neutrino. The other track is thought to be that of a gamma ray.

OPPOSITE. The Brookhaven solar neutrino detector. The tank, 20 feet in diameter and 48 feet long, contains 100,000 gallons of perchloroethylene. It is located in the Homestake Gold Mine at Lead, South Dakota, 4850 feet underground, in order for cosmic rays and all other particles except neutrinos to be shielded out by the earth. This detector was designed to observe the solar neutrino flux by the capture of neutrinos to form radioactive argon 37 by the reaction $\nu + {}^{37}\text{Cl} \rightarrow {}^{37}\text{Ar} + e^-$.

(Brookhaven National Laboratory)

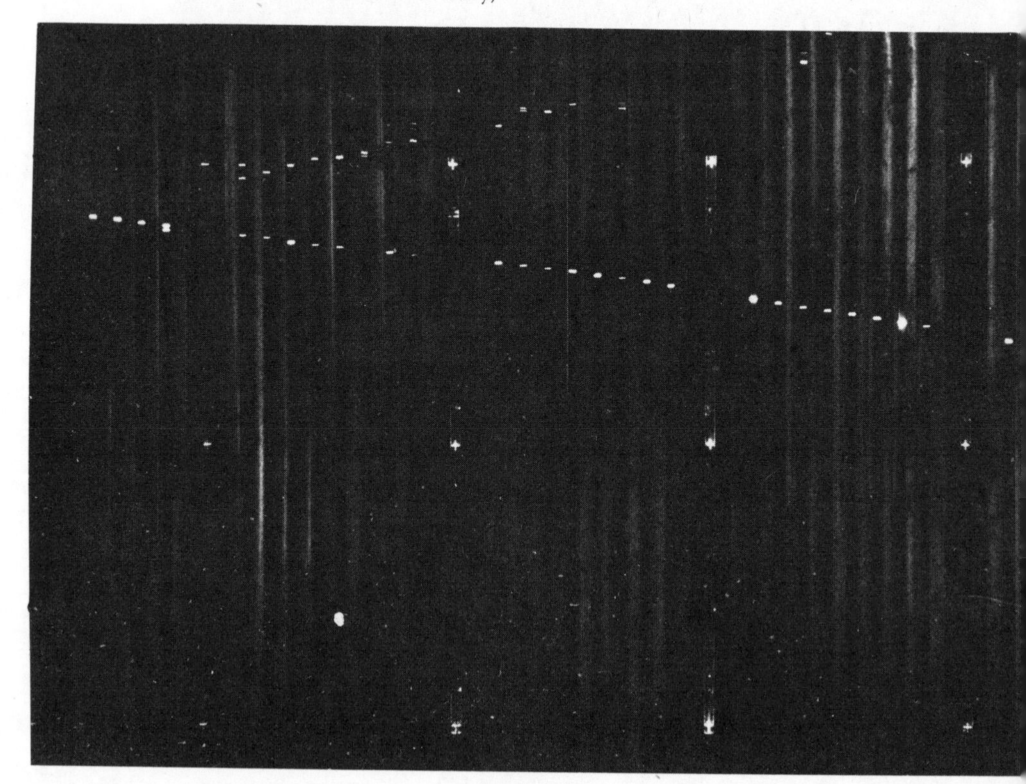

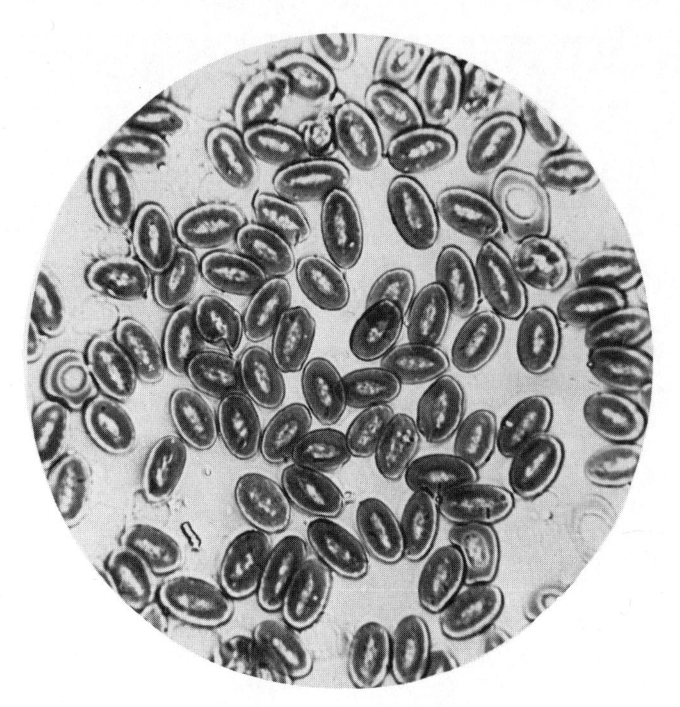

One of the earliest photographs of cells taken with a microscope. This photomicrograph, made by J. J. Woodward, U.S. Army surgeon, in 1871, shows cells in the blood of a pigeon. (Armed Forces Institute of Pathology)

Effects of ionizing radiation on chromosomes.

BELOW, LEFT. A normal plant cell with chromosomes divided into two groups; RIGHT, the same type of cell after X-ray exposure, showing broken fragments and bridges between groups, typical abnormalities induced by radiation. (Brookhaven National Laboratory)

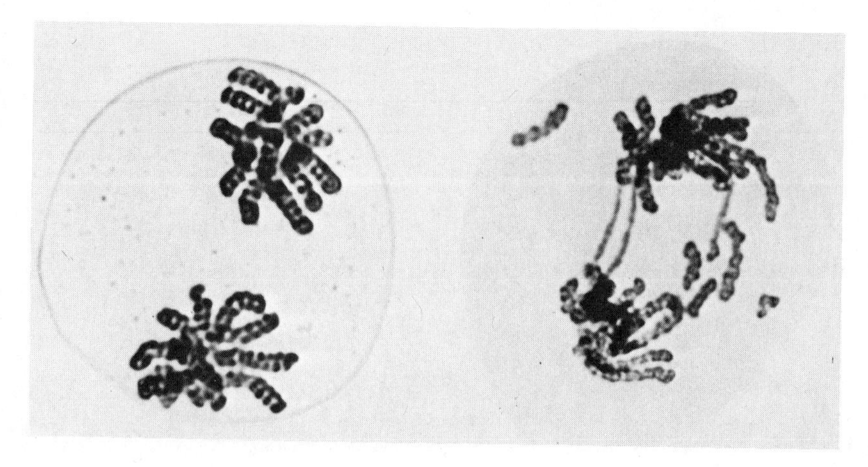

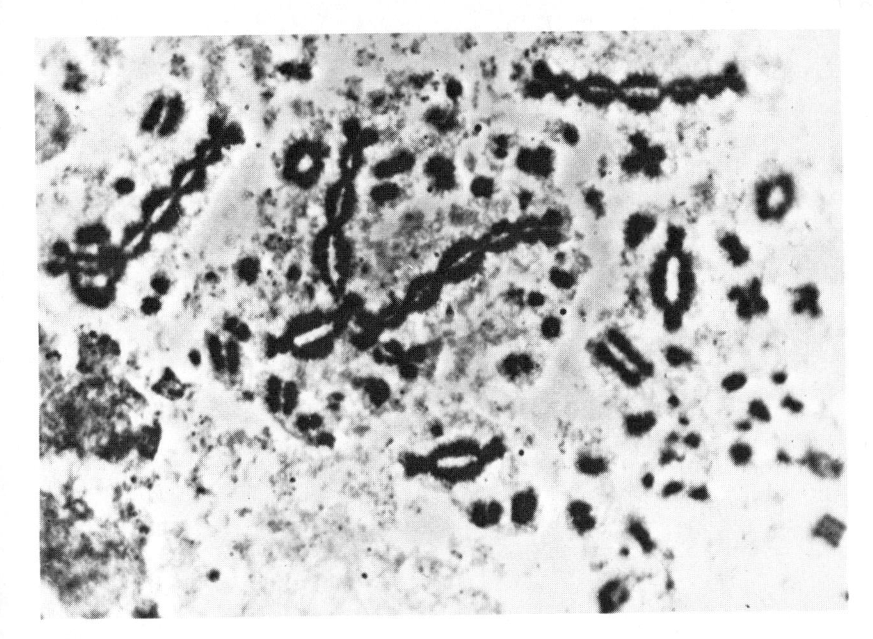

A cell nucleus, enlarged 275 times, showing chromosomes that have abnormally doubled, tripled, and so on, because of irradiation. Such a cell could easily become cancerous. (Dr. C. K. Yu, Argonne National Laboratory)

This giant cell nucleus, enlarged 50 times, has grown abnormally due to irradiation; it now has more than 700 chromosomes and is about 10 times the size of a normal cell. (Dr. C. K. Yu, Argonne National Laboratory)

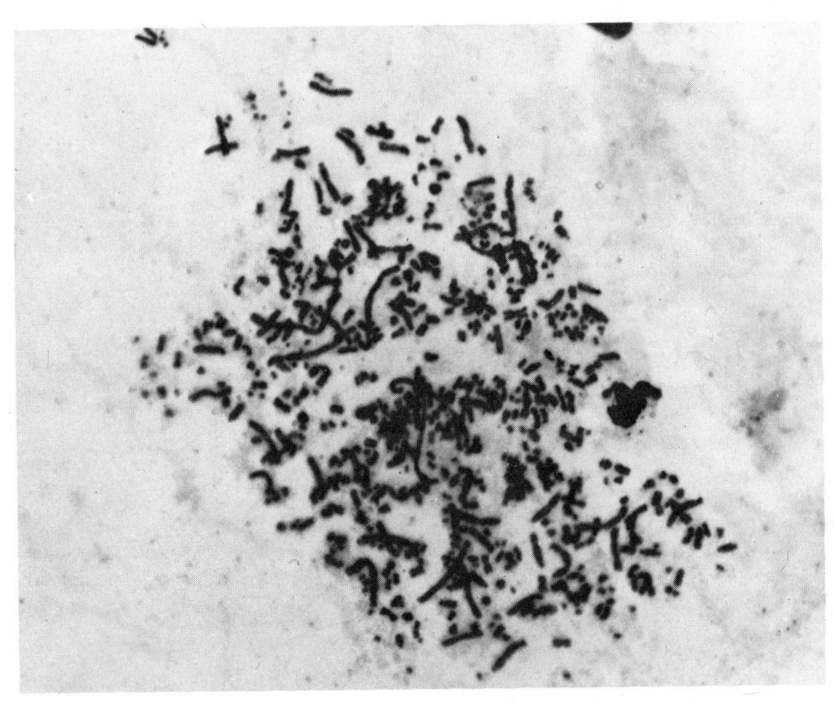

Experiments in X-radiation of fruit flies (*Drosophila*) performed at the California Institute of Technology caused gene mutation and chromosome rearrangements with resulting abnormalities in thorax and wings. (A) normal male; (B) four-winged male with a double thorax; (C and D) three-winged flies with partially doubled thoraxes. (U.S. Atomic Energy Commission)

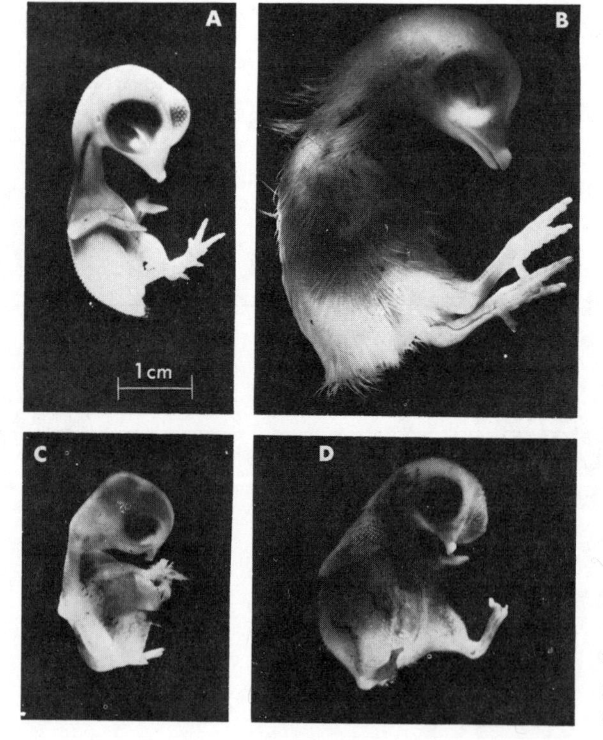

Effects on chick embryos of irradiation with cobalt 60 gamma rays. (A) Normal embryo 10 days after fertilization; (C) 10-day embryo irradiated on the sixth day after fertilization; note deformities of beak and toes and generalized hemorrhage and swelling. (B) Normal embryo 13 days after fertilization; (D) 13-day embryo irradiated on the sixth day. In addition to the defects in C, there is serious retardation in growth. (U.S. Atomic Energy Commission)

RIGHT. Of two originally identical groups of 14-month-old mice, the top group was untreated; that below received a large but not fatal dose of radiation as young adults. Only three members of the treated group have survived, and they are gray and senile; the mice in the untreated group are all normal, healthy, and active. (Brookhaven National Laboratory)

A patient with Cushing's disease, a pituitary gland disorder, (LEFT) before treatment and (RIGHT) 8 months afterward. She received 8500 rads of 910 MeV alpha particles, delivered to the pituitary over an 11-day period. Five years later she was still apparently free of the disease. (Lawrence Radiation Laboratory)

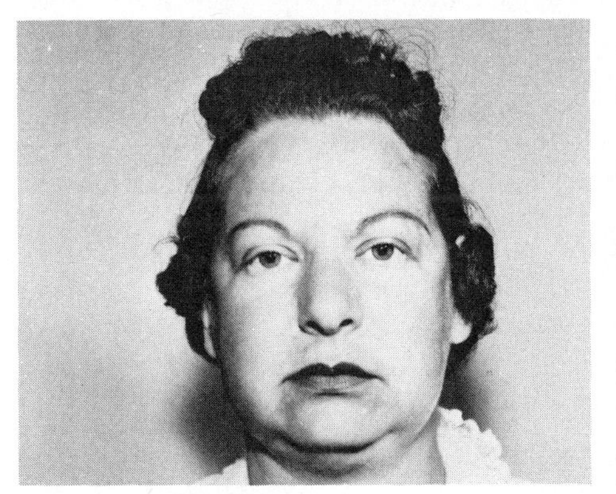

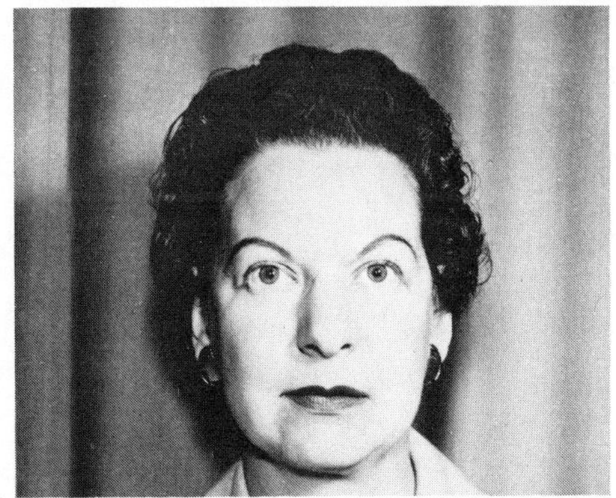

The 100-pound Hanford miniature swine at right, compared here with a 600-pound Poulouse swine, was developed for its similarities with man in size and physiology. (U.S. Atomic Energy Commission)

OPPOSITE. Early advertisement for radium preparations (Dr. Charles E. Miller, Argonne National Laboratory)

"STANDARD"

RADIUM PREPARATIONS

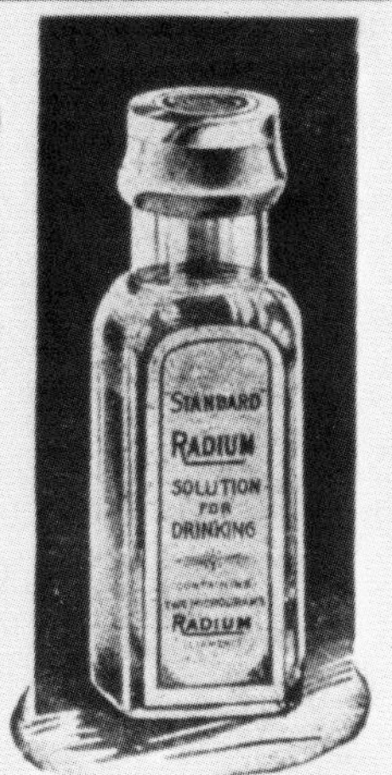

"Standard" Radium Solution for Drinking

—

Each bottle contains two micrograms radium element in 60 cc. aqua dist.

—

Maximum-equilibrium constant of radium emanation, 5400 mache units.

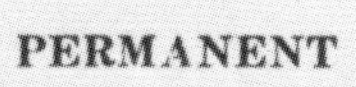

PERMANENT

"Standard" Radium Solution for Intravenous Use.

—

In Ampulles of 2 cc. N. P. S. S. containing 5, 10, 25, 50, or 100 micrograms radium element.

PERMANENT

"Standard" Radium Compress

—

A means of applying radium locally for the relief of pain.

—

A flexible pad of standardized, guaranteed radium element content.

PERMANENT RADIO-ACTIVITY

INDICATIONS

Subacute and Chronic Joint and Muscular Conditions.
High Blood Pressure. Nephritis.
The Simple and Pernicious Anemias.

"The value of radium is unquestionably established in chronic and subacute arthritis of all kinds (luetic and tuberculous excepted) acute, subacute and chronic joint and muscular rheumatism (so called) in gout, sciatica, neuralgia, polyneuritis, lumbago and the lancinating pain of tabes."—Rowntree and Baetjer, Journal A. M. A. Oct. 18, 1913.

For Descriptive and Clinical Literature Address

RADIUM CHEMICAL COMPANY

PITTSBURGH

New York
C. Everett Field, M. D.,
50 E. 41st St.

Boston
Samuel Delano, M. D.,
39 Newbury St.

Chicago
C. W. Hanford, M. D.,
749 1st Nat'l Bank Bldg.

San Francisco
Fred I. Lackenbach
Biologic Depot
908 Butler Bldg

Enrico Fermi, leader of the team of scientists that initiated the first man-made nuclear chain reaction on December 2, 1942 (Argonne National Laboratory)

Sketch of the first atomic pile (Argonne National Laboratory)

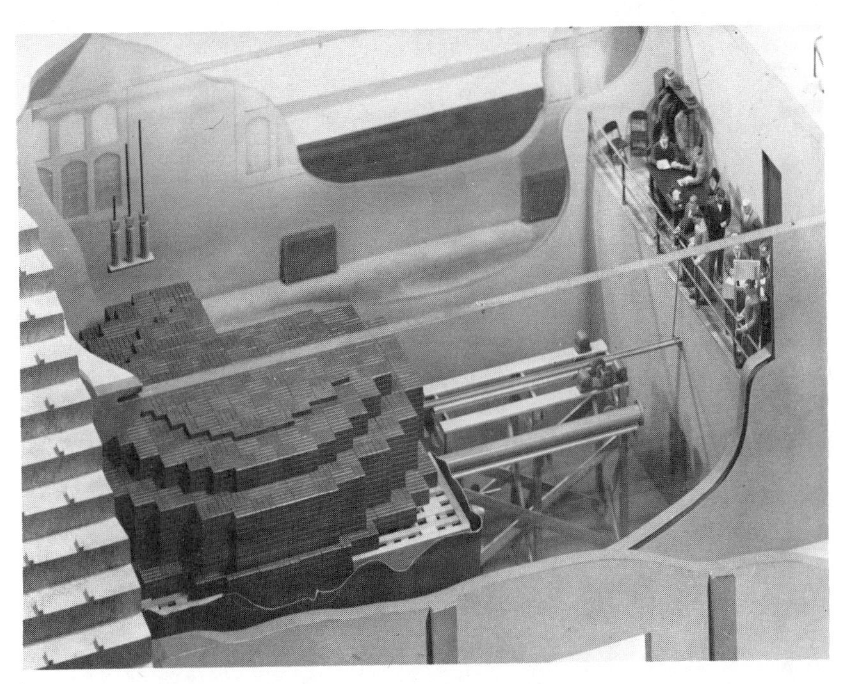

Cutaway scale model of the West Stands of Stagg Field at the University of Chicago, showing the first pile in the squash court beneath. The apparatus for withdrawing the emergency control rod is in the center, and the rope attached to the rod is tied to the rail of the balcony. (Argonne National Laboratory)

The West Stands during World War II (Argonne National Laboratory)

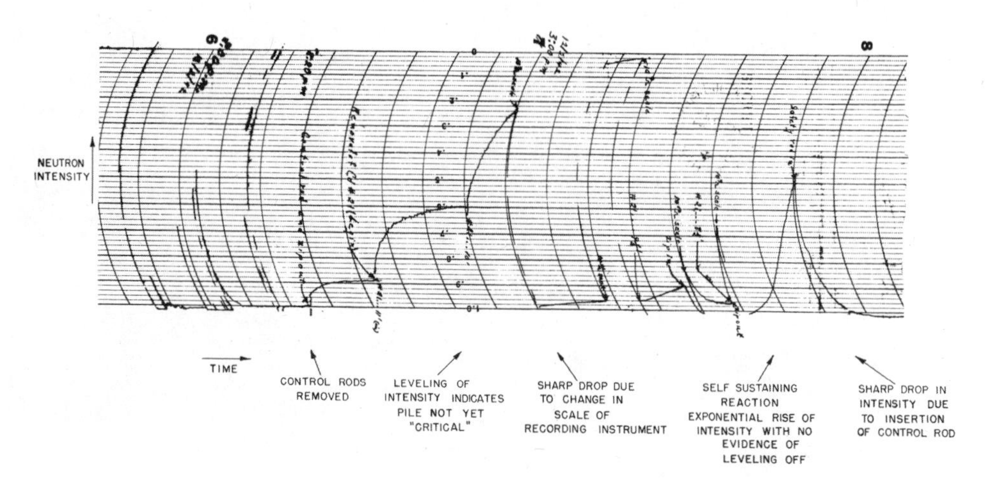

The "birth certificate" of the atomic age. The graph indicates the neutron intensity as recorded by a galvanometer during the start-up of the first self-sustaining chain reaction on December 2, 1942. (Argonne National Laboratory)

Nuclear Energy by Henry Moore commemorates the twenty-fifth anniversary of the first chain reaction at the University of Chicago. (Sandi Kronquist, University of Chicago)

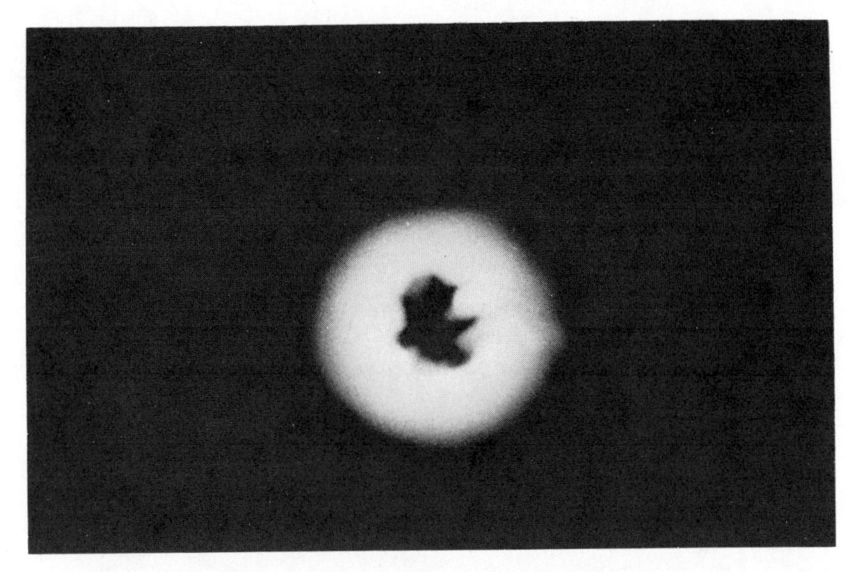

Glowing hot plasma of ionized gas in a linear pinch tube. The plasma is being pushed outward by an internal magnetic field as instabilities grow on its internal surface. Photographed by fast-shutter photography permitting sequences at intervals of 3 to 5 millionths of a second. (Gulf Oil Corp.)

The Juggernaut research reactor at Argonne National Laboratory. The device within the L-shaped wall next to the reactor is used for neutron radiography; the neutron beam emerges from the reactor core through the circular beam port. (Argonne National Laboratory)

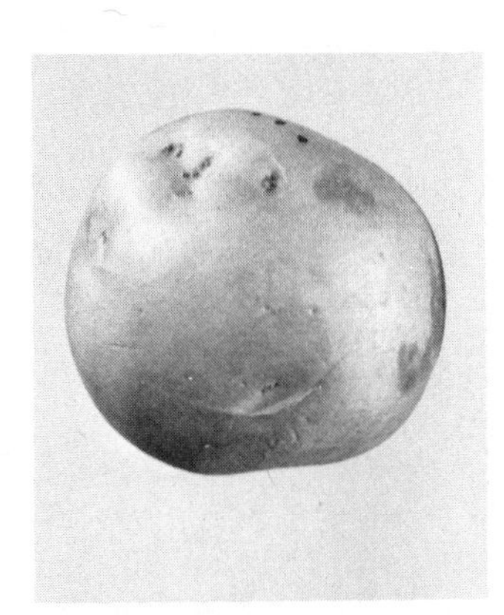

The potatoes shown here illustrate the effect of radiation in preserving foods. The one at left was exposed to 20,000 rads of gamma radiation; the other was not treated. The photographs were taken after both potatoes had been stored for 15½ months. The irradiated potato had no sprouts and was still firm, fresh-looking, and edible. (Brookhaven National Laboratory)

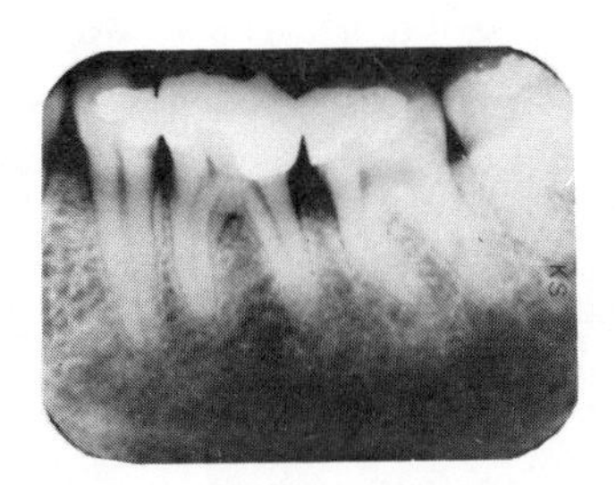

Dental X ray. Gums appear darker than teeth because soft tissues are less absorbing for the X-ray beam. Fillings, which contain heavy materials such as silver, are very absorbing and appear as light areas. If there had been a cavity, it would have appeared dark within the shadow of the tooth because there is less material to absorb the radiation.

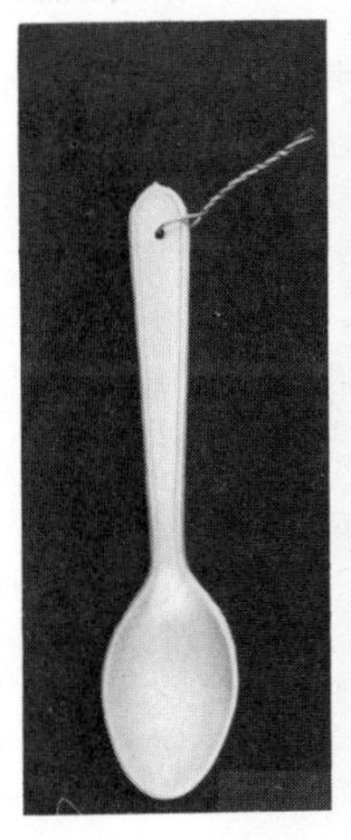
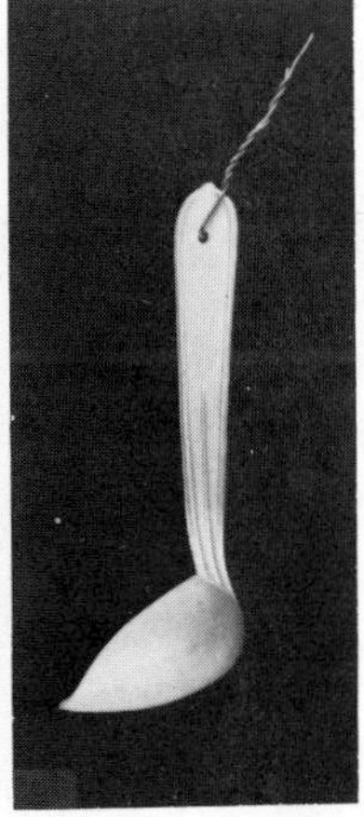

RIGHT. The spoon at left is made from irradiated plastic; unlike the ordinary plastic ones it does not become misshapen in hot water. (Salt Lake Pipe Line Company)

OPPOSITE: *Swords into Plowshares* (Isaiah 2:4), sculptured in bronze by Moissaye Marrans, symbolizes the Plowshare program for peaceful uses of nuclear explosives. The statue, 14½ feet tall, is on the façade of the Community Church of New York. (The National Sculpture Society and the Community Church of New York)

A 14,000-year-old burial site in the area of the Aswan Reservoir in Sudan. To determine the age of such ancient remains, archaeologists search for any scrap of associated wood or charcoal that could be used for age measurement by carbon 14, one of the "nuclear clocks." (Department of Anthropology, Southern Methodist University)

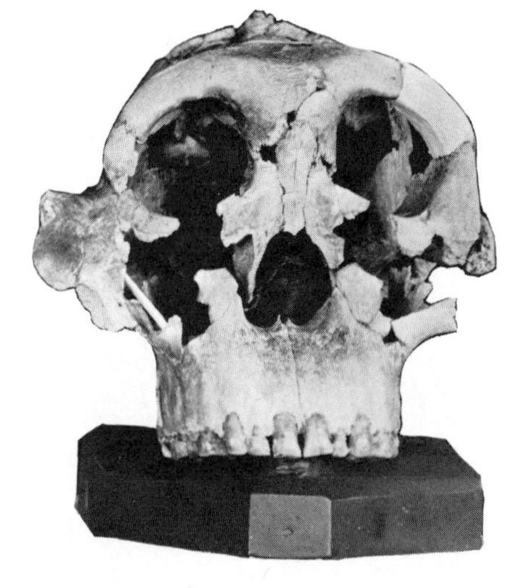

Fossil skull of *Zinjanthropus,* discovered in Olduvai Gorge, Tanzania, in 1959 by Louis and Mary Leakey. Dating by the potassium-argon method showed it to be nearly 2 million years old. (Jen and Des Bartlett, Bruce Coleman, Inc.)

PART III

THE ATOM AND THE ENVIRONMENT

ONCE scientists were bent on "conquering" nature. In recent years they have concluded that nature is not to be conquered, and that it is more sensible to try to live within nature's laws.

Man, ever multiplying his numbers and his wants, modifies his environment, sometimes in unpredictable ways. It has become imperative to understand the processes and balances of our environment, and to become sensitive to changes and what they may imply.

Atomic energy added a new dimension to environmental alteration. It is fortunate that the subject of environmental pollution has come to the fore before the problems added by atomic energy have become irremediable. These problems must be solved within the context of environmental problems as a whole—which includes learning how to manage the world's limited natural resources while there is still time.

12 Space Radiation

EVERY second of your life you are pierced from head to toe by a score of minute projectiles. You cannot see, hear, or feel these subatomic bullets, but they bathe every square inch of the earth's surface with an invisible, unrelenting rain. Collectively, these penetrating particles are called cosmic rays. They come from the distant stars, and some began their journey even before the earth was born. They are the first kind of space radiation.

A second kind is encountered only by the astronauts as their capsules penetrate the great belts of protons and electrons that the earth has captured in its magnetic trap. Named after their discoverer, James A. Van Allen, the Van Allen belts enshroud the earth and visibly announce their presence only when they dump some of their contents into the polar atmospheres and help kindle the auroras that light the winter skies.

Unmanned satellites and space probes radio back to us news of still a third kind of space radiation—solar plasma. Only when instrument carriers and astronauts break through an invisible, world-enveloping, teardrop-shaped barrier called the magnetopause can solar plasma be detected directly. Beyond this protective shell lies the open interplanetary sea, swept by a steady solar wind and lashed frequently by colossal tongues of plasma that the sun spews out from convulsed regions on its surface. How did we learn about this celestial sea, and how can we travel to the planets through its storms?

Man was long oblivious to space radiation because of its invisibility. Cosmic rays, the Van Allen belts, and the solar plasma were all great surprises when they were disclosed to a world long effectively sheltered from interplanetary weather by our thick atmosphere. Now, however, the scents of each of these three quarries lure us along trails as intellectually fascinating and physically adventurous as any Antarctic expedition or voyage into the sea's depths.

When a deep-space rocket, its fuel nearly gone, breaks through the earth's magnetopause into interplanetary space, it is like leaving the refuge of a good harbor and venturing out onto the open sea. The noun sea brings to mind steady trade winds, rolling ground swells, fickle currents, and storms. The analogy is indeed apt, for deep space has just these characteristics. On the earth's surface, we detect only a few subtle hints about what transpires beyond our three protecting breakwaters: the atmosphere, the ionosphere, and the earth's magnetic field.

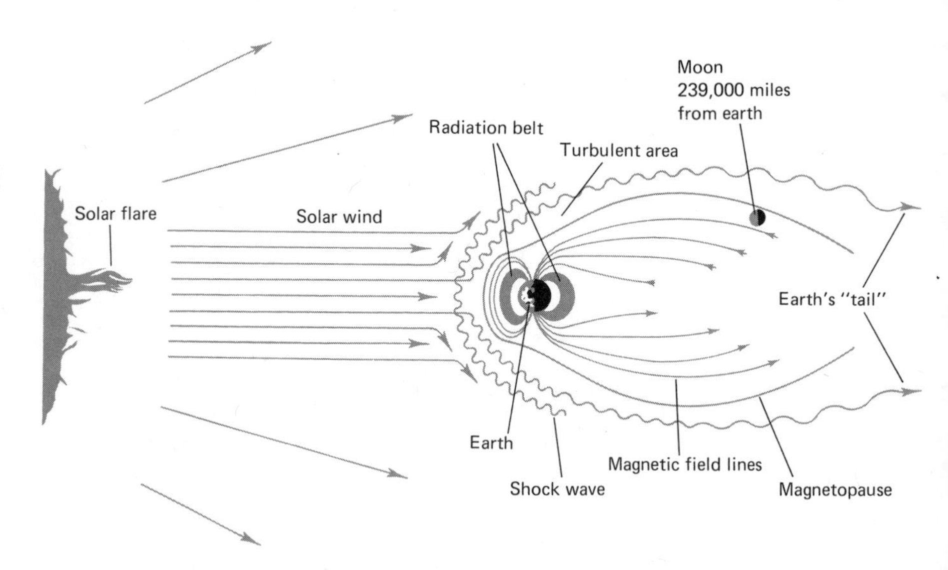

The sun is the major weather maker in outer space. Solar plasma tongues and cosmic rays are born during solar-flare eruptions. In addition, a steady "solar wind" is boiled off the sun's hot atmosphere. Galactic cosmic rays (not shown) come from the stars and arrive at the earth uniformly from all directions. (Distances on diagram are not to scale.)

The innermost barrier, the atmosphere, is an effective shield against both electromagnetic and particulate radiations. In weight, the atmosphere above us is equivalent to a layer of water about 34 feet deep. Most of it hugs the surface. A handy rule of thumb states that air density drops by a factor of ten for each 10 miles (16 kilometers) of altitude, up to 60 miles (100 km), where the air density is only about one-millionth that at sea level. The thick, bottommost part of the atmosphere is the strongest part of the shield, since it is mainly collisions with the air's nitrogen and oxygen molecules that prevent the incoming photons and charged particles from reaching the surface. At 9000 feet, for example, one is pierced three times more frequently by cosmic rays than at sea level.

One of three things can happen when a charged particle or photon of energy enters the upper atmosphere from above. Collision with an air molecule may turn aside the incoming projectile, the molecule perhaps absorbing some of its energy in the process. This process is termed scattering; and if it occurs often enough a charged particle from outer space ultimately may lose most of its energy and settle down to become a permanent resident of the atmosphere. This is the fate of most of the low-energy particles hitting our atmosphere. The kinetic energy they lose in collisions is often revealed as visible light in the upper atmosphere. The brilliant auroras, for example, may be due to stimulation of air molecules and atoms by extraterrestrial radiation.

Charged particles with somewhat higher energies may be absorbed rather than scattered by the nuclei of atoms in the atmosphere. Some particles in the cosmic-ray stream or flux are so energetic that they smash nitrogen and oxygen nuclei into many subatomic pieces, which then go on to do their own atom smashing. The result of a very energetic primary cosmic ray, then, is a shower or avalanche of secondary cosmic rays (see diagram, page 218).

Photons from the sun and stars are also absorbed and scattered by molecules in the atmosphere. The blue sky is a familiar reminder that blue light is scattered by atoms in the upper at-

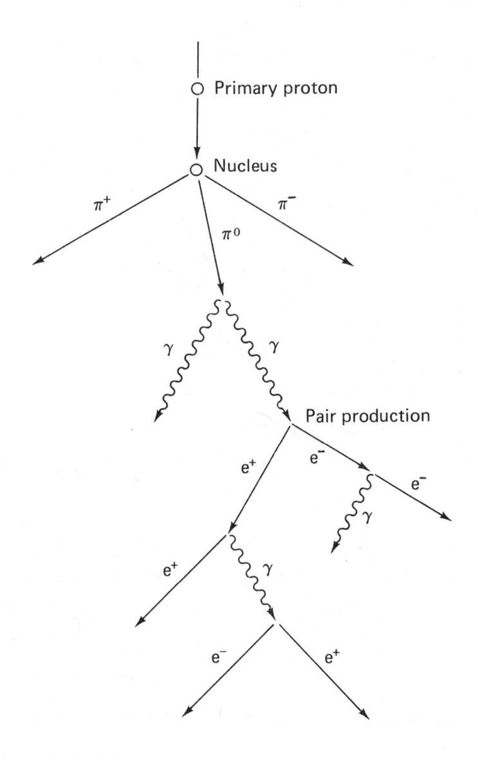

A cosmic-ray avalanche. A high-energy primary proton interacting with a nucleus produces dozens of secondary particles, which may produce other collisions. The secondary particles decay, are involved in pair production (of electrons and positrons), are annihilated, or produce gamma-ray photons, which may create new pairs of particles over and over again. Finally the photons and electrons begin to lose energy to the atmosphere by scattering, ionization, and photoelectric processes. A single event such as is shown may produce several million new particles during a period of a few millionths of a second.

mosphere. While most solar photons eventually reach the earth's surface in the form of visible sunlight, many short-wavelength and long-wavelength photons never make it. Infrared photons, with longer wavelengths than visible light, are readily absorbed in the high atmosphere by molecules, particularly those of water vapor. Some of the ultraviolet photons, which possess short wavelengths, collide with and are absorbed by oxygen (O_2) and nitrogen (N_2) molecules. The photon energy goes into the dissociation or splitting of diatomic molecules and the ionization of single atoms, as:

$$N_2 + \text{photon} = N + N$$
$$O_2 + \text{photon} = O + O$$
$$O + \text{photon} = O^+ + \text{electron}^-$$

The overall effect of the atmosphere is insulation of the earth's surface from all electromagnetic radiation arriving from space, save for some at certain wavelengths that enters through atmospheric windows—that is, sections of the electromagnetic spectrum where photons are not absorbed in ionization, dissociation, and other processes.

It is fortunate that ultraviolet photons from the sun are stopped before they reach the earth's surface, because unfiltered.solar ultraviolet light would be deadly to most terrestrial life. Astronauts must be protected from this ultraviolet flux just as carefully as they are shielded from the hard vacuum of outer space. (36)

To put it differently, if the atmosphere were transparent to ultraviolet light, life would have had to develop in a form capable of surviving in the presence of ultraviolet. Human eyes able to see ultraviolet light might have evolved.

The short-wavelength radiation stopped high in the atmosphere creates the second of the earth's breakwaters. The ionization of oxygen and nitrogen atoms results in layers of unattached, free electrons, between 35 and 200 miles in altitude, which act as mirrors to low-frequency radio signals. Collectively, these layers are termed the ionosphere (diagram, page 220). When a radio wave from either the earth or outer space penetrates the ionosphere, it is bent away from its direction of propagation as the free electrons change the velocity of its wavefront. The lower the frequency and the more abundant the free electrons, the more sharply the radio waves are bent. Some are bent so far that they are reflected back in the direction from which they came. Without the reflecting power of the ionosphere, we would not have long distance transmission of radio signals below 20 megacycles per second (20 MHz).

Although the ionosphere also reflects radio waves from deep space that are below 20 MHz and acts as a shield in this

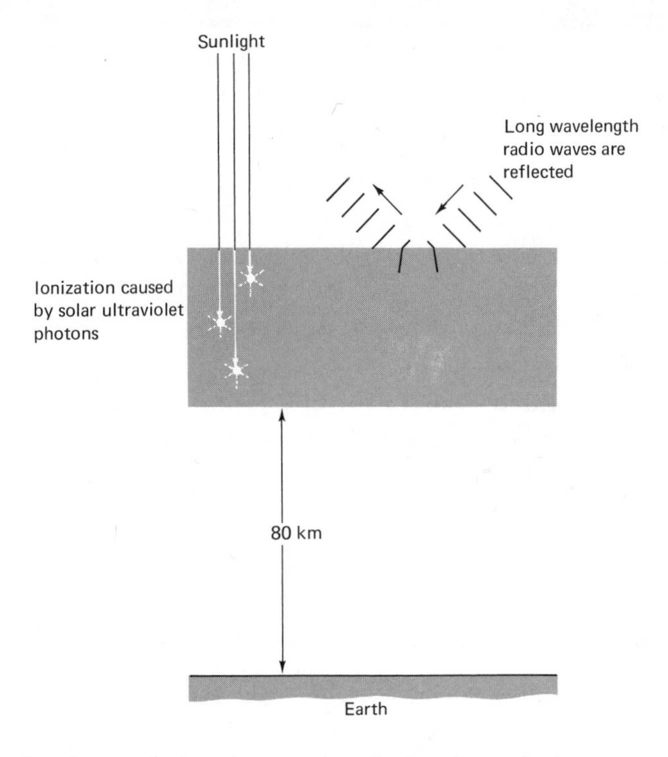

The ionosphere is created when short-wavelength solar photons ionize oxygen and nitrogen high in the atmosphere. Long-wavelength radiowaves speed up when they penetrate these layers containing many free electrons. Consequently, the wavefronts bend away from the initial direction of travel and the radiowaves may be reflected.

sense, we hardly need it for protection. The ionosphere, however innocuous it may be, is a distinct nuisance to radio astronomers because it blinds them to all the tantalizing low-frequency radio waves emitted by the stars and galaxies, and narrows the radio window through which we can "see" the universe. Thus the ionosphere wraps the earth in yet another muffling layer of insulation.

The third and last barrier that separates terrestrials from the sights and sounds of the interplanetary expanses, the earth's magnetic field, has none of the substance of the layers of free electrons in the ionosphere or of the material atmosphere beneath. The earth's magnetic field originates inside our planet's rocky mantle, perhaps through the dynamo action of immense subterranean electrical currents, and extends thousands of miles out into space. The external field can be pictured by imagining lines of force that emerge from one magnetic pole and curve around the earth to reenter at the opposite magnetic pole (see diagram). Before the space age began, the external magnetic field was visualized as one that would result from a bar magnet imbedded in the solid earth, but displaced some 11° from the axis of rotation. The field was thought to be symmetric around the magnetic poles and it was assumed it extended to infinity. As is explained later, data obtained by satellites have forced drastic revision of this idealized portrait.

The ideal dipole field shown in the diagram is nevertheless

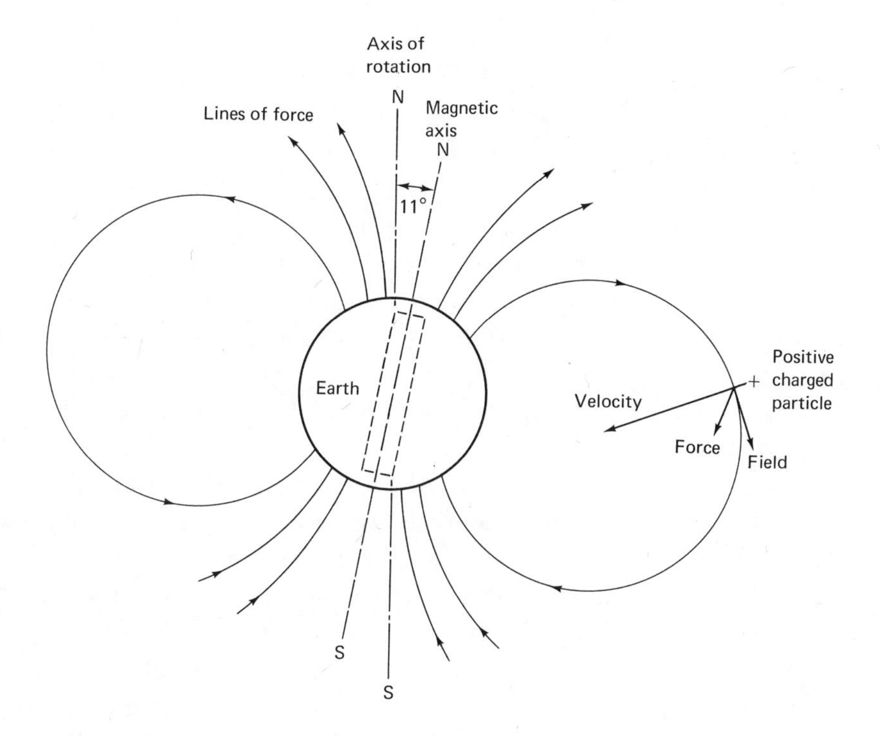

The magnetic field of the earth is canted about 11° with respect to the axis of rotation. The precise positions of the magnetic poles vary with time. An incoming charged particle will be forced off a straight trajectory as shown.

close enough to reality to support the magnetic shielding story that follows. The key to understanding magnetic shielding is recognition that the path of an electrically charged particle is bent at right angles to both the direction of the magnetic field and the direction in which the particle travels. The bending force F is given by a simple equation:

$$F = qvB \sin(v,B)$$

where q = the electrical charge
v = the particle velocity
B = the magnetic field strength
$\sin(v,B)$ = the sine of the angle between the velocity of the particle and the direction of the magnetic field

The resulting force F is perpendicular to both v and B.

A charged particle approaching the earth from space will tend to be turned away if it is aimed at the equator. On the other hand, it would not be deflected at all if it arrived on a path parallel to the earth's magnetic axis at the poles. The earth, then, is enclosed within a magnetic shield that is open at the poles and strongest in the equatorial regions. The solar wind never even reaches the earth's upper atmosphere because its weak (but not innocuous) charged particles are easily turned away by the magnetic lines of force. Energetic primary cosmic rays, however, zip right through as if there were no field around us at all.

But all this is getting ahead of the story. At the turn of this century, had you suggested at a scientific meeting that the earth was under intense bombardment by cosmic projectiles you would have been laughed out of the room. For the science of 1900, like the earth, also was well insulated from all forms of space radiation by the three barriers just described. There was a conviction among scientists that space radiation did not and could not exist. Still, a few wispy clues penetrated this insulation and there were some scientists who could not resist following them to the tops of mountains, the bottoms of lakes, and high into the stratosphere.

Before 1900, the idea that minute solid objects could pass through one's body and leave the shadows of one's bones on a photographic plate was incredible—perhaps as unbelievable as the dream of heavier-than-air flying machines. Nonetheless, Wilhelm Roentgen had discovered his invisible, penetrating X rays in 1895, and Henri Becquerel had found radioactivity, with its subtle, unseen emanations, a year later. Scientists began to realize that a whole new world lay just below the threshold of human senses; detecting radiation was like grappling with ghosts.

The ghostly radiation registered on only two instruments common in turn-of-the-century laboratories: photographic film and the electroscope. Charged particles and photons plowing through a photographic emulsion left trails of excited molecules behind them. Development of the film made these tracks visible. When charged particles and penetrating photons pass into an electroscope chamber, they ionize the enclosed gas and allow the electroscope to discharge (see diagram, page 104). The quantity of radiation can be crudely measured by watching the rate at which the gold leaves collapse. Electroscopes do not record the passage of individual particles; rather they register the effects of many particles over a relatively long period of time.

Early experimenters with radiation were troubled by the perplexing tendency of electroscopes to discharge slowly in the absence of any known source of radiation. In other words, the air inside the electroscopes was always slightly ionized. Even when the ions in the electroscope were neutralized intentionally, new ones continually formed, indicating a constant source of radiation from somewhere. Heavy shields around the electroscopes reduced somewhat the rate at which the electroscopes discharged. But the unknown radiation was much more penetrating than that from X-ray tubes or natural radioactivity. For many years, scientists believed that this mysterious flux of photons or particles came from the earth itself, the laboratory walls, and even the materials used in constructing the electroscope. Almost everyone looked down instead of up.

While most physicists were satisfied with terrestrial explanations of their newly found penetrating radiation, others detected a few extraterrestrial currents slipping through the earth's triple ring of breakwaters. When rough lodestones had been replaced by more sensitive compass needles, navigators on land and sea quickly noted that even the best needles were never still. They wandered slightly and quivered; sometimes a needle would be relatively quiet for days, then some unseen force would shake it violently. As far back as 1759, John Canton, a London schoolmaster, had discovered that the more energetic excursions of his compass needle occurred when the northern lights were conspicuous. It seemed likely that the same phenomenon that stimulated the auroras also shook compass needles—whatever that phenomenon might be.

Almost a hundred years passed before scientists realized that the sun might be reaching across 93 million miles and jostling the earth. In 1852, British Major General Edward Sabine found from magnetic studies in Canada that magnetic storms were more frequent when there were lots of sunspots. Subsequent studies showed that auroras, sunspots, and nervous compass needles were synchronized. This was the first strong hint that some unseen radiation from the sun was affecting the earth.

The polar auroras with their flickering flames and glowing draperies have always been known by man. In spite of the fact that satellites fly above them and sounding rockets continually pierce them from below, the auroras have still not yielded to detailed explanation. A few things seem sure: The auroras are associated with solar and geomagnetic activity and present us with the only visible (and sometimes audible) manifestations of the great radiation belts that surround our planet.

One has only to look at the shape of the earth's dipole field to see that the weakest chinks in the magnetic armor exist at the magnetic poles where the lines of force intersect the earth's surface. The auroras occur most often between magnetic (not geographical) latitudes of 65° and 70°; that is, in a ring-shaped band about each magnetic pole. Whatever causes the auroras obviously is strongly affected by magnetic fields; and only charged particles have this response. Coupling this observation with the strong correlation between auroras, the 11-year sunspot cycle, and the frequency of solar flares, it seems inescapable that eruptions of charged particles from the sun have something to do with igniting the auroras.

Alas, science seems so easy looking backward in time, but in the early 1900s no one had yet put these pieces of the puzzle together.

In 1907 the Canadian scientist John C. McLennan carried his electroscopes out onto the thick winter ice of Lake Ontario. The expectation was that the electroscopes would no longer discharge because the natural radioactivity in the rocks and soil would be screened by hundreds of feet of nonradioactive water. McLennan's measurements did show a definite decrease in the discharge rate, but the electrometers persisted in discharging slightly. McLennan was forced to conclude that although some of the invisible radiation came from rocks and soil, another component originated either in the atmosphere or outer space. (36)

So scientists went up to look—in balloons.

The Austrian-American physicist Victor F. Hess (1883–1964) carried electroscopes far higher in the atmosphere than had any of his airborne predecessors. His memorable flight of August 1912 produced data that shocked his earthbound comrades. As Hess's balloon rose, the rate of electroscope discharge decreased. But above 2000 feet, his electroscopes began to discharge faster, indicating an increase in the radiation level around the balloon. At 16,000 feet the radiation level was four times that measured on the ground.

Hess interpreted his results in this way: The initial reduction in radiation level was due to leaving terrestrial radioactivity behind, and the marked increase above 2000 feet was due to extraterrestrial radiation.

It was subsequently shown that cosmic rays were high-speed charged particles and that most primary cosmic rays could penetrate a full meter of lead at sea level. This fact implied that cosmic-ray energies were in the billion-electron-volt range.

The groundwork for the next step in cosmic-ray research had been laid by a Norwegian geophysicist back in the days when physicists were still puzzled by electroscopes that discharged for no apparent reason. Fredrik Carl Störmer (1874–1957) was trying to explain the aurora as a manifestation of terrestrial bombardment by low-energy, sun-emitted particles. It turned out that Störmer's work was more appropriate to cosmic rays than to auroras.

Störmer's approach was this: Suppose that a charged particle is heading toward the earth from a great distance. At the turn of the century, everyone believed that the earth's dipole magnetic field extended deep into interplanetary space, so Störmer thought the magnetic forces tended to bend the particles off their initially straight courses. The mathematics was complicated and tedious, but he was able to compute a great many trajectories that demonstrated (theoretically at least) how charged particles would approach the earth.

Two important conclusions emerged: (1) Charged particles should be deflected away from the earth's equator toward the poles. Thus, radiation detectors at the same elevations but different latitudes should record more cosmic rays near the poles. (2) Forbidden zones appeared that extraterrestrial charged particles in theory could not penetrate. Furthermore, particles occupying these zones could not escape, assuming they could somehow invade them. The forbidden zones are, of course, the zones of trapped radiation that were to surprise James Van Allen in early 1958. To Störmer these zones were of no interest because he supposed particles could not enter them. In the context of understanding cosmic rays, the important point of his work was the predicted latitude effect, that is, that higher cosmic-ray levels would be found near the poles.

By the late 1930s, balloons and aircraft had carried detection instruments up to 10 miles. Beyond was unknown territory. The ideal vehicle for exploring the regions above the limits of balloons and aircraft was the rocket.

Rockets were not new to warfare or to upper atmosphere research. The Chinese reputedly employed war rockets as far back as 1000 A.D. The American rocket pioneer Robert Goddard (1882–1945) installed instruments on some of his primitive liquid-fueled rockets as early as July 17, 1929. On this date a Goddard rocket transported a recording thermometer and barometer to the "tremendous" height of 90 feet; this was the first sounding rocket. The contribution of World War II to rocket research was the development of large rockets capable of breaching the atmosphere completely and, for the first time, reaching into outer space itself. (36)

In the late 1940s cosmic-ray experiments were put aboard rockets. One group of experimentors was led by James Van Allen.

Van Allen and his colleagues analyzed the telemetry data from their counters and concluded that considerable soft (or low-energy) radiation existed in the regions over the geomagnetic poles. The radiation was assumed to be low-energy electrons. The desire to explore further this unexpectedly high concentration of electrons was one of the moving forces behind Project Vanguard, the first U.S. satellite program.

Just what did the scientists expect to find with satellite radiation detectors? In the main, they looked for nothing significantly different from the data balloons and rockets had already placed on the record books. By the end of 1957, however, there was a considerable body of theoretical work supporting the existence of zones about the earth where charged particles could be trapped for long periods of time.

Thus, on the day Sputnik 1 was launched, it was well known that charged particles could be magnetically trapped in zones around the earth, but there was no direct evidence that such particles were really out there. (Sputnik 1, launched by the USSR, was the first man-made satellite. The date, October 4, 1957, opened the space age.) Despite all the preparation, final confirmation of the reality of the Van Allen belts came as quite a surprise. Moreover, the trapped zones were larger and more heavily populated with high-energy particles than anyone had suspected.

Sputnik 1 was the first man-made contrivance to reach orbit, but if it carried radiation detectors the Russians have never disclosed the fact. In about 4 weeks, on November 3, 1957, Sputnik 2 followed Sputnik 1 into orbit. This time the spacecraft carried a doomed dog and also some radiation detectors. In 1958 the Russian scientist Sergei Nikolaevich Vernov reported that Sputnik 2 had registered considerable soft cosmic radiation at high altitudes, similar to what Van Allen had found. If Sputnik 2 had been within radio range of Russian telemetry stations when it reached its highest altitudes, Vernov would have had the distinction of reporting the discovery of the radiation belts. Sputnik 2 was just too low as it passed over the Soviet Union beeping out its instrument readings.

Meanwhile, the shock of the Sputnik successes had stimulated the United States into organizing a second satellite effort to back up the Vanguard program. Explorer 1, the first satellite in this program, propelled one of Van Allen's GM (Geiger-Müller) tubes into a 1584-by-234-mile orbit on January 31, 1958.

The telemetry signals received from Explorer 1 indicated radiation levels were pretty much as expected for cosmic rays, as long as the satellite remained below 370 miles (600 km). Above 500 miles (800 km) the radiation levels increased dramatically. In fact, the radiation was so intense at high altitudes that the GM tube was saturated, that is, so many particles pierced the tube that it discharged continuously. As a result the telemetry showed no GM counts at all at very high altitudes. Equipment malfunction was suspected, but because the tube started counting again as the satellite reached lower altitudes, saturation was a more reasonable explanation. In order to jam the Explorer 1 GM counter, the radiation levels had to be some fifteen thousand times that expected from primary cosmic rays.

Van Allen concluded that the earth was surrounded by intense belts of trapped radiation. The second of the three major kinds of space radiation was disclosed. And again, the discovery had been generally unexpected.

The first map of the Van Allen belts was constructed with the help of data from the Explorer 4 satellite and the space probe Pioneer 3. The map displayed contours of equal counting rates that were symmetric around the earth's geomagnetic axis. Two zones of especially intense radiation were found: an inner, kidney-shaped belt and an outer crescent-shaped belt. "Horns" poked down into the earth's atmosphere in the vicinity of the auroral zones.

Continued mapping of the Van Allen belts has made it clear that the two belts shown in early maps really merge into one vast earth-centered doughnut of trapped protons and electrons with an inner zone of high-energy protons and an outer

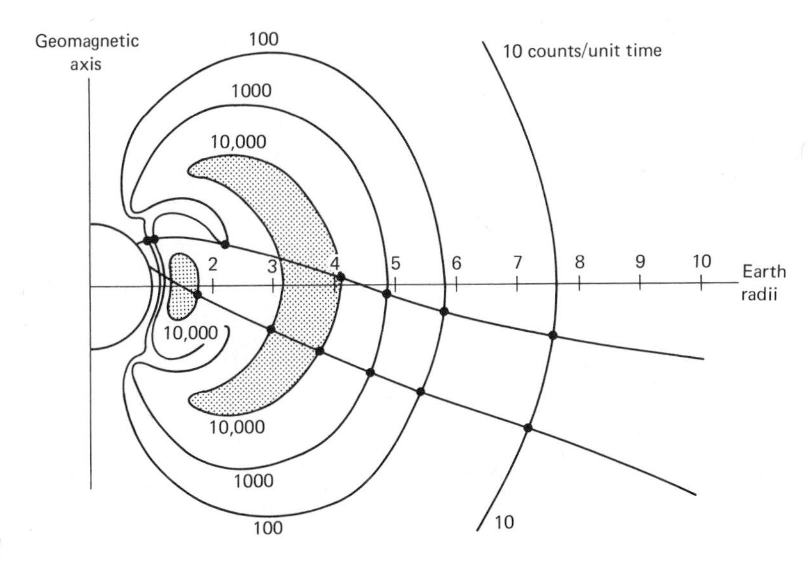

James A. Van Allen's map of the radiation belts derived from the GM tubes carried by Explorer 4 and Pioneer 3. The contours indicate regions of equal counting rates.

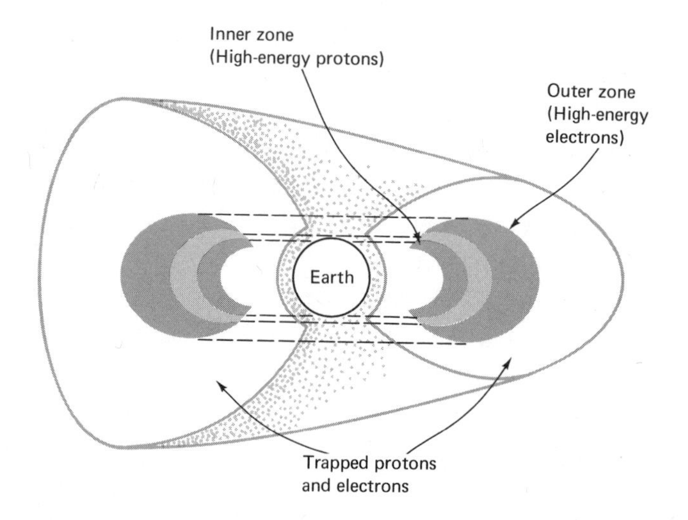

Section through the geomagnetic axis showing the present view of the structure of the zones of trapped radiation. These zones vary in shape and intensity with the solar cycle. The impact of the solar wind flattens the "doughnut" on the sunward side.

zone of high-energy electrons. In between is a slot where trapped particles with relatively low energies abound.

Like people, occupants of the radiation belts are born, live a while in a crazy, oscillating world, and then die—in the sense that they finally escape their magnetic trap.

The population of the inner zone of high-energy protons is relatively stable in time, whereas the high-energy electron flux in the outer zone may increase a thousandfold soon after a solar flare is seen on the sun. Therefore two different birth processes exist.

Theoretical progress has been made, but the birth processes require much more research before they can be fully explained. Death, on the other hand, is easier to understand. Collisions of electrons and protons with atoms in the polar atmosphere are the most likely cause. Electrical neutralization of a belt particle by collision with such atoms manifestly opens the door because the magnetic field can no longer contain it. Collisions can also scatter particles out of the belt by changing their directions of travel and their energies. Just as light is not reflected well from a dirty mirror, belt particles do not always bounce back from the spongy atmosphere.

Between birth and death, a trapped particle zips back and forth along spiral paths between geomagnetic poles; in addition, it drifts westward or eastward around the earth.

Why don't the particles just keep on going until they strike the atmosphere and are absorbed? A few do, of course, but under certain velocity conditions, a particle may be turned around (reflected) before it hits the atmosphere. For these particles, the earth's converging field acts like a magnetic mirror that reflects the particles from pole to pole. (Artificial magnetic mirrors are often employed in thermonuclear power experiments to create magnetic bottles that can confine ionized gases, or plasmas, at temperatures of millions of degrees. See chapter 19.) Calculations indicate that a trapped proton in the inner zone may spiral from pole to pole once every few seconds over a life-span as long as several hundred years. Trapped protons obviously lead active lives.

Beyond the Van Allen belts stands the third and last breakwater sheltering the earth from much that transpires in interplanetary space. As with cosmic rays and the radiation belts, a few suggestive observations and surmises preceded the actual discovery of the magnetopause and the current of solar wind that breaks against it.

Before there were satellites and space probes, the only hint that a steady solar wind sweeps the entire solar system came from the observation that comet tails almost always point away from the sun, regardless of the comet's direction of travel. In the early 1950s, the German physicist Ludwig F. Biermann showed that a steady current of high-speed protons and electrons boiling off the sun could account for the blowing comet tails. Biermann's hypothesis was supported in 1958 by Eugene N. Parker, a young physicist at the University of Chicago. Parker showed theoretically that some electrons and protons in the sun's million-degree corona were so hot (that is, had such a high speed) that they would escape the sun's gravitational field and fill interplanetary space with a steady "wind" consisting of several protons per cubic centimeter and blowing at several hundred miles per second. (Although this speed seems high, the particle energies are relatively low for space radiation; for example, a proton travelling at 500 km per second has an energy of only 1300 electron volts compared with the millions of electron volts possessed by trapped particles.) The theoretical stage was thus set for the discovery of this third, very low-energy component of space radiation.

Still another hint—though not a very conclusive one—came from the observation in 1937 by Scott E. Forbush, an American physicist, that the intensity of primary cosmic rays decreased soon after the eruption of a solar flare. The Forbush effect could be due either to the flare material itself reaching out toward the earth or to fluctuations in the earth's magnetic field caused by the flare. Something associated with solar flares was shielding the earth from primary cosmic rays.

It is important to note that just about all cosmic rays were thought to have their origins well out in our galaxy, beyond the solar system, until the great solar flare of February 23, 1956. On that date a huge flare appeared on the sun's face, and in a few minutes cosmic-ray detectors all over the earth began registering many times their normal count of cosmic rays. In the preceding 20 years only four cases of solar cosmic rays had been definitely recorded. The 1956 event convinced everyone that the sun indeed did generate cosmic rays—in copious amounts at times. This discovery underscored the intimacy of the sun-earth relationship, despite the distance of 93 million miles separating the two bodies.

This modest groundwork did not adequately prepare the scientific community for the rather bizarre pictures drawn by the Pioneer deep-space probes and satellites, such as Explorers 18 and 21, which had long eccentric orbits that took them tens of thousands of miles into space.

From theory, one would expect that the earth's magnetic field strength would decrease with the cube of the distance from the earth. The first magnetometer-carrying probes and satellites found these expectations fulfilled, up to a point. Beyond a distance of about 8 earth radii, however, the magnetic field showed extreme disorder and then a sudden drop to the level of the solar magnetic field. Subsequent mapping confirmed that the earth was surrounded by a rather sharp transition region, with the earth's field bottled up inside and the sun's field excluded outside.

This transition region is called the magnetopause, and the volume inside it, containing the earth's field, is the magnetosphere. It turned out that the magnetosphere was far from spherical; it was in reality shaped like a teardrop with a long wake streaming out behind it like an invisible comet tail. What causes this streamlined shape? The satellite-borne detectors of low-energy protons and electrons (plasma probes) that penetrated the magnetopause discovered the solar wind that had been hypothesized by Biermann and Parker. Sunward, this solar wind was ramming up against the earth's magnetic field and compressing it into a shock front, flowing around it, and creating the teardrop shape one expects to see as fluid rushes past a spherical obstruction. Leeward of the earth, the earth's magnetic lines of force stretched out for hundreds of thousands of miles forming the earth's tail.

The earth thus presents a strange apparition to a viewer seeing it through magnetometers and particle detectors. As our planet swings about its orbit, it is sheltered from the solar wind by an elongated magnetic hull that always presents a blunt prow toward the sun.

It is a rare sea that is devoid of storms. The interplanetary sea is no exception. The sun, always restless, creates interplanetary storms when a center of activity on its surface, say, a sunspot, spews forth a cloud of plasma at jet-plane velocity (600–1200 mph). The plasma sweeps along the sun's outwardly spiraling lines of force toward the planets, piling up a shock front consisting of the slower travelling solar wind. The most popular view of this plasma cloud presents it as a tongue that

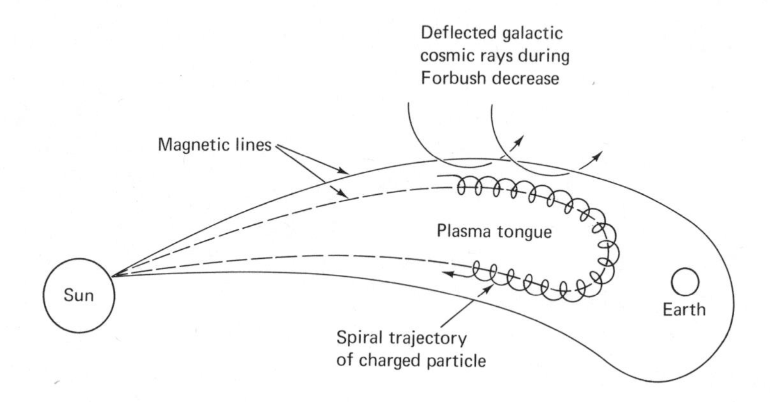

Hypothetical plasma tongue engulfing the earth and causing a Forbush decrease in the intensity of primary cosmic rays

carries part of the sun's magnetic field along with it, as electrical conductors such as plasmas are wont to do (see diagram). The tongue, in this view, is an elongated magnetic bottle confining the hot plasma ejected by the sun.

When the onrushing plasma tongue smashes into the earth's magnetopause, the turbulence causes a series of magnetic storms. Brilliant auroras may also be seen, possibly because the storm helps precipitate electrons from the Van Allen belts, though this is only surmise at present. The solar plasma tongue easily engulfs the entire earth and shields it with its magnetic bottle. The Forbush decrease in primary cosmic-ray intensity occurs during this brief protective period.

Plasma tongues and the solar wind are, of course, not things we can see or feel unaided by instruments. The 1962 Venus probe Mariner 2 showed that the plasma in a solar tongue has a density of only 10–20 protons per cubic centimeter—a degree of rarefaction unobtainable even in our best vacuum chambers on earth. Like all other space radiation, solar plasma is sparse and invisible, though perhaps not innocuous.

Is Space Travel Safe?

Persons who already felt that space travel could not succeed found satisfaction in the discovery of the Van Allen belts and the solar storm radiation, for surely, they reasoned, these would be lethal to all who ventured beyond the protecting breakwaters of the atmosphere, ionosphere, and magnetopause. But the many astronauts who have returned safely from orbit since 1961 show how inflated these fears were. Space radiation isn't harmless, but astronaut protection is really an engineering problem, with reasonable, available solutions.

To begin with, complete protection from space radiation is impossible, even for people on earth. The human race has survived primary cosmic rays for millions of years without artificial protection and without disastrous effects. It is a question of how much additional exposure can be absorbed, for space radiation is little different in its effects from that emitted by dental and medical X-ray machines.

There are three important sources of potentially dangerous space radiation: the Van Allen belts, galactic cosmic rays, and the intense bursts of solar cosmic rays emitted during solar flares. The solar wind and the plasma within the solar-plasma tongues constitute no threat to humans in space, because the protons and electrons are too weak to penetrate spaceship hulls or even space suits.

American and Russian astronauts usually orbit well below the intense inner Van Allen belt. Even though they are above the atmospheric shield, the additional primary cosmic rays and the few trapped particles on the fringes of the Van Allen belts amount to only a few millirads per day, much less than the exposure permitted workers in atomic energy installations. Spacecraft could orbit indefinitely beneath the Van Allen belts without fear of overexposing the crews.

The use of massive shields to protect astronauts against primary cosmic radiation is hardly needed. Even extremely thick slabs of heavy metals would hardly diminish the flux of primary cosmic rays; besides, in penetrating such dense matter a few high-energy protons would create avalanches of secondary cosmic rays that could be more dangerous than the primary ones that caused them. Fortunately, the bulk of the primary cosmic rays are protons and helium nuclei that pose little danger in small quantities. However, if one of the larger nuclei, such as iron, which make up 2%–3% of the primary cosmic rays, were to be absorbed and release its energy in a vital spot, like the brain, enough cells might be destroyed to incapacitate an astronaut. The probability of this happening is considered so minute that space mission planners ignore it.

The most intense radiation levels in the Van Allen belts are high enough to kill an unprotected man within a few days. The easiest solution to the problem this poses is to program orbital flights so that they avoid the Van Allen belts altogether, as they have in the past. In practice, this means that the satellites must avoid the regions where the belts bend down toward the atmosphere and, in addition, remain below an altitude of roughly 500 miles. Spacecraft aimed at the moon and the planets can ascend right through the belts without danger, however, because the time of their transit of the belts will be measured only in minutes—far too short to cause important biological damage.

The solar cosmic rays emitted during solar flares are considered the most dangerous kind of space radiation. Perhaps a half dozen dangerous flares occur each year, more during the peaks of the 11-year sunspot cycle. Because a large flare might increase cosmic-ray intensity in the vicinity of the earth by a hundred times and maintain high levels for several days, exposed astronauts might receive lethal doses before they could reach the haven of our magnetosphere and atmosphere.

Space flight beyond the magnetopause can be made safe in two ways: spacecraft shielding and timing the launches to occur during a quiet period on the sun. (36)

What would be the exposure to the crew and passengers of supersonic transport that is to fly at 70,000–80,000 feet? It doesn't look too bad for an hour's flight under normal conditions. However, during a solar flare the sun spits out an intense burst of protons. There is a good possibility that the pilot will have to take evasive action and drop to 40,000 feet. Otherwise he will subject his passengers to as much in 1 hour as we now receive on the ground in a month.

What happens at the moon? Radiation conditions there ought to be similar to those experienced by our supersonic transport planes near the top of our atmosphere. Cosmic rays will come barreling in from space (with no atmosphere to shield the astronaut), smash into the moon's soil, and generate neutrons near the surface, which will come boiling back out to pester the poor humans who have the gall to settle on the surface. There is one important difference, however, between the top of our atmosphere and the moon's topsoil. There is no nitrogen to gobble up thermal neutrons (and generate ^{14}C). Thus the flux or intensity of thermal neutrons could very well be orders of magnitude higher than on earth. As thermal or very slow neutrons are easily captured by stable elements, making them radioactive, this environment may prove a real hazard to spacemen who wish to stay on the moon for any length of time.

An additional but less important contribution to man's radiation environment on the moon is the lunar radioactivity. Like the terrestrial, it comes about mainly from basalt and granite regions containing potassium, thorium, and uranium. The lunar soil composition was established in a preliminary way by the Russian experiments on their satellite Luna 10, which orbited the moon in March 1966.

Since then, of course, man has managed the colossal achievement of bringing back samples of the moon's surface. These were checked for their gamma-ray emission in a specially designed radon-free room, 15 meters below the ground. As a result of these measurements, we now know that the concentration of thorium and uranium in these moon samples is much like that of similar rock on earth. The radioactive potassium concentration in the Apollo 11 samples, however, proved to be much lower in the lunar surface material than for earth rocks or even meteorites. Furthermore, there were definite amounts of radioactive aluminum, sodium, scandium, and other radioactive elements, which are not naturally present on earth.

Their presence on the moon is explainable on the basis that the lunar surface material has been exposed to cosmic radiation for at least several million years, thus generating these unusual radioisotopes, which are only artificially available on earth.

An interesting sidelight to the moon's continual bombardment by radiation from space is the very reasonable suggestion that the energy stored during bombardment on the night (and cold) side of the moon is released by the sun's rapid heating in the form of visible light. It is well known that many imperfect and impure crystalline materials, such as quartz or fluorite, will trap (at impurity centers) the electrons resulting from ionizing radiation and later, when heated rapidly, release these electrons to recombine with ions and produce light. The light intensity is proportional to the total radiation exposure. This phenomenon is called thermoluminescence.

A thrilling prospect for a very practical application of the natural lunar radiation background is the possibility of using the energy trapped in the moon's soil during the lunar night. Moon settlers might well tap a vast storehouse of energy on the dark side of the moon. By the end of the lunar night, the bombarded material possesses 6 kilowatt-hours per pound, or more energy than is stored in coal. The dust could be fed into an insulated chamber to generate heat for a lunar base or to produce electricity by heating a thermoelectric generator. (14)

13 ◊ Down to Earth

THE environment in which we live is recognizable as a single complex, composed of many subenvironments—land, oceans, atmosphere, and the space beyond our envelope of air. The deer in the forest, the lizard in the desert burrow, and the peavine in the meadow are different kinds of organisms living in situations that are seemingly unalike. Each creature is part of its environment and a contributor to it, but it also is part of the total biosphere; the living world, the sum of all living, interacting organisms. All creatures are linked to each other, however remotely, in their dependence on limited environments that together form the whole of nature.

We know much about the life of the earth, but there is far more that we do not know. Understanding of the large cyclical forces has continued to elude us. We do not even yet grasp the small and seemingly random biological relations between individual organisms—relations involving predator and prey, for instance, and those among species and families—such as exist together in symbiotic harmony and interdependence. Through centuries of observation we have gained a store of information. We are left, however, with a still unsatisfied curiosity about the reach and strength of the tenuous biological cords that bind together the lives of the deer, the lizard, and the peavine.

Life on earth evolved amid constant exposure to ionizing radiation, from the earth itself and from space, known as background radiation. Therefore environmental studies must be conducted in relation to, and with understanding of, background radioactivity.

Of some 340 kinds of atoms that have been found in nature, about 70 are radioactive. Three families of radioactive isotopes—the uranium, thorium, and actinium series—produce a large proportion of the natural radiation. Other radionuclides occur singly rather than in families, and some of them, such as potassium 40 and carbon 14, are major contributors of natural radioactivity. Traces of natural radioactivity can be found, in fact, in all substances on earth.

When man began experimenting with atomic fusion and fission, he placed in his environment—across vast landscapes, in the oceans, and in the atmosphere—measurable additional amounts of radioactivity. These additions were composed of the longer-lived members of some two hundred kinds of atomic radiation. Although the additions constituted but a fraction of the background burden, they represented the first alteration of the radiological balance that had existed since the early ages of the planet. Thus it became necessary to determine what the impact of such a change might be. In the process of inquiry, these ideas emerged:

1. The addition of man-made radioactivity presents the possibility of delayed or cumulative effects. Long-term studies geared to the assessment of biological effects from extremely low radioactivity are essential.

2. The addition of radioactivity makes possible broad-gauged studies to trace the movement and concentration of radionuclides in the environment. These studies, in turn, can disclose new information on biological complexes and mechanisms.

The quantities of low-level long-lived radioactivity already released into our environment will provide materials for future studies covering decades. Further, because radioisotopes are chemically similar to nonradioactive forms, observation of their biological fate will provide clues to the transport, concentration, dilution, or elimination of many other kinds of man-made toxic agents and contaminants of the environment.

Each environment presents its own sets of conditions and

unknowns. It is important to appreciate those that are characteristic of water, land, and atmosphere.

The oceans are the basins into which are poured all the nutrients or wastes transported from the land by rivers and winds. An ocean, a river, or a lake is an area of constant physical and biological motion and change. In the ocean the surface waters form a theater of kaleidoscopic, and frequently violent, action. The presence of man-made radioactivity in water has made it possible to follow the disposition of nutrients and wastes in the restless aquatic ecosystem.

In a water environment the minerals necessary to life are held in solution or lie in bottom sediments. They become available to animal life after being absorbed by plants, both large floating or rooted plants and tiny floating ones called phytoplankton; because the phytoplankton are found everywhere in the sea, they play a larger role. The phytoplankton concentrate minerals and become food for filter-feeding fish and other creatures, including the smaller zooplankton, floating one-celled animals, which in turn are food for other organisms. Thus the minerals enter extremely complex food chains. The cycles of nutrition are completed when fish and plants die and decomposition again makes the minerals available to the phytoplankton.

Some radionuclides that are introduced into an aquatic environment enter the food chains exactly as do the stable minerals essential to life, because the radionuclides are merely radioactive forms of the nutrients. Elements such as copper, zinc, and iron are less plentiful in the water environment than hydrogen, carbon, or oxygen, for example, but are concentrated by phytoplankton because they are necessary for life. Such elements are in short supply but in constant demand; thus, when their radioactive forms are deposited in water, they are immediately taken up by aquatic plants and begin to move through the food chains. Fission products such as strontium 90, for which there is little or no metabolic demand, are taken up by aquatic food chains to only a minor extent.

The precise paths of radioelements through aquatic ecosystems are almost unknown. In addition to their movement in food chains, radioelements also may be moved physically from place to place in the tissues of fish or other creatures. Some radionuclides for which there is no biological demand may sink into bottom sediments and remain there until they have lost their radioactivity. Or radioactivity actually may be transported uphill, from water to land, as when birds that feed on fish containing radioactivity leave their excretions at nesting areas. The routes and modes of transport seem numberless.

The surface waters of the seas, down to depths of 200 meters, are areas of rapid mixing in which temperature, density, and salinity are almost uniform. Below the surface water is a zone in which temperature decreases and density and salinity increase with depth. This zone, known as the thermocline, may reach a depth of 1000 meters. Because density is increasing here, vertical motion is reduced, and exchanges between the surface and the deep waters are impeded. Physical conditions affect the rates of physical movement of radioactivity in the mixed (surface) layer, the degree to which radionuclides are held at the thermocline, and the processes by which radionuclides pass the thermocline and enter the deep-water cycles and upwellings.

Plankton take up large amounts of radioactivity. Planktonic forms, in fact, proved to be the most sensitive indicators of the presence of radioactivity in the marine environment. Further, the daily vertical migrations of plankton—down in response to sunlight and up at night—seemed a part of the process by which radionuclides move from the upper waters to the deeps.

Strontium 90 and cesium 137, important in fallout on land, enter the marine cycles only in minute amounts. Practically no fission products are found in fish. Since strontium 90 is not concentrated strongly by marine organisms, the question of what happens to it in the ocean remains unanswered. Studies have suggested, however, that strontium moves in solution and thus indicates the movement of water. If this is true, strontium 90 may be contained in the deep currents and eventually will be brought again to the surface. Some observers believe this process has begun.

The freshwater environment differs from the marine in the greater variety of its minerals, among other things. Rivers vary greatly in character and change radically from season to season because of rainfall and other factors. General understanding of their biological workings is difficult to formulate. But rivers are the routes by which minerals and wastes are transported toward the sea, and estuaries are significant because of the many forms of life that flourish there.

Studies of radioactivity in rivers and estuaries usually have been made in relation to the fate of effluents from nuclear plants. Concentration factors have been established for significant radionuclides in phytoplankton, algae, insects, and fish, and typical patterns of dilution and dispersion have been plotted.

Radiobiologists are studying biological distribution, while oceanographers are using the trace amounts of effluent radioactivity to verify the patterns of dispersion of river waters in the ocean. (4)

The word fallout, destined for the vocabulary of all men, was coined in 1945. It refers to the radioactive debris that settles to the surface of the earth following nuclear explosions. Since most organisms living on earth can now be exposed to detectable radiation from a single nuclear explosion, there naturally has been great interest in the possible effects of fallout on living creatures, including man.

Apprehension and a degree of controversy have resulted from lack of knowledge of the nature of fallout, from its association with nuclear armaments, and the involvement of personal convictions.

In a nuclear explosion tremendous quantities of heat are produced in a relatively small quantity of matter within a few thousandths of a second. The casing and other structural parts of the exploded device, as well as the fission products and surrounding air, are raised to a temperature approaching that in the center of the sun, probably several million degrees. In such heat all materials are vaporized during the first few thousandths of a second. The high temperature creates what is known as a fireball, which expands rapidly, heating material in its vicinity, and then rises above the explosion point. Thus, right at the outset a nuclear explosion creates a mixutre of gases, melted nuclear fuel, and perhaps some partly melted environmental materials. As the fireball cools the melted materials begin to solidify, and the gaseous materials condense and solidify. This produces the debris that is destined to fall back to earth.

The characteristics of fallout resulting from a specific explosion are determined by two main factors: the height of the burst and the size or power of the explosion. (9)

When a nuclear weapon is detonated a number of radioactive nuclides are produced that may slowly come to earth far from the site of the detonation. The nuclides of chief interest biologically are plutonium 239, cesium 137, strontium 90, strontium 89, and iodine 131. Also, the neutrons produced may be absorbed by the nitrogen 14 of the air to give carbon 14, according to the equation

$$^{14}N + n \rightarrow {}^{14}C + p$$

One atom of nitrogen 14, in other words, captures a neutron, then liberates a proton to become carbon 14.

The average dose from fallout in the United States is less than 15 millirads per year—less than 10% of the natural background. The chief danger from fallout lies not so much in its total dose as in the fact that many of the nuclides resemble normal body elements in their chemical behavior. Thus, each may find its way to a critical organ or tissue and exert an effect out of proportion to its average dose. ^{89}Sr and ^{90}Sr resemble calcium and so will find their way to the bones. ^{131}I will behave like stable iodine and concentrate in the thyroid and salivary glands. ^{239}Pu finds its way to certain sensitive sites in bone. ^{14}C, following normal carbon, may lodge within the critical DNA molecules of our cell nuclei. (39)

Natural radionuclides find their way into plants' metabolic processes. Man-made radionuclides also are so incorporated—even some, such as uranium or radium, that have no known metabolic role. The man-made nuclides, whether they reach the earth in fallout or by other means, mix with the stable nuclides to which they are chemically related, increasing by small fractions the total amount of each element available to participate in plant growth cycles. Because artificial radionuclides behave so typically, they present, on the one hand, a possible long-term hazard and, on the other, the expectation that their detectability will reveal much about the biological courses of minerals and nutrients.

The disposition of man-made radioactivity on land is determined in part by such factors as topography and the presence or absence of water. Topography may influence the distribution by setting patterns of drainage and exposure of surface soils to wind and rain. Water may affect dilution, or it may leach radionuclides out of surface soils and thus remove them from the level in which plants are rooted. The leaching may carry radionuclides elsewhere, however, possibly causing mild contamination of the water table.

Plants take up radionuclides through their roots or through their foliage. But the role of soils is significant. Some radionuclides are bound as ions to clays and thus are withheld in large measure from entry into the plant system. Cesium 137, for example, is held so tightly by soils that uptake through plant roots is slight, and thus a more significant mode of entry of cesium 137 into food chains is by direct deposit on plant leaves. Variables are introduced by the physical configuration of the plant itself, by seasonal differences in plant metabolism, and by the effects of rain and snow. In the case of iodine 131, a short half-life—8 days—virtually precludes the possibility of extensive uptake through plant roots. But the half-life is not too short to prevent grazing cattle from ingesting radioiodine deposited in fallout and thus allow the appearance of radioiodine in milk. (4)

Not long before scheduled nuclear tests in March and April 1962, scientists measured radioiodine in the thyroids of California and Colorado deer. When the tests were completed, they found a marked radioiodine increase in the thyroids of the animals in both states. One experiment revealed that cattle and sheep grazing an open range accumulated 10,000 times as much radioiodine as penned cattle and sheep fed on stored fodder.

Cesium 137 has been detected in the bodies of Alaskan caribou and in Alaskan Eskimos and Indians who dine on caribou meat. Caribou browse on lichens (simple plants) during the winter. Lichens easily absorb cesium 137. Caribou hunted by the Eskimos in early spring contain cesium 137 from lichens. And, researchers found, cesium 137 from caribou meat accumulates in the Eskimos' bodies in considerable amounts in late spring and summer. (1)

Much attention has been devoted to strontium 90 and to its availability to man by deposit on plants and soils. Soils, however, present confusing factors. Experiments and fallout observations show that strontium 90 does not penetrate soils deeply. In typical instances it remains in the upper inch or two of the soil surface, where its availability to root systems is as variable as the conditions of mixing, leaching, and plant growth. Experiments have shown that plant uptake of strontium from soils can be reduced by introduction of calcium (to which it bears a close chemical relation) in available form into the soil.

Radiobiological developments on land result from combinations of environmental influences. Studies in the Rocky Mountains show that ecological conditions above the timberline, particularly in areas where snowbanks accumulate, are efficient in concentrating fallout radionuclides. Concentrations thus take place in the snow-packed heights that are the sources of mountain streams flowing to the plains far below.

The environment of the earth is a product of weather—of the transport of moisture, of the actions between winds and oceans, of the cycling of energy through biotic systems. Understanding of biological potentials of atmospheric factors involves understanding of atmospheric motions affecting transport and mixing of contaminants and the processes of deposition of radionuclides from atmosphere to earth.

At some thousands of feet above the earth's surface—at 30,000–40,000 feet in the middle and polar latitudes and at 50,000–60,000 feet in the tropics—there is a level, the tropopause, at which air temperature, rather than decreasing, becomes constant or increases with height. Below this level is the troposphere, the turbulent zone of clouds, rain, and fog. Above it is the stratosphere, where there is no turbulence and only a slow mixing of dry and cloudless air. The stratosphere continues to a height of about 100,000 feet. Investigators have noted the importance of rain or snow in washing fallout particles from the air in the troposphere. There is disagreement on the precise modes of distribution of radioactive materials projected into the stratosphere. (4)

Brazil Nuts and Cereals—Internal Radiation Exposure

WHENEVER scientists bring up the question of internal exposure due to natural radioactivity in food they immediately think of Brazil. Not only because of the very high gamma-ray activity of the soil, which exists in certain areas of Brazil as well as India, but also because of the Brazil nut with its extraordinarily high radioactivity—some 14,000 times that of common fruits. Of course the Brazil nut is exceptional and by no means characteristic of nuts in general. Cereals are also relatively high—

perhaps as much as 500–600 times that of fruits, which have the lowest concentrations of natural radioactivity. The following table points out pretty clearly that it pays to keep away from rich foods for more than one reason.

Actually the table discusses only the alpha activity of foods and this tells nothing about the radioactive potassium contribution. All muscle-building foods contain potassium and we must eat to live. Recent surveys indicate that for most populations the radioactive potassium in our muscles provides 10–20 times the internal exposure of any of the other incorporated radioactive materials, such as radium or carbon 14.

RELATIVE ALPHA ACTIVITY OF FOODS

From *Proceedings of the Second United Nations International Conference on the Peaceful Uses of Atomic Energy,* September 1–13, 1958, Geneva, Switzerland, United Nations Publication, vol. 23, "Experience in Radiological Protection," p. 153, W. V. Mayneord. (14)

FOODSTUFF	RELATIVE ACTIVITY	FOODSTUFF	RELATIVE ACTIVITY
Brazil nuts	1400	Biscuits	2
Cereals	60	Milk (evaporated)	1–2
Tea	40	Fish	1–2
Liver and kidney	15	Cheese and eggs	0.9
Flour	14	Vegetables	0.7
Peanuts and peanut butter	12	Meats	0.5
Chocolates	8	Fruits	0.1

It is usually futile to try to reduce radioactivity intake significantly by alterations in diet. An excellent example is presented by milk, which has been widely publicized as a carrier of strontium 90. Paradoxically, even though milk is the largest single source of ^{90}Sr in the diet, reducing milk intake tends to increase the body burden of ^{90}Sr. This is because milk calcium is always less contaminated with ^{90}Sr than plant calcium, and if milk intake is reduced the body gets proportionally more of its calcium from plant sources.

But the most important consideration is the adverse nutritional effect than can result from misguided action. (9)

In the midwest there is a study in progress that began with the discovery, 20 years ago, of the high natural radioactivity in the drinking water of several municipal water supplies. In 1955 the radioactivity was found to result from radium, which had been leached out of the soil along with the other more common materials, such as calcium and magnesium.

The water from Maine's wells has 3000 times the radium concentration as the Potomac River. Of course, Maine's radioactivity is mild indeed when compared with that of some springs in Kansas, Colorado, or Jachymov, Czechoslovakia, where the concentration is 10,000 times greater.

Amazingly enough it has been less than 50 years since such quantities of radium in water were considered a great asset. Cures for all sorts of ills were ascribed to radium or thorium. An advertisement (reproduced in the photograph section) that appeared at the time of World War I is unequivocal in its support of radium as a medication. Fortunately it only took a generation for most of the medical profession to become aware that radium chemically resembles natural calcium so much that it tends to pile up in our bones and do serious damage. There are still places in the world that tout the advantages of bathing in radioactive mineral springs. Very recently someone ordered mineral water in Milan and noticed that the label on the bottle boasted about the radioactive contents and the merits of such radioactivity in curing a host of gruesome diseases.

In certain parts of southwest India (Travancore-Cochin State) the coastline is characterized by patches of radioactive sand with the highest thorium content in the world—as much as 33%. Thus conditions are particularly favorable for studies of radiation-induced mutations. The gamma-ray backgrounds are often such that the native population receives more exposure than our U.S. governmental agencies will allow a worker whose job necessitates handling radioactivity.

Similar conditions exist in the state of Minas Gerais near Rio de Janeiro in Brazil. In 1964 a genetic cell study of the population of the village of Guarapari (population 6000) was started by the Institute of Biophysics, University of Brazil, and the Institute of Environmental Medicine, New York University. Radiation dosimeters were given to selected individuals. The dose rate these individuals received in moving through the variable gamma-ray field over a long period of time was thus accumulated. To induce these people to wear the packets continuously, they were disguised as religious medallions and distributed by the principal scientists, Jesuit Fathers Thomas L. Cullen and the late Francisco X. Roser. The average gamma-ray dose rate for the exposed population has been established as about six times the world average with some exposures as high as thirty times this amount. (14)

There are radiation differences not only in different parts of the world but in different houses on the block.

There may be disadvantages in having a brick and concrete or granite home. Dr. William Spiers of Leeds, England, one of the world's foremost experts on natural radiation, pointed out that the intensity of gamma rays from radium in the brick may provide as much as three times the exposure one might expect in a wooden house. (14)

And differences inside the house.

It recently has been shown that cigarette smoke contains a naturally occurring radioactive element, polonium 210. Studies have demonstrated that more of this radioelement is found in the lung and other soft tissues of cigarette smokers than those of nonsmokers.

Another naturally occurring radioactive isotope, lead 210, which undergoes radioactive decay to become polonium 210, also is present in cigarette smoke. In addition, lead 210 was found in lung tissue and in the bones of cigarette smokers in twice the concentrations normally found in the non-smoking population.

Lead 210 in the smoke is the primary source of the increased concentrations of both these radioelements in body tissues. These higher concentrations, in turn, may increase by 8–30% the internal radiation dose to the bones of cigarette smokers over that received by mankind as a whole.

Recently, laws have been passed that force fairly stringent controls on exposure. There is now a flurry of activity to build instruments to measure both the external and internal exposure hazards. The principal radiation exposure in uranium mines is caused by the presence of radon gas and its solid, particulate radioactive daughters. The particulates are likely to be inhaled by the miner and a portion of the particles may be retained in the respiratory tree. The alpha particles emitted from these daughters are capable of producing tissue damage.

Not all mines exhibit a high level of radioactivity. In fact, quite the contrary. Mines of common table salt are often very free of radioactive impurities and deep enough to shield out most of the extraterrestrial radiation. Such mines in Ohio and Texas have been used by researchers to do experiments that require a radiation-free environment. They mean it when they say "Back to the salt mines."

A delicate experiment to determine the stability of the proton is being carried out 2 miles below the earth's surface in a gold mine in South Africa. So far, it has been established that if the proton is unstable at all, its average lifetime is at least 10^{25} years! (14)

Thus, we see that an internal background also exists, since each of the naturally radioactive nuclides finds its way into the body. The most important, of course, are ^{40}K, ^{14}C, and ^{3}H, since our bodies contain relatively high percentages of potassium, carbon, and hydrogen. Thorium and uranium also enter the body and, as they decay, pass through a number of radioactive daughter elements before eventually becoming lead. Like Th and U themselves most of these daughters pass through the body rapidly. But at least one, radium, Ra, chemically resembles natural calcium so much that it tends to pile up in our bones. In this way it too adds noticeably to our internal background.

DOSE RATES DUE TO EXTERNAL AND INTERNAL IRRADIATION FROM NATURAL SOURCES IN NORMAL AREAS (14)

SOURCE	DOSE RATES (MRAD/YR.)*
External irradiation	
Cosmic rays at sea level	0
Ionizing component	28
Neutrons	0.7
Terrestrial radiation	50
Cosmic rays at 20,000 feet	1500 (= 1.5 rad/yr.)
Cosmic rays near top of atmosphere	30 rad/yr.
Internal irradiation	
Potassium 40	20
Rubidium 87	0.3
Carbon 14	1
Radium 226, 228	1
Hydrogen 3 (tritium)	2
Average total dose to body	100

* Rad (*r*adiation *a*bsorbed *d*ose) is the basic unit of absorbed dose of ionizing radiation. A dose of 1 rad means the absorption of 100 ergs of radiation energy per gram of absorbing material. 1 millirad = 0.001 rad. (A roentgen of gamma rays will deposit almost 1 rad in tissue.)

All in all, these sources bring our natural radiation dose to about 0.15 rad per year or less, for a lifetime dose of about 10 rads. A few places on earth, notably in India and Brazil, have soils so rich in Th and U that the natural background may be as high as 2 rads per year. Also, air tends to act as a shield against cosmic-ray dose so that dwellers in high places (such as mountainous regions) may receive a little more radiation than those at sea level. But 0.15 rad per year is a safe maximum figure for most of us.

Artificial Backgrounds

A number of objects of our daily experience constitute sources of noticeable radiation. A television set, for example, is simply a low voltage X-ray machine. Electrons from the picture tube filament are speeded up by electrical voltage until they collide with the phosphor coating of the tube. This is precisely the mode of operation of an X-ray tube. However, the low voltages employed, the poor efficiency of the phosphor for X-ray production, the glass and plastic shielding through which the X rays must pass and the distance of the average viewer from the screen all combine to lower the received dose. On the average this source contributes less than 1 millirad per year to one's total dose.

NONOCCUPATIONAL ARTIFICIAL EXPOSURES (14)

SOURCE	DOSE OR RATE
Wristwatch dial, approx. 1 microgram of radium, gamma rays	1 mr/hr
Airplane instruments—pilot position	1 mr/hr
Shoe fitting (20 sec)	10 r
Diagnostic X ray	
14 × 17 chest plate	0.1 r
Photofluorographic chest	1 r
Extremities	0.5 r
G1 series (per plate)	1 r
Pregnancy	9 r
Fluoroscopy	15 r/min
Dental (per film)	0.5 r
Radiation safety guide for population	0.5 r/yr.
Radiation safety guide for radiation worker	
Extremities	75 r/yr.
More sensitive body organs	5 r/yr.
Skin and thyroid	30 r/yr.

For many years various radium-containing paints have been used to render watch and clock dials luminous. [The radiation emitted by the radium caused the phosphorescent paint to emit light.] Since radium and its daughters emit both gamma and beta rays energetic enough to penetrate a watch glass, 25 millirads per year per person from this source is probably a good average figure. In recent years ^{3}H has come into vogue for these self-luminescent paints. It emits only beta rays of energies low enough to be completely stopped by glass or plastic, so the dose from a watch with this material on the face is about nil.

A few other scattered sources of radiation turn up at times: the bathtub sets glazed with a uranium pigment that afforded the users a dose rate of over 100 millirads per hour or the houses built of radioactive stone that bombarded occupants with 10 millirads per hour.

The average American has about four dental X rays and one medical X ray every 10 years. With modern equipment and practice this will result in perhaps 1 rad per year of local dose, but only 50 millirads or so of whole body dose. All in all, radiation from medical sources exceeds the natural background for only a few of us. (39)

14 ◊ Atoms in Agriculture

To know what questions to put to Nature—that is 95 percent of scientific research.

—Alfred North Whitehead

If man's existence on the earth is compared to a calendar year, then he began farming in the very early morning of December 30 and began applying systematic knowledge to agriculture at 10:15 P.M. on December 31.

The first traces of man on the earth are dated at about 1.75 million years ago. Plant life then was very much like plant life today, but the animal population was quite different. Man became a producer of plants and animals instead of merely a gatherer and hunter about 8000 years ago. He has applied systematic study to cultivated plants and animals for only 300 years. (5)

Estimated crop losses to weeds, insects, and diseases are many billions of dollars each year. These losses can be reduced through research.

How fast do roots grow? How deep? How soon does water get to them after a rain?

How far will pine pollen travel on the wind?

To answer these questions, scientists need some kind of miniature genie, one who will shout "I'm here!" when the root has reached the fertilizer or the water has reached the root. Such a helpful genie exists as the radioactive atom.

One way to see how valuable radioactive tracers are is to compare them to standard chemical techniques. A sensitive chemical test can perceive molecules as dilute as 10^{-7}; that is, it can detect a molecule surrounded by 10 million molecules of another kind. A good radioactive tracer technique can distinguish concentrations of 10^{-11}; that is, it can trace 1 in 100 billion.

In other words, by the chemical test you could find a person in metropolitan New York with a secret tattoo on the roof of his mouth. By the tracer method you could find this same person anywhere in the world, even if the world population were multiplied fiftyfold.

In the chemical test you could distinguish the equivalent of one kernel of corn in one-tenth of a boxcar load; in the tracer, one kernel in 850 boxcars.

Most studies of plant nutrition and metabolism pertain to the following questions: What do plants need for their best growth? How do they take in the materials they need? What things are absorbed by roots and what things by foliage? How does the plant turn water and other simple compounds into carbohydrates and proteins?

Early research indicated that only 10–12 percent of phosphorus fertilizers was taken up by plants in the first year; the rest was locked into the soil or washed away. With radioactive phosphorus 32 scientists found that as much as 50–70 percent of the phosphorus in a plant came from the fertilizer during the first two or three weeks of growth.

Fertilizer applied to soil is largely wasted because it is either bound by soil particles or is washed out of the root zone. If chemical elements could go directly into leaves and bypass the wastefulness of soils, a tremendous saving would result.

Botanists have learned in recent years that the foliage of plants can take in some nutrients much as roots can. With tracers they discovered that many nutrients are readily taken up by foliage, including bark of dormant trees, even at temperatures below freezing. As shown by isotopic tracers, elements such as phosphorus, nitrogen, and potassium move both up and down from the point of application at rates similar to those following root absorption. Urea (a nitrogen compound) is now used as a nutrient foliar spray for many fruit and vegetable crops in this country.

Even before the use of tracers, agronomists realized the inefficiency of spreading fertilizer uniformly over a seedbed. They know the fertilizer should be placed somewhere near the seed, but where? Above? Below? Beside? Below and beside? How far away? They had conducted some research, but the methods were slow and tedious.

Using tracers, the researchers confirmed earlier findings that roots within 2 or 3 days reached fertilizer placed less than 2 inches directly below seeds, but the roots tended to congregate there. When the fertilizer was 2 inches below and 2 inches to the side, roots reached it within a week and a better root system developed. With 3 inches between seeds and fertilizer, the desired seedling "boost" was delayed 3 or 4 weeks.

The movement of radioactive phosphorus from root to leaf was found to be remarkably fast, sometimes requiring less than 20 minutes.

Some plants take in chemicals that the plant probably cannot use: for example, the so-called locoweeds accumulate enormous amounts of selenium. With tracer techniques, we can see that the root uptake process has poor powers of discrimination.

Tracer experiments reveal that roots cannot distinguish potassium (needed in large amounts) from other elements which are chemically similar but quite different in size. Once inside the plant, only potassium can be metabolized and similar but heavier elements (rubidium, cesium) are useless. This is like an absentminded builder who buys brick, boulders, and gravel indiscriminately for his wall and then finds he can use only part of his materials.

The process called photosynthesis whereby green plants use energy from the sun to convert simple compounds from air and soil into complex, energy-rich substances has been termed the most important chemical reaction in the world. It is the basis for man's entire food supply and, except for nuclear energy, all significant fuel as well. Tracer techniques have multiplied the research efforts on photosynthesis tremendously.

When only chemical tests were available, food manufacturing in green leaves had to progress for hours before scientists could measure the products. But with tracers and other new techniques they have narrowed the experimental time to minutes and finally to seconds. Today they know that a green leaf has formed sugars more complex than fructose, fruit sugar, after exposure to light for only 1 second.

Radioactive isotopes have been used to study insects, their life cycles, dispersion, mating and feeding habits, parasites, and predators. Several hundred such studies have been made on dozens of insect species.

As one example, nearly half a million mosquito larvae were tagged with radioactive phosphorus in Canada. Some of the adults from these larvae were later found as far as 7 miles away, but most were recovered within ⅛ mile.

In a companion study grasshoppers were labeled with the same isotope. Their average rate of movement was only 21 feet per hour, and after 7 days their position was based entirely on random motions plus prevailing winds. It seems that grasshoppers have no ability to move toward food.

How far do insects carry pollen? This question is of practical importance in knowing how far to separate seed fields to maintain pure varieties of plants. In the past it was studied by the laborious method of growing a plant having a dominant marker gene for some visible trait surrounded by plants without the marker. Seeds from plants at various distances from the marked plant were grown the following year to see how far the genetically marked pollen had been carried. It was difficult to obtain strains genetically pure for presence or absence of the marker gene. Also, considerable testing and bookkeeping were involved.

With tracers the answer may be found in a few days. A plant is injected with radioactive phosphorus; after a few days its pollen is highly radioactive. Flowers at various distances from the tagged plant may be checked daily for radioactivity. In one study with alfalfa, radioactive pollen was carried as far as 30 feet by bees, but more than one-third was deposited on plants adjacent to the labeled one.

Two characteristics of soils besides fertility are vitally important and difficult to measure. These characteristics are moisture and density. Moisture must be determined frequently for efficient irrigation. Density controls the pore space available for water and oxygen; the possible damage to the soil from tillage and harvesting machines is revealed by before-and-after tests of density.

Both soil moisture and density were formerly determined by laboratory methods, which had two drawbacks: the methods were laborious, and they tested soil in an unnatural state. Today a sort of double-barreled radiation method can be used to measure these two soil characteristics.

Neutrons are readily scattered by water but not by soil; gamma rays are absorbed by both soil and water. In practice the experimenter drills two holes in the soil a few feet apart. Into one he puts a gamma-ray source; into the other, a radiation detector. The reading on his detector dial tells him the amount of gamma rays absorbed by both soil and water. Replacing the gamma-ray source with a neutron source, he obtains a reading on absorption by water only. The difference between the two readings is ascribed to the density, or degree of compaction, of that soil in its native state.

With insect pests, as with plant diseases, biological control is more economical than artificial control. The use of insecticides too often results in destruction of helpful insects along with pests. Limited success has been achieved in breeding certain plants for resistance to insects.

Two important uses of biological control in agriculture have been made in recent years: importing an insect from Australia to eradicate a weed in California and disseminating ladybird beetles to control certain scale insects.

Helpful parasites and predators must first be identified

before they can be used. In the case of small or nocturnal insects, this can be exceedingly troublesome. Tagging the pests with radioisotopes in order to identify the predators which consume them is much simpler because the most efficient predators contain the most activity.

Radioactive labeling is also valuable in studying helpful insects. In one case the indolence of drone bees was indicated by finding that even with adequate syrup in their cage they still received identical syrup from worker bees in an adjoining cage. (5)

One of the most important agricultural uses of radiation is to eliminate insect pests that injure and kill farm animals. The success of radiation was demonstrated dramatically against the screwworm fly.

The female of this insect—twice the size of a housefly—lays eggs in open wounds of many warm-blooded animals, including man. Even a tiny scratch in the hide of a cow, for instance, serves as a hatchery for this insect's eggs—and each female deposits about 250 eggs at a time.

The eggs develop quickly into larvae that look like tiny screws, hence the fly's name. About the size of a paper clip, these larvae burrow deep into the flesh. They can kill a full-grown steer or a sheep in a little more than a week.

Screwworm fly larvae once caused an estimated $20 million a year livestock damage in the Southeast. The southwestern states suffered an annual loss in livestock of up to $100 million. So during the 1950s the U.S. Department of Agriculture began a program that set fly against fly.

Radiation impairs the ability of the male fly to reproduce. Male screwworm flies were subjected to heavy radiation from a cobalt 60 isotope source. Then, in a dramatic test, hordes of the irradiated flies were released on the island of Curaçao in the Caribbean, where screwworm flies infested most of the livestock. Five months later there were no screwworm flies left on Curaçao. How were they exterminated?

The irradiated males had mated with native female flies. Because of the irradiation, the eggs produced by the mating were not fertile. No young flies appeared. The native flies died off, and, isolated by the ocean, Curaçao was saved from the insect menace.

In a follow-up, 7 billion irradiated males were released in the southern part of the United States by the mid-1960s. The screwworm fly now has been all but eliminated from this country.

Radiation has been especially effective against diseases carried by insects. Among the most serious of these is malaria, which is caused by a tiny protozoan, transmitted from person to person by the bite of the female *Anopheles* mosquito.

Anopheles lays its eggs in stagnant water. When the eggs hatch, young wigglers skitter across the surface of the water. Later, they mature, change to adult mosquitoes, and fly off.

Scientists need to understand the movement of *Anopheles* to control malaria in tropical countries. They need to know, for example, how far insects can fly from their breeding water.

Radioisotopes help keep tab on them. The young can be tagged for life with a bit of radioisotope placed in waters where the insects breed. Once they leave the breeding area, the mosquitoes may be detected wherever they go by instruments that record the amount of radiation given off by the radioactive tracer.

Pesticides—poisons such as DDT—are still our main weapon for insect control. Many health officials worry that man has spread too much insect poison into the environment. They fear, for example, that the poisons may contaminate human food. It is important, therefore, to know how much and how fast insecticides accumulate in various animals' bodies. (1)

To obtain this information many experiments have been performed. In one of these Ohio State University scientists tagged DDT with a radioisotope and studied how much pesticide was absorbed by marsh animals and how long it took to be absorbed.

15 ⸎ *Nuclear Power and the Environment*

Increasing concern is being expressed about the environmental effects of electrical generating plants, both conventional and nuclear.

Electric power requirements in this country have been doubling about every 10 years. Future expansion is expected to continue in much the same pattern. Steam electric power plants, whether fossil fired or nuclear, must be relied upon in the main to meet these ever increasing power needs. There are relatively few economical sites available for hydroelectric plants. Efforts to develop acceptable alternate systems for meeting our bulk power needs are unlikely to prove successful in the foreseeable future.

The use of steam electric power plants will inevitably have an impact on the environment. The fossil fuel plants accelerate the exhaustion of irreplaceable resources; add heat to the air and water; consume oxygen and add carbon dioxide, sulfur dioxide, and other gaseous and particulate matter to the environment. Nuclear power reactors also add waste heat to the air and water and, in addition, add low levels of radioactivity to the environment.

As for nuclear accidents, no accidents of any type affecting the general public have occurred in any civilian nuclear power plant in the United States. Since the beginning of the atomic energy program in 1943, seven U.S. workers have died in radiation-connected accidents. Of these deaths, three occurred in an AEC-owned experimental reactor [SL-1] at a remote testing station in Idaho, two were from criticality accidents in the weapons program, and two occurred in nuclear fuel processing plants. This record compares most favorably with similar development and industrial activities.

The increasing demand for electric power can be attributed to a number of factors. The population growth, of course, has been important but it is only part of the story. Electric power usage per person has been increasing at a much faster rate than the population. Industry usage has grown. Electricity has been used in many new areas such as residential air conditioning and space heating.

Total consumption of electrical energy in the United States quadrupled between 1950 and 1968, while the population increased by one-third. The consumption per capita rose in that period from 2000 to 6500 kilowatt hours per year. The estimated per capita consumption in 1980 is some 11,500 kilowatt hours and about 25,000 kilowatt hours by the year 2000.

U.S. ELECTRIC UTILITY POWER STATISTICS
RELATING TO POPULATION AND CONSUMPTION

	1950	1968	1980 (EST.)	2000 (INTERPOLATED PROJECTION)
Population (millions)	152	202	235	320
Total power capacity (millions of kilowatts)	85	290	600	1352
Kw capacity/person	0.6	1.4	2½	~4¼
Power consumed per person per year (kilowatt-hours)	2000	6500	11,500	~25,000
Total consumption (billion kilowatt-hours)	325	1300	2700	~8000
Nuclear power capacity (% of total)	0	<1%	25%	~69%

The projected growth of generating capacity in the eleven northeastern states illustrates these mounting electric power demands. A report to the Federal Power Commission (*Electric Power in the Northeast 1970–1980,* A Report to the Federal Power Commission, prepared by the Northeast Regional Advisory Committee, December 2, 1968) indicates that between now and 1990 the power industry in these eleven states must build about four times as much electrical generating capacity as the industry has provided thus far in its 80-year history.

This same report concludes that nuclear power will account for about 60% of the total generation in the Northeast by 1980 and more than 80% by 1990. Reasons for the choice of nuclear power, particularly in the New England–New York areas, are the low fuel cost, the low fuel transportation cost, and the virtual absence of atmospheric pollutants from nuclear fuels. (20)

Men and Ashes

An archaeologist kneels in the Mexican valley of Tehuacan, gently brushing the dirt from crude artifacts discarded thousands of years before. Fragments of earthenware and chipped stones in the ashes of ancient fires tell their story. They are wastes from the first human culture on the North American continent.

Today, in the wheat fields of Kansas, tons of straw left behind by threshing machines must be disposed of each year by baling for later use or by burning. Wastes of straw may never be seen by archaeologists of the future, but other wastes of our time will intrigue them—piles of slag from steel mills, city dumps under tons of soil, and mountains of rusted automobiles.

Almost every act of man leads to wastes of one sort or another—hunters' ejected cartridges, smoke from power-plant chimneys, wastepaper in schoolrooms, or the curls of childhood lost forever on the barbershop floor.

The nuclear industry is no exception. Managers of nuclear

installations are concerned with the handling, processing, and disposal of wastes. In many ways this activity is comparable to other disposal processes, but it is different in a few important respects.

The most significant difference between nuclear wastes and those of other industries is that nuclear wastes are radioactive, and this accounts for the special methods required for their disposal. The highly radioactive wastes from spent nuclear fuels, in particular, pose a challenging problem. They cannot be piled in open fields or dumped into rivers or the ocean. Another problem exists because of the millions of gallons of water containing traces of radioactivity that must be disposed of each year. (28)

Nuclear energy is discussed in part IV. The management of nuclear wastes is an environmental problem.

The management of radioactive waste material in the growing nuclear energy industry can be classified under two general categories. The first is the treatment and disposal of materials with low levels of radioactivity. These materials are the low-activity gaseous, liquid, and solid wastes produced by reactors and other nuclear facilities such as fuel fabrication plants. The second category involves the treatment and permanent storage of much smaller volumes of wastes with high levels of radioactivity.

Neither the reprocessing of used fuel nor the disposal of high level wastes is conducted at the sites of the nuclear power generating stations. After the used fuel is removed from the reactor, it is securely packaged and shipped to the reprocessing plant. After reprocessing, the high level wastes are concentrated and stored in tanks under controlled conditions at the site of the reprocessing plant. Only a few reprocessing plants will be required within the next decade to handle the used fuel from civilian nuclear power plants.

More than 20 years of experience has shown that underground tank storage is a safe and practical means of interim handling of high level wastes. Tank storage, however, does not provide a long term solution to the problem. Accordingly, using technology developed by the AEC, these liquid wastes are to be further concentrated and changed into solid form. These solids will then be transferred to a federal site, such as an abandoned salt mine, for final storage. These salt mines have a long history of geologic stability, are impervious to water, and are not associated with usable ground water resources.

Thermal Effects

ALL steam-electric generating plants must release heat to the environment as an inevitable consequence of producing useful electricity. Heat from the combustion of fossil fuel in a boiler or from the fission of nuclear fuel in a reactor is used to produce high temperature and pressure steam which in turn drives a turbine connected to a generator. When the thermal energy in the steam has been converted to mechanical energy in the turbine, the spent steam is converted back into water in a condenser.

Condensation is accomplished by passing large amounts of cooling water through the condenser. In the least costly and most widely used method, the cooling water is taken directly from nearby rivers, lakes, estuaries, or the ocean. The cooling water is heated 10°–30° F—depending on plant design and operation—and then usually returned to the same source. Thermal effects is a term used to describe the impact that the heated water may have on the source body of water.

The thermal effects may be detrimental, beneficial, or insignificant, depending on many factors such as the manner in which the heated water is returned to the source water, the amount of source water available, the ecology of the source water, and its desired use. The addition of the heated water from the plant condenser to the source body of water does raise its temperature and this increased temperature can affect fish and other aquatic life.

In some situations, cooling methods other than the once-through method described above may be employed. Artificial ponds can be constructed to provide a source of water for circulation through the condensers. Cooling towers—either of the wet or dry type—can be used in other instances. Combinations of cooling methods can also be used effectively in many situations.

In wet cooling tower systems, the cooling water is brought in direct contact with a flow of air and the heat is dissipated principally by evaporation. The flow of air through the cooling tower can be provided by either mechanical means or natural draft, and makeup water must be added to replace evaporative losses.

In dry cooling tower systems the cooling water is carried through pipes over which air is passed and the heat is dissipated by conduction and convection rather than by evaporation. Because of the larger surface area required for heat transfer and the larger volume of air that must be circulated, dry cooling towers are substantially more expensive than wet cooling towers and hence seldom used.

It is important to emphasize that although these alternatives may offer relief from a potential thermal effects problem, their use can involve other environmental effects and economic penalties. Whatever method of cooling is chosen, the waste heat—from both fossil and nuclear plants—still must eventually be dissipated into the environment.

The light water power reactors currently being marketed operate at a lower efficiency and therefore reject more heat than the most modern of today's fossil fuel plants of the same generating capacity. For this reason and because about 10 percent of the heat from fossil fuel plants is discharged directly into the atmosphere through the stack, modern fossil fuel plants currently discharge approximately one-third less waste heat to cooling water than do nuclear plants. With the advanced reactors now under development, however, the difference in the amounts of heat released to the cooling water by nuclear and fossil plants will be greatly reduced.

Because so much of today's power comes from conventional fossil fuel plants, they are the major contributors of waste heat to the environment today. In 1968, nuclear power contributed only about one percent of the waste heat. However, at estimated rates of growth, 30 percent of the heat wasted by steam generating plants in 1980 will come from nuclear plants. By 1995, the contributions from both sources are expected to be about equal.

The scope and depth of thermal effects research conducted for individual nuclear power stations have increased considerably in recent years. One example of such research is the comprehensive study of the fish life, ecology, and hydrology of the lower Connecticut River in the vicinity of Connecticut Yankee's nuclear power plant at Haddam Neck. The study, being carried out by a group of independent scientists through financing made available by Connecticut Yankee, was initiated in the fall of 1964.

The study covers six major areas of investigation: hydrology; studies of organisms on the river bottom; fish studies, including both resident and migrating (shad) populations; bacteriology, microbiology, and algae studies; radiological surveys; and thermal studies. The thermal study work has included temperature distribution predictions and measurements using a variety of techniques such as flow measurements and airborne infrared temperature surveys.

The shad tagging program, now in its sixth year, has shown that the pattern of upstream migration is substantially the same as that before the plant went into operation. The radiological survey of river water, fish, shell fish, plankton, and bottom sediment has shown only a negligible increase in radioactivity since the plant has been in operation. The Connecticut River Study Staff and cooperating agencies as yet have found no significant change in the ecology of the Connecticut River resulting from the discharge of heated water from the Connecticut Yankee Atomic Power Plant after more than 2½ years of operation.

The Columbia River is a large, cold, clean river which supports runs of salmon, steelhead, and shad. There were six reactors operating during 1944–55, and eight from 1955–64. In 1964, the river had nine nuclear reactors in a short stretch through the Hanford reservation, but the number operating has since decreased to three. It is worth noting how salmon, which require cold water, have responded to the reactor operation.

A 1968 report summarizes some of the results of work which has been under way for about 20 years at the Hanford site (*Biological Effects of Hanford Heat on Columbia River Fishes—A Review,* R. Nakatani, presented to the Isaac Walton League, Portland, Ore., February 17, 1968). During the period of the study, all but 44 miles of the salmon spawning area on the Columbia has been inundated by water backed up by a series of dams. The only spawning areas left are from Richland up to the Priest Rapids Dam. Much of this fast water lies in the Hanford Reservation in the vicinity of the reactors. The heat discharged by the reactors has had no apparent effects upon salmon eggs or fry, probably because the salmon spawn while the natural river temperature is low (mid-October through November).

The question of whether the heated water from the reactors interferes with the passage of fish is also being studied to determine if the fish trying to migrate up to tributaries of the Columbia above Hanford will be prevented from passing. The Bureau of Commercial Fisheries and Battelle Northwest scientists have been observing the movement of adult salmon tagged with sonic emitters. Results indicate the fish avoid the warmer water on the reactor side of the river, but the important point is that their progress past the reactor site is not impeded. The fish migrate along the same shoreline above the reactor discharge so factors such as current velocities may also be important in determining their path.

Perhaps the best evidence of the absence of any harmful effects from the reactors on Columbia River salmon is the increase in nesting sites on the Hanford reservation.

Although a considerable amount of thermal effects research has been conducted, more attention should be given to beneficial uses of heated water. For example, it has been proposed that heated water be used to irrigate crops and to warm the soil, thereby possibly extending the growing season or protecting crops from freezing. Aquaculture also is a distinct possibility since productivity of commercial species (such as catfish) can probably be greatly increased, particularly in winter months.

Use of by-product heat to delay freezing and thereby to extend the shipping season in northern waterways has also been suggested. Obviously, more can be done along these lines. Heated water is an energy source, which suggests that ingenuity can find beneficial uses for this energy.

Another alternative exists in the sea where water is plentiful and where the temperature structure in many areas is especially promising for beneficial modification by thermal discharges. Oceanographers have asked if there is some way to use the excess energy from a steam power plant to induce upwelling in the ocean and thus enhance the biological productivity. The ocean is relatively stable, with warm, less dense water overlying cold denser water. Separating the two is a layer, usually at a depth of 300–2200 feet, where there is a sharp temperature change (the thermocline, mentioned earlier). In the summer, particularly in mid latitudes where seasonal temperature variation is greatest, a thermocline develops much nearer the surface. Water below this thermocline is not only colder than surface water, but also much richer in nutrients.

Growing plants are sparse or absent in the deeper water, because there is insufficient light. Above the thermocline light is usually abundant, but nutrients such as nitrate and phosphate are in short supply. If the nutrients of the colder, deeper water could be transported upward into the warmer upper layers, biological productivity could be increased. A vastly more productive and valuable fishery might develop in the vicinity of the nutrient-rich water.

On the other hand, changes in the temperature regime may favor growth of undesirable species, and organisms commonly used by man may be crowded out. Perhaps both views are correct, depending on geographical location. In our present knowledge of the probable response of marine ecosystems to artificially induced upwelling and limited heating, feasibility studies seem worth pursuing. (20)

Reduction of power, or even of the rate of power increase, is to be avoided if possible, for it implies a reduction in the standard of living, or a stagnation at the present not very equitable level throughout the world.

Where Are We Now?

RADIOBIOLOGICAL studies that are environmental in scope became, with the release of atomic energy, a mandate on the twentieth century.

The atom as a tool of the environmental radiobiologist has, of itself, solved few problems. Its significance is that it has speeded up—to a degree still not fully tested—our ability to study ecosystems and their relations to each other.

The first decades of the atomic age have comprised a period of swift maturity. Much has been done to gain perspective. Atomic energy as a potential force for destruction has not been controlled. But there is a surer knowledge of the hazard inherent in the absence of control and a rational hope that the new power will be directed toward peaceful objectives. We know that:

The uninhibited release of nuclear products into the environment of the earth will create problems—fundamentally biological problems—of long duration and of still unassessed ultimate effect.

Use of atomic weapons in war could have a biological cost beyond calculation. Yet, in terms of constructive employment of atomic resources, we also know that:

Atomic energy may help solve the very problems that the new age presents.

Careful and controlled development of atomic forces will provide the reservoirs of energy that will be needed to sustain the world's populations of the next century and beyond.

In whatever case, the solutions lie in the direction of environmental knowledge.

Man, the human animal, will live in the environment he has the intelligence to understand and to preserve. (4)

PART IV

THE ATOM AT WORK

ONLY a few events in history have materially altered civilization. One of these, surely, was learning how to extract energy from the nucleus of the atom. The nuclear reactor has been compared to the steam engine and the automobile in its import for social change.

In this part we open with the history of atomic energy. In the three decades since World War II we may have lost sight of why the atomic age was blasted in with two bombs. Although social questions have mostly not been dealt with in this book, the social responsibility of scientists has been a sensitive subject, especially in connection with the Vietnam War, and will undoubtedly remain one for some time to come. It is important, therefore, that the background to the construction of the atomic bomb be understood.

The next several chapters, which together with the opening one make up almost half of part IV, are devoted to various aspects of the production of nuclear power: modern reactors, power plants, nuclear fuels, and thermonuclear fusion. The last chapters are about peacetime applications of the atom other than energy production. Some applications have already been discussed in connection with medicine, biology, and agriculture, but there are many more. And the future, no doubt, holds still more.

16 ◊ The First Reactor

THE story of the first primitive reactor is an account of the birth of a new era. To understand atomic energy as a force that shaped that era, and to give meaning to the present and perspective to the future, it is interesting and rewarding to know how the era began. This chapter tells the story in the words of two men whose job it was to report an event in which others, more renowned and more directly concerned, participated.

The First Pile

ON December 2, 1942, man first initiated a self-sustaining nuclear chain reaction, and controlled it.

Beneath the West Stands of Stagg Field, the University of Chicago athletic stadium, late in the afternoon of that day, a small group of scientists witnessed the advent of a new era in science. History was made in what had been a squash-rackets court. (See photograph section).

Precisely at 3:25 P.M., Chicago time, scientist George Weil withdrew the cadmium-plated control rod and by his action man unleashed and controlled the energy of the atom.

As those who witnessed the experiment became aware of what had happened, smiles spread over their faces and a quiet ripple of applause could be heard. It was a tribute to Enrico Fermi, to whom, more than to any other person, the success of the experiment was due.

Fermi had been working with uranium for many years. In

1934 he bombarded uranium with neutrons and produced what appeared to be element 93 and element 94. However, after closer examination it seemed as if nature had gone wild; several other elements were present, but none could be fitted into the periodic table near uranium—where Fermi knew they should have fitted if they had been the transuranic elements 93 and 94. It was not until five years later that anyone, Fermi included, realized he had actually caused fission of the uranium and that these unexplained elements belonged back in the middle part of the periodic table.

Fermi was awarded the Nobel prize in 1938 for his work on transuranic elements. He and his family went to Sweden to receive the prize. The Italian Fascist press severely criticized him for not wearing a Fascist uniform and failing to give the Fascist salute when he received the award. The Fermis never returned to Italy.

From Sweden, having taken most of his personal possessions with him, Fermi proceeded to London and thence to America.

The modern Italian explorer of the unknown was in Chicago that cold December day in 1942. An outsider looking into the squash court where Fermi was working would have been greeted by a strange sight. In the center of the 30×60-foot room, shrouded on all but one side by a gray balloon cloth envelope, was a pile of black bricks and wooden timbers, square at the bottom and a flattened sphere on top. Up to half of its height, its sides were straight. The top half was domed, like a beehive. During the construction of this crude-appearing but complex pile (the name which has since been applied to all such devices) the standing joke among the scientists working on it was: "If people could see what we're doing with a million-and-a-half of their dollars, they'd think we are crazy. If they knew why we are doing it, they'd be sure we are."

Three years before the December 2 experiment, it had been discovered that when an atom of uranium was bombarded by neutrons, the uranium atom sometimes was split, or fissioned. Later, it had been found that when an atom of uranium fissioned, additional neutrons were emitted and became available for further reaction with other uranium atoms. These facts implied the possibility of a chain reaction. The facts further indicated that if a sufficient quantity of uranium could be brought together under the proper conditions, a self-sustaining chain reaction would result. This quantity of uranium necessary for a chain reaction under given conditions is known as the critical mass, or more commonly, the critical size of the particular pile.

Years of scientific effort and study lay behind this demonstration of the first self-sustaining nuclear chain reaction. The story goes back at least to the fall of 1938 when two German scientists, Otto Hahn and Fritz Strassman, working at the Kaiser Wilhelm Institute in Berlin, found barium in the residue material from an experiment in which they had bombarded uranium with neutrons from a radium-beryllium source. This discovery caused tremendous excitement in the laboratory because of the difference in atomic mass between the barium and the uranium. Previously, in residue material from similar experiments, elements other than uranium had been found, but they differed from the uranium by only 1 or 2 units of mass. The barium differed by approximately 98 units of mass. The question was, where did this element come from? It appeared that the uranium atom when bombarded by a neutron had split into two different elements, each of approximately half the mass of the uranium.

Before publishing their work in the German scientific journal *Die Naturwissenschaften,* Hahn and Strassman communicated with their colleague Lise Meitner, who, having fled the Nazi-controlled Reich, was working with Niels Bohr in Copenhagen, Denmark.

Dr. Meitner was very much interested in this phenomenon and immediately attempted to analyze mathematically the results of the experiment. She reasoned that the barium and the other residual elements were the result of a fission, or breaking, of the uranium atom. But when she added the atomic masses of the residual elements, she found this total was less than the atomic mass of uranium.

There was but one explanation: The uranium fissioned or split, forming two elements each of approximately half of its original mass, but not exactly half. Some of the mass of the uranium had disappeared. Miss Meitner and her nephew Otto R. Frisch suggested that the mass which disappeared was converted into energy. According to the theory advanced in 1905 by Albert Einstein in which the relationship of mass to energy was stated by the equation $E = mc^2$, this energy release would be of the order of 200 MeV for each atom fissioned. [As was stated before, this is an enormous amount of energy for an elementary particle.]

Einstein himself, nearly 35 years before, had said this theory might be proved by further study of radioactive elements. Bohr was planning a trip to America to discuss other problems with Einstein, who had found a haven at Princeton's Institute for Advanced Study. Bohr came to America, but the principal item he discussed with Einstein was the report of Meitner and Frisch. Bohr arrived at Princeton on January 16, 1939. He talked to Einstein and John A. Wheeler, who had once been his student. From Princeton the news spread by word of mouth to neighboring physicists, including Enrico Fermi at Columbia. Fermi and his associates immediately began work to find the heavy pulse of ionization which could be expected from the fission and consequent release of energy.

Fermi and Bohr exchanged information and discussed the problem of fission. Fermi mentioned the possibility that neutrons might be emitted in the process. In this conversation, their ideas of the possibility of a chain reaction began to crystallize.

Within days, experimental confirmation of Meitner and Frisch's deduction was obtained from four laboratories in the United States. Later it was learned that similar confirmatory experiments had been made by Frisch and Meitner. Frédéric Joliot-Curie in France, too, confirmed the results.

On February 27, 1939, the Canadian-born Walter H. Zinn and Leo Szilard, a Hungarian, both working at Columbia University, began their experiments to find the number of neutrons emitted by the fissioning uranium. At the same time, Fermi and his associates, Herbert L. Anderson and H. B. Hanstein, commenced their investigation of the same problem. The results of these experiments showed that a chain reaction might be possible since the uranium emitted additional neutrons when it fissioned. (10)

It was evident that a powerful weapon might be developed. Some of the foreign-born physicists, notably Leo Szilard and Eugene Wigner, felt strongly that the government should begin intensive work on the problem, particularly since news began to leak that the Germans were doing research relating to a fission bomb. A letter to President Roosevelt was drafted and Albert Einstein, a pacifist, was persuaded to sign it, as he was the most distinguished scientist in the world. The letter was delivered by Alexander Sachs, an economic consultant to the White House:

Albert Einstein
Old Grove Rd.
Nassau Point
Peconic, Long Island

August 2nd, 1939

F. D. Roosevelt
President of the United States
White House
Washington, D.C.

Sir:

Some recent work by E. Fermi and L. Szilard, which has been communicated to me in manuscript, leads me to expect that the element uranium may be turned into a new and important source of energy in the immediate future. Certain aspects of the situation which has arisen seem to call for watchfulness and, if necessary, quick action on the part of the Administration. I believe therefore that it is my duty to bring to your attention the following facts and recommendations:

In the course of the last four months it has been made probable—through the work of Joliot in France as well as Fermi and Szilard in America—that it may become possible to set up a nuclear chain reaction in a large mass of uranium, by which vast amounts of power and large quantities of new radium-like elements would be generated. Now it appears almost certain that this could be achieved in the immediate future.

This new phenomenon would also lead to the construction of bombs, and it is conceivable—though much less certain—that extremely powerful bombs of a new type may thus be constructed. A single bomb of this type, carried by boat and exploded in a port, might very well destroy the whole port together with some of the surrounding territory. However, such bombs might very well prove to be too heavy for transportation by air.

The United States has only very poor ores of uranium in moderate quantities. There is some good ore in Canada and the former Czechoslovakia, while the most important source of uranium is Belgian Congo.

In view of this situation you may think it desirable to have some permanent contact maintained between the Administration and the group of physicists working on chain reactions in America. One possible way of achieving this might be for you to entrust with this task a person who has your confidence and who could perhaps serve in an inofficial capacity. His task might comprise the following:

a) to approach Government Departments, keep them informed of the further development, and put forward recommendations for Government action, giving particular attention to the problem of securing a supply of uranium ore for the United States;

b) to speed up the experimental work, which is at present being carried on within the limits of the budgets of University laboratories, by providing funds, if such funds be required, through his contacts with private persons who are willing to make contributions for this cause, and perhaps also by obtaining the co-operation of industrial laboratories which have the necessary equipment.

I understand that Germany has actually stopped the sale of uranium from the Czechoslovakian mines which she has taken over. That she should have taken such early action might perhaps be understood on the ground that the son of the German Under-Secretary of State, von Weizsäcker, is attached to the Kaiser-Wilhelm-Institut in Berlin where some of the American work on uranium is now being repeated.

Yours very truly,
Albert Einstein

President Roosevelt subsequently appointed the Advisory Committee on Uranium, which on November 1 reported that a chain reaction was possible and that the government should support a complete investigation.

Further impetus to the work on a uranium reactor was given by the discovery of plutonium at the Radiation Laboratory, Berkeley, California, in March 1940. This element, unknown in nature, was formed by uranium 238 capturing a neutron, and thence undergoing two successive changes in atomic structure with the emission of beta particles. Plutonium, it was believed, would undergo fission as did the rare isotope of uranium, ^{235}U.

Meanwhile, at Columbia, Fermi and Zinn and their associates were working to determine operationally possible designs of a uranium chain reactor. Among other things, they had to find a suitable moderating material to slow down the neutrons traveling at relatively high velocities. In July 1941, experiments with uranium were started to obtain measurements of the reproduction factor (called k), which was the key to the problem of a chain reaction. (10)

The k factor is simply the number of effective neutrons produced per decay. (Fermi's description of a chain reaction appears later in this chapter.)

If this factor could be made sufficiently greater than 1, a chain reaction could be made to take place in a mass of material of practical dimensions. If it were less than 1, no chain reaction could occur.

Since impurities in the uranium and in the moderator would capture neutrons and make them unavailable for further reactions, and since neutrons would escape from the pile without encountering uranium 235 atoms, it was not known whether a value for k greater than 1 could ever be obtained.

Fortunate it was that the obtaining of a reproduction factor greater than 1 was a complex and difficult problem. If Hitler's scientists had discovered the secret of controlling the neutrons and had obtained a working value of k, they would have been well on the way toward producing an atomic bomb for the Nazis.

At Chicago, by July 1942, measurements obtained from experimental piles had gone far enough to permit a choice of design for a test pile of critical size.

It was necessary to use uranium oxides because metallic uranium of the desired degree of purity did not exist. Although several manufacturers were attempting to produce the uranium metal, it was not until November that any appreciable amount was available.

Although the dies for the pressing of the uranium oxides were designed in July, additional measurements were necessary to obtain information about controlling the reaction, to revise estimates as to the final critical size of the pile, and to develop other data. Thirty experimental subcritical piles were constructed before the final pile was completed.

Meantime, the Manhattan Engineer District was created, and in September 1942, Major General L. R. Groves assumed command. (The Atomic Energy Commission, or AEC, a civilian agency, succeeded the Manhattan Engineer District as the governmental organization to control atomic energy on January 1, 1947.)

Construction of the main pile at Chicago started in November. The project gained momentum, with machining of the graphite moderator blocks (see chapter 17), pressing of the uranium oxide pellets, and the design of instruments. Fermi's two construction crews worked almost around the clock.

Original estimates as to the critical size of the pile were pessimistic. As a further precaution, it was decided to enclose the pile in a balloon cloth bag which could be evacuated to remove the neutron-capturing air.

This balloon cloth bag was constructed by Goodyear Tire and Rubber Company. Specialists in designing gasbags for lighter-than-air craft, the company's engineers were a bit puzzled about the aerodynamics of a square balloon. Security regulations forbade informing Goodyear of the purpose of the envelope and so the army's new square balloon was the butt of much joking.

The bag was hung with one side left open; in the center of the floor a circular layer of graphite bricks was placed. This and each succeeding layer of the pile was braced by a wooden frame. Alternate layers contained the uranium. By this layer-on-layer construction a roughly spherical pile of uranium and graphite was formed.

Facilities for the machining of graphite bricks were installed in the West Stands. Week after week this shop turned out graphite bricks.

Before the structure was half complete, measurements indicated that the critical size at which the pile would become self-sustaining was somewhat less than had been anticipated in the design.

Day after day the pile grew toward its final shape. And as the size of the pile increased, so did the nervous tension of the men working on it. Logically and scientifically they knew this pile would become self-sustaining. It had to. All the measurements indicated that it would. But still the demonstration had to be made. As the eagerly awaited moment drew nearer, the scientists gave greater and greater attention to details, the accuracy of measurements, and exactness of their construction work.

Guiding the entire pile construction and design was the nimble-brained Fermi, whose associates described him as completely self-confident but wholly without conceit.

So exact were Fermi's calculations, based on the measurements taken from the partially finished pile, that days before its completion and demonstration on December 2, he was able to predict almost to the exact brick the point at which the reactor would become self-sustaining.

At Chicago during the early afternoon of December 1, tests indicated that critical size was rapidly being approached. At 4:00 P.M. Zinn's group was relieved by the men working under Anderson. Shortly afterward, the last layer of graphite and uranium bricks was placed on the pile. Zinn, who remained, and Anderson made several measurements of the activity within the pile. They were certain that when the control rods were withdrawn, the pile would become self-sustaining. Both had agreed, however, that should measurements indicate the reaction would become self-sustaining when the rods were withdrawn, they would not start the pile operating until Fermi and the rest of the group could be present. Consequently, the control rods were locked and further work was postponed until the following day.

That night the word was passed to the men who had worked on the pile that the trial run was due the next morning.

About 8:30 on the morning of Wednesday, December 2, the group began to assemble in the squash court.

At the north end of the squash court was a balcony about 10 feet above the floor of the court. Fermi, Zinn, Anderson, and Arthur H. Compton were grouped around instruments at the east end of the balcony. The remainder of the observers crowded the little balcony. One of the young scientists who worked on the pile put it this way: "The control cabinet was surrounded by the 'big wheels'; the 'little wheels' had to stand back."

On the floor of the squash court, just beneath the balcony, stood George Weil, whose duty it was to handle the final control rod. In the pile were three sets of control rods. One set was automatic and could be controlled from the balcony. Another was an emergency safety rod. Attached to one end of this rod was a rope running through the pile and weighted heavily on the opposite end. The rod was withdrawn from the pile and tied by another rope to the balcony. Norman Hilberry was ready to cut this rope with an axe should something unexpected happen, or in case the automatic safety rods failed. The third rod, operated by Weil, was the one which actually held the reaction in check until withdrawn the proper distance.

Since this demonstration was new and different from anything ever done before, complete reliance was not placed on mechanically operated control rods. Therefore, a liquid-control squad stood on a platform above the pile. They were prepared to flood the pile with cadmium-salt solution in case of mechanical failure of the control rods.

Each group rehearsed its part of the experiment.

At 9:45 Fermi ordered the electrically operated control rods withdrawn. The man at the controls threw the switch to withdraw them. A small motor whined. All eyes watched the lights which indicated the rods' position.

But quickly, the balcony group turned to watch the counters, whose clicking stepped up after the rods were out. The indicators of these counters resembled the face of a clock, with hands to indicate neutron count. Nearby was a recorder, whose quivering pen traced the neutron activity within the pile.

Shortly after 10:00, Fermi ordered the emergency rod, called Zip, pulled out and tied.

"Zip out," said Fermi. Zinn withdrew Zip by hand and tied it to the balcony rail. Weil stood ready by the vernier control rod which was marked to show the number of feet and inches which remained within the pile.

At 10:37 Fermi, without taking his eyes off the instruments, said quietly:

"Pull it to thirteen feet, George." The counters clicked faster. The graph pen moved up. All the instruments were studied, and computations were made.

"This is not it," said Fermi. "The trace will go to this point and level off." He indicated a spot on the graph. In a few minutes the pen came to the indicated point and did not go above that point. Seven minutes later Fermi ordered the rod out another foot.

Again the counters stepped up their clicking, the graph pen edged upward. But the clicking was irregular. Soon it leveled off, as did the thin line of the pen. The pile was not self-sustaining—yet.

At 11:00, the rod came out another 6 inches; the result was the same: an increase in rate, followed by the leveling off.

Fifteen minutes later, the rod was further withdrawn and at 11:25 was moved again. Each time the counters speeded up, the pen climbed a few points. Fermi predicted correctly every movement of the indicators. He knew the time was near. He wanted to check everything again. The automatic control rod was reinserted without waiting for its automatic feature to operate. The graph line took a drop, the counters slowed abruptly.

At 11:35, the automatic safety rod was withdrawn and set. The control rod was adjusted and Zip was withdrawn. Up went the counters, clicking, clicking, faster and faster. It was the clickety-click of a fast train over the rails. The graph pen started to climb. Tensely, the little group watched, and waited, entranced by the climbing needle.

Whrrrump! As if by a thunder clap, the spell was broken. Every man froze—then breathed a sigh of relief when he realized the automatic rod had slammed home. The safety point at which the rod operated automatically had been set too low.

"I'm hungry," said Fermi. "Let's go to lunch."

Perhaps like a great coach, Fermi knew when his men needed a break.

It was a strange between halves respite. They got no pep talk. They talked about everything else but the game. The redoubtable Fermi, who never said much, had even less to say. But he appeared supremely confident. His team was back on the squash court at 2:00 P.M. Twenty minutes later, the automatic rod was reset and Weil stood ready at the control rod.

"All right, George," called Fermi, and Weil moved the rod to a predetermined point. The spectators resumed their watching and waiting, watching the counters spin, watching the graph, waiting for the settling down and computing the rate of rise of reaction from the indicators.

At 2:50 the control rod came out another foot. The counters nearly jammed, the pen headed off the graph paper. But this was not it. Counting ratios and the graph scale had to be changed.

"Move it six inches," said Fermi at 3:20. Again the change—but again the leveling off. Five minutes later, Fermi called: "Pull it out another foot."

Weil withdrew the rod.

"This is going to do it," Fermi said to Compton, standing at his side. "Now it will become self-sustaining. The trace will climb and continue to climb. It will not level off."

Fermi computed the rate of rise of the neutron counts over a 1-minute period. He silently, grim-faced, ran through some calculations on his slide rule.

In about a minute he again computed the rate of rise. If the rate was constant and remained so, he would know the reaction was self-sustaining. His fingers operated the slide rule with lightning speed. Characteristically, he turned the rule over and jotted down some figures on its ivory back.

Three minutes later he again computed the rate of rise in neutron count. The group on the balcony had by now crowded in to get an eye on the instruments, those behind craning their necks to be sure they would know the very instant history was made. By this time the click of the counters was too fast for the human ear. The clickety-click was now a steady brrrrr. Fermi, unmoved, unruffled, continued his computations.

"I couldn't see the instruments," said Weil. "I had to watch Fermi every second, waiting for orders. His face was motionless. His eyes darted from one dial to another. His expression was so calm it was hard. But suddenly, his whole face broke into a broad smile."

Fermi closed his slide rule—

"The reaction is self-sustaining," he announced quietly, happily. "The curve is exponential."

The group tensely watched for 28 minutes while the world's first nuclear chain reactor operated.

The upward movement of the pen was leaving a straight line. There was no change to indicate a leveling off. This was it.

"O. K., Zip in," called Fermi to Zinn, who controlled that rod. The time was 3:53 P.M. Abruptly, the counters slowed down, the pen slid down across the paper. It was all over.

Man had initiated a self-sustaining nuclear reaction—and then stopped it. He had released the energy of the atom's nucleus and controlled that energy.

(For a reproduction of the graph of December 2, 1942, see photograph section.)

Right after Fermi ordered the reaction stopped, the Hungarian-born theoretical physicist Eugene Wigner presented him with a bottle of Chianti. All through the experiment Wigner had kept this wine hidden behind his back.

Fermi uncorked the wine bottle and sent out for paper cups so all could drink. He poured a little wine in all the cups, and silently, solemnly, without toasts, the scientists raised the cups to their lips—the Canadian Zinn, the Hungarians Szilard and Wigner, the Italian Fermi, the Americans Compton, Anderson, Hilberry, and a score of others. They drank to success—and to the hope they were the first to succeed.

A small crew was left to straighten up, lock controls, and check all apparatus. As the group filed from the West Stands, one of the guards asked Zinn: "What's going on, doctor, something happen in there?"

The guard did not hear the message which Arthur Compton was giving James B. Conant at Harvard, by long-distance telephone. Their code was not prearranged.

"The Italian navigator has landed in the New World," said Compton.

"How were the natives?" asked Conant.

"Very friendly." (10)

(Report signed by Corwin Allardice and Edward R. Trapnell, November 17, 1949)

On the tenth anniversary of the successful first pile experiment, Fermi himself wrote the story for the *Chicago Sun-Times* in a matter-of-fact, undramatic way:

When Hahn achieved fission, it occurred to many scientists that this fact opened the possibility of a form of nuclear (atomic) energy.

The year was 1939. A world war was about to start. The new possibilities appeared likely to be important, not only for peace but also for war.

A group of physicists in the United States—including Leo Szilard, Walter Zinn, now director of Argonne National Laboratory, Herbert Anderson, and myself—agreed privately to delay further publications of findings in this field.

We were afraid these findings might help the Nazis. Our action, of course, represented a break with scientific tradition and was not taken lightly. Subsequently, when the government became interested in the atom bomb project, secrecy became compulsory.

Here it may be well to define what is meant by the chain reaction which was to constitute our next objective in the search for a method of utilizing atomic energy.

An atomic chain reaction may be compared to the burning of a rubbish pile from spontaneous combustion. In such a fire, minute parts of the pile start to burn and in turn ignite other tiny fragments. When sufficient numbers of these fractional parts are heated to the kindling points, the entire heap bursts into flames.

A similar process takes place in an atomic pile such as was constructed under the West Stands of Stagg Field at the University of Chicago in 1942.

The pile itself was constructed of uranium, a material that is embedded in a matrix of graphite. With sufficient uranium in the pile, the few neutrons emitted in a single fission that may accidentally occur strike neighboring atoms, which in turn undergo fission and produce more neutrons.

These bombard other atoms and so on at an increasing rate until the atomic "fire" is going full blast.

The atomic pile is controlled and prevented from burning itself to complete destruction by cadmium rods which absorb neutrons and stop the bombardment process. The same effect might be achieved by running a pipe of cold water through a rubbish heap; by keeping the temperature low the pipe would prevent the spontaneous burning.

The first atomic chain reaction experiment was designed to proceed at a slow rate. In this sense it differed from the atomic bomb, which was designed to proceed at as fast a rate as was possible. Otherwise, the basic process is similar to that of the atomic bomb.

The further development of atomic energy during the next three years of the war was, of course, focused on the main objective of producing an effective weapon.

At the same time we all hoped that with the end of the war emphasis would be shifted decidedly from the weapon to the peaceful aspects of atomic energy.

We hoped that perhaps the building of power plants, production of radioactive elements for science and medicine would become the paramount objectives.

The problems posed by this world situation are not for the scientist alone but for all people to resolve. Perhaps a time will come when all scientific and technical progress will be hailed for the advantages that it may bring to man, and never feared on account of its destructive possibilities. (10)

The woman's view of the first reactor is presented in Laura Fermi's book *Atoms in the Family* (Chicago: University of Chicago Press, 1954). Knowing nothing of her husband's secret work, she gave a party the evening of the day of the first chain reaction. Several guests, upon arrival, greeted their host with "Congratulations!" "For what?" Mrs. Fermi kept asking. She received no answer until 2½ years later, when her husband handed her the official Smyth report on atomic energy.

17 ◊ Present-Day Reactors

THE first nuclear reactor by today's standards was a fairly crude apparatus—essentially an assembly of uranium and graphite bricks about 24½ feet on a side and 19 feet high. The method of assembly, which was simply to place one brick on top of another, gave rise to the name atomic pile; nuclear reactor is now the preferred term.

A nuclear reactor is simply a device for starting and controlling a self-sustaining fission chain reaction. For reasons that will become evident, it could as well be called a neutron machine.

Nuclear reactors are used in several ways:

(1) to supply intense fields or beams of neutrons for scientific experiments; (2) to produce new elements or materials by neutron irradiation; (3) to furnish heat for electric power generation, propulsion, industrial processes, or other applications.

The basic parts of a nuclear reactor are: a core of fuel; a neutron moderator, which is a material that aids the fission process by slowing down the neutrons; a means of regulating the number of free neutrons and thereby controlling the rate of fission; a coolant, which removes the heat generated in the core; and radiation shielding. (23)

The fission process requires a heavy element, such as uranium or plutonium, as a basic material. Let us consider ura-

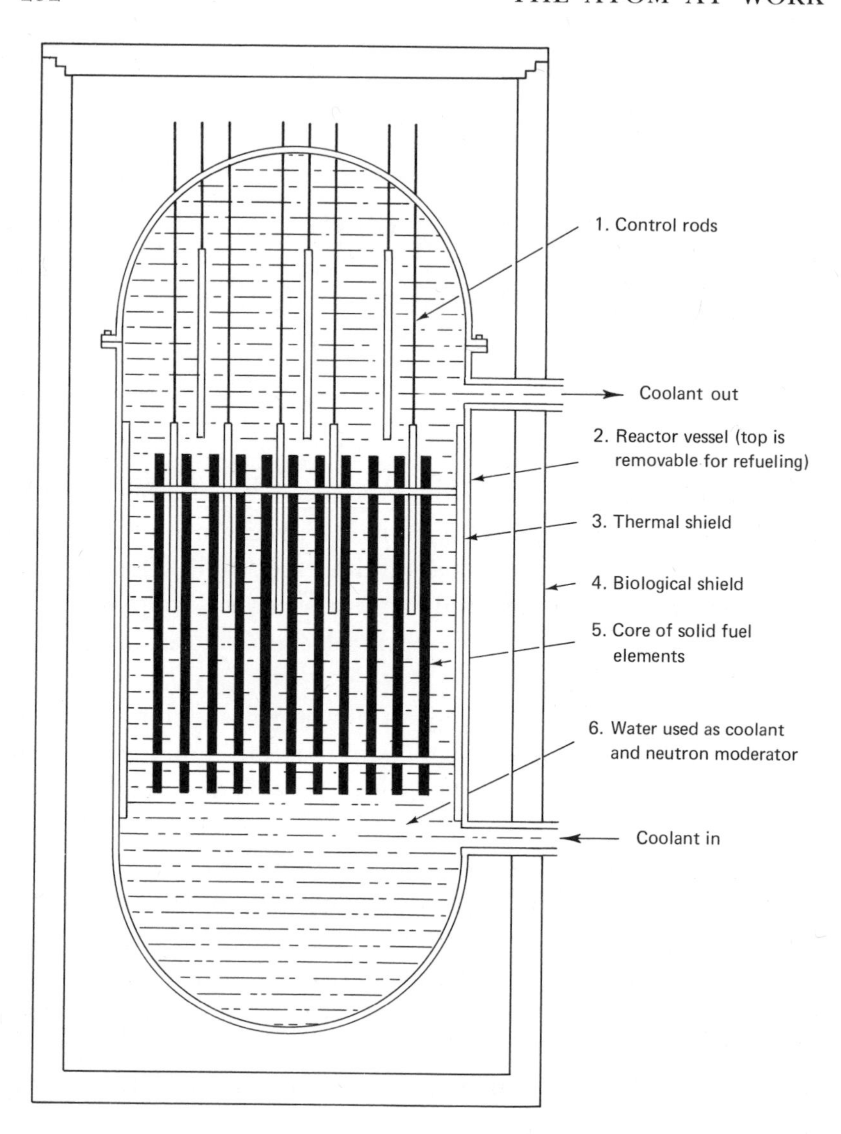

Vertical cross section of a pressurized-water nuclear reactor

nium. Natural uranium is a mixture of three isotopes. An atom of one of these isotopes, uranium 235, can readily undergo fission when a free neutron strikes its heavy central nucleus. The nucleus breaks into two pieces that fly apart at high speed; in

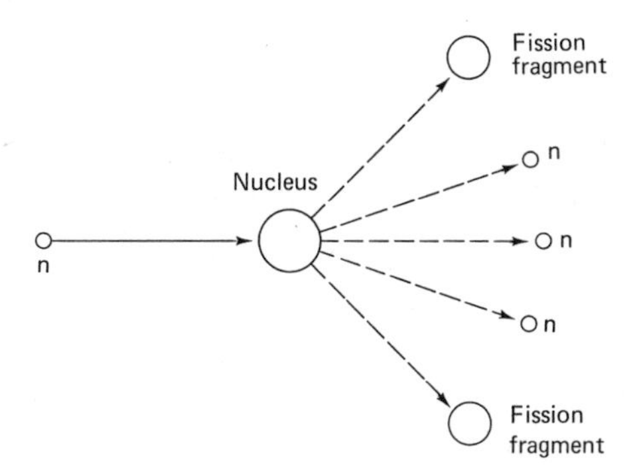

Typical fission reaction

addition, two or three new neutrons are released. (See diagram above.) The kinetic energy of the flying fission fragments is converted to heat when they collide with surrounding atoms, and the released neutrons cause a chain reaction by initiating new fissions in other ^{235}U atoms.

Sustaining the chain reaction is important because more than 30 billion fissions must occur in 1 second to release each watt of energy. If the chain reaction is to be useful, the fissions must occur at a desired rate, and the heat that is generated by the process must be removed. The job of the nuclear reactor, then, is to provide an environment in which fission reactions can be initiated, sustained, and controlled, and to make possible recovery of the resultant heat. (21)

If you bring a few pounds of ^{235}U together very rapidly, you can create a nuclear explosion—an uncontrolled release of energy from fissioning ^{235}U. The rate at which fission occurs in ^{235}U or in any other fissionable isotope depends upon how the reactor's neutron economy is managed. Neutrons are the medium of exchange in a nuclear reactor economy. When a single ^{235}U nucleus fissions spontaneously, two or more neutrons are released, in addition to a substantial amount of energy. Collectively, the two released neutrons can cause more than one additional fission in the surrounding uranium in less than one thousandth of a second. Each new fission can repeat the process. Therefore, if an average of only 1.2 secondary fissions occurred as a result of each initial fission, 1.2^{1000}, or 10^{79}, fissions would (theoretically) occur in 1 second. The energy release would be immense. The essence of reactor control is: To keep the power level in a nuclear reactor steady, the neutrons released in each fission should go on to cause precisely one more fission. When this occurs, the reactor is self-sustaining or critical. The reactor power output may be raised or lowered by permitting slightly more or slightly less than one additional fission to occur until the desired power level is achieved. The just critical condition can then be reestablished by control-element adjustments.

Neutron economy, like dollar economy, is controlled by balancing income and outgo. Three things can happen to each fission-generated neutron: (1) It can go on to cause another fission and, in the process, release more than one new neutron (profit). (2) It can be absorbed in a nonfission reaction with atoms in the coolant, the structure, or even uranium itself (loss). (3) It can bounce (scatter) off atoms in the reactor without being absorbed and ultimately escape from the fuel region altogether (loss). (34)

The reader may well be visualizing a nuclear reactor as a kind of three-dimensional and very high-speed pinball game played with neutrons in a box of fuel and moderator atoms, with an adjustable plunger for control and a fan for cooling. What does a reactor look like? The answer is that many basically different reactor designs have been worked out and many more are possible.

There are several reasons for the multiplicity of reactor designs. First, the designer has a wide choice of reactor materials. Second, there is a broad spectrum of reactor uses. Third, different reactor designers often have different ideas as to the best way of designing a reactor for a given purpose. (23)

Here are some details on various uses of reactors.

Research reactors are a uniquely versatile source of atomic radiation for experimental purposes. Some examples of the ways in which they can serve subject areas of science are:

Nuclear physics. Studying nuclear reactions by irradiating target materials.

Solid-state physics. Determining the crystal structure of materials by neutron diffraction techniques.

Radiation chemistry. Studying the effects of radiation on chemical reactions and on the properties of materials such as plastics.

Analytical chemistry. Identifying trace impurities in materials by activation analysis techniques. (Every species of radioactive atom has a distinctive pattern of radioactive decay. In activation analysis, a sample is made radioactive by neutron activation. By analyzing the resulting radioactivity with sensitive detection instruments, the identity of substances present in the sample is determined.)

Biology. Inducing genetic mutations in plant species by seed irradiation.

Medicine. Experimental treatment of certain brain cancers by a technique known as neutron capture therapy.

Other. Production of radioisotopes for use in laboratory programs. In some experiments, materials are inserted in the reactor for irradiation; in others, experimental apparatus is set up in the path of neutron beams emanating from openings (ports) in the reactor shield.

There are several basically different research reactor designs. The two most commonly used are pool reactors and tank reactors. In the former, the reactor core is suspended in a deep, open pool of water, which serves as coolant, moderator, and radiation shield. This arrangement affords flexibility, since the position of the core can easily be shifted and experimental apparatus can readily be positioned; also it permits direct observation of the proceedings.

In tank reactors, the reactor core is held in a fixed position inside a closed tank. The coolant most often used is ordinary water, but some installations use heavy water. Tank reactors generally operate at higher power levels than pool reactors and therefore as a rule provide a higher neutron flux. (23)

Production reactors are designed to produce elements and isotopes not usually found in nature. The most frequently produced element is plutonium, which can later be used as fuel in other reactors or in nuclear weapons. Plutonium reactors are designed so that excess neutrons produced are captured within the core in uranium 238, which then decays radioactively to become plutonium. Reactors designed to make radioactive isotopes by bombarding stable elements with neutrons are also called production reactors.

There is a distinction between research reactors and experimental reactors. A research reactor is designed to produce neutrons for research. An experimental reactor is itself an experiment. Experimental reactors test the design of new types of reactors and seldom are used for any external research. Experimental reactors usually try out new designs for power reactors. (33)

Power reactors are used in the following way: In conventional steam-electric power plants, a fossil fuel (coal, oil, or natural gas) is burned in a boiler and the resulting heat is used to generate steam. The steam is used, in turn, to drive a turbogenerator, thereby producing electricity. In a nuclear power plant, a nuclear reactor furnishes the heat; the reactor thus substitutes for the conventional boiler. (23)

Among the different types of nuclear reactors there are wide differences in their net consumption of nuclear fuel. On one end of the scale, there are reactors that have a high net fuel consumption; these are used in most of the commercial nuclear power plants operating in the United States today. Next come reactors with a low, but positive, net fuel consumption. The ultimate reactors, insofar as fuel conservation is concerned, are those that have a negative net fuel consumption, which means that they produce more fuel than they use. (Actually, they produce new fissionable material that can be processed for use as fuel.) These are known as breeder reactors and will be popular for central station nuclear power plants that begin operation in, say, 10–20 years. The breeding principle has proved workable, and economically attractive reactors must now be developed so that breeder plants can be built.

Several reactors have a potential for breeding—that is, for producing more nuclear fuel than they consume—because of the materials, or combinations of materials, that are used to build them.

How does a breeder work? As you recall, a uranium 235 atom can fission when its nucleus absorbs a neutron. The fission reaction releases free neutrons that may, in turn, initiate other fissions. All the neutrons released, however, are not absorbed by fissionable material; some are absorbed in the structural material of the reactor, the control elements, or the coolant; some escape from the reactor and are absorbed by shielding; and some are absorbed by fertile material. When the nucleus of an atom of fertile material absorbs a neutron, the fertile atom can be transformed into an atom of a fissionable material—the substance that forms the basis for the nuclear chain reaction. By careful selection and arrangement of materials in the reactor—including, of course, fissionable and fertile isotopes—the neutrons not needed to sustain the fission chain reaction can fairly effectively convert fertile material into fissionable material. The breeder reactor improves the efficiency of the neutron process both by increasing the number of free neutrons released in fission and by decreasing the number of neutrons wasted, thereby making a larger number available for absorption in fertile material. If, for each atom of fissionable material that is consumed, more than one atom of fertile material becomes fissionable material, the reactor is said to be breeding. One fertile material is uranium 238, which is always found in nature with fissionable uranium 235. When uranium 238 absorbs neutrons it is converted to fissionable plutonium 239.

Another fertile material is thorium 232, which can be converted to fissionable uranium 233. (21)

We turn now to nuclear power plants, also known as atomic power plants.

It should be understood at the outset that it is physically impossible for an atomic power plant to behave like an atomic bomb. In the latter, pieces of essentially pure fissionable material are rapidly compressed into a dense mass which is forcibly held together for an instant of time to enable the chain reaction to spread through it. These conditions do not and cannot exist in the reactors used in atomic power plants. They employ relatively dilute fuel; they are designed along different principles; and they operate differently. (3)

However, they use radioactive materials, and special precautions must be taken. Here is how fission products are controlled during operation of a nuclear power plant.

The discussion will be based on plants employing water-cooled reactors, which are the most widely used at the present time; however, the principles behind the plant features and operating procedures described apply to central-station atomic power plants in general.

A large water-cooled reactor contains 50–100 tons of fuel. The fuel material most commonly used today is slightly enriched uranium dioxide (UO_2) in the form of small cylindrical pellets. The pellets are placed in thin-walled metal tubes to form fuel rods, a number of which are bundled together in a long metal can to make up a fuel element. A number of these are positioned in a grid to make up the reactor core. The core is contained in a massively constructed steel tank, known as the reactor vessel, through which cooling water flows.

The inventory of fission products in the plant, after several months of operation, amounts to several hundred pounds. The fission products are, of course, formed inside the fuel. On a weight basis, in excess of 99.99% of the fission product inventory of the plant normally remains confined within the fuel elements. As this fact indicates, it is difficult for the fission products to escape the fuel. There are two reasons. First and most important, it is the nature of uranium dioxide to hold tenaciously onto the fission products. Second, fission products which manage to break the grip of the uranium dioxide must find a way to get past the fuel cladding (that is, the metal tubes) in order to get out. Those that do get out of the fuel enter the coolant (see below).

When it comes time to refuel the plant, which is done at intervals of a year or longer, the reactor is shut down and the top of the reactor vessel is removed. A crane is used to lift out the spent fuel elements and move them to a storage vault or pool. There they are left for several months to allow for the shorter-lived radioactivity to subside. By the end of this cooling period, nearly all of the gaseous fission products have lost their radioactivity. The fuel elements are then loaded into ruggedly-built lead-shielded steel containers for shipment via truck, rail, or barge, to a plant where they will be chemically processed to recover their unused fuel content for future use. It is at the processing plant that the fission products contained in the fuel elements are removed, concentrated and stored.

Thus all but an extremely small fraction of the fission products formed during the operation of an atomic power plant are normally held captive in the heart of the reactor or in spent-fuel storage, and leave the premises when the spent fuel is shipped away. (3)

Two factors determine the amount of fuel burnup that can be achieved in a power reactor. The first is radiation damage to the fuel material, one cause of which is the bombardment the material receives from fission fragments. The result is physical distortion of the fuel, leading in time to failure of the cladding and radioactive contamination of the reactor coolant. The second factor is that fission products lower the reactivity of the fuel by soaking up neutrons. An excessive accumulation of these nuclear ashes would make it impossible to keep the reactor running.

Because of these effects—and either may be the limiting

factor—the fuel must be replaced when only partially consumed. In fact, in most of the reactors being used today for civilian power generation, the fuel must be replaced when only 1%–2% of the fuel atoms have been used up. (2)

There are two basic types of water-cooled reactors—pressurized water reactors and boiling water reactors. In both systems the primary coolant circulates within a closed equipment circuit and is completely cut off from its original source (river, lake, or ocean). Indeed, in all commercial atomic power plants essentially the only water that goes from a waterway into the plant and then empties back directly into the waterway is that which is used to cool the turbine condensers. This water does not flow through the reactor. Its function is merely to carry nonusable heat away from the plant.

To maintain the purity of the water and to limit the amount of radioactivity in the primary cooling system, the reactor coolant is purified. This is done by drawing off a portion of the primary coolant flow, passing it through purification equipment, and then returning it to the system.

In addition to processing a portion of the primary coolant flow, the coolant purification system may also handle water collected from other points in the reactor installation (for example, water that has been used to clean out equipment during maintenance operations).

All but a small fraction of the solid or liquid radioactive substances removed during the purification process are collected as waste concentrates, which are temporarily stored. The balance, averaging a few millionths of a gram per day during routine operation, is discharged to the waterway serving the plant in a dilute waste stream generally so feebly radioactive that it meets Atomic Energy Commission standards for drinking water. Further dilution occurs as the waste stream is dispersed in the waterway.

The radioactive gases removed during the purification process average a few hundred thousandths of a gram per day during routine operation. This material is released to the atmosphere through a tall chimney on a controlled basis to assure that there is sufficient dilution and atmospheric dispersion of the radioactivity to meet AEC regulations, which are based on the annual radiation exposure that might be received by persons living at the plant boundary.

The radioactive waste concentrates from the purification process, together with other miscellaneous solid wastes, are encased in concrete in steel barrels. When a sufficient number of barrels accumulate, they are shipped from the plant to an AEC-approved site for burial or long-term storage.

Most of us know someone who is contrary in the sense that the harder we try to get him to do something, the more reluctant he is to do it. We usually attribute this to a stubborn streak in his nature. Reactors designed for central-station service have a similar streak in their nature when it comes to nuclear excursions—accidental increases in the rate of the fission chain reaction—for they are so designed that when an excursion begins, their natural tendency is to slow themselves down. Several factors contribute to this inherent stubbornness. The most important is a complex phenomenon to describe, but the gist of it is that, as the temperature of the fuel rises, the proportion of neutrons captured by nonfissioning atoms increases and the rate of fission therefore tends to slow down. The effect is not only automatic but instantaneous, and so offers immediate resistance to any increase in reactor power level.

A second factor is that as the fuel becomes hotter its density decreases slightly, which also acts to lower its reactivity.

Thirdly, in water-cooled reactors, the water that flows through the reactor core, besides carrying away the heat, serves also to moderate the neutrons and thereby encourage the fission chain reaction. Just as the fuel density decreases with increasing temperature, so does the density of the water and with the same effect—that is, lowering of reactivity.

In these and other ways, accidental nuclear excursions

tend to be self-correcting. Thus a runaway reaction can only occur if there is an accidental addition of reactivity so large as to override the losses of reactivity which accompany the excursion. Various design safeguards are provided to prevent this from happening.

To understand how reactors are controlled it is necessary to explain what is known as excess reactivity. If a reactor were loaded with the bare minimum of fuel needed to initiate a fission chain reaction, it could not operate for more than a split second. Why? Because as soon as the fission reaction is started, some fuel would be consumed and the reactor's fuel inventory would fall below the bare minimum needed. Also, fission products would begin to be formed and these substances would absorb some of the neutrons needed to sustain the chain reaction. For the latter reason, even if enough fuel were added to replace exactly the amount that had been consumed, the reactor still could not operate. Before the reactor could be started up again, one would have to add a little extra fuel to make up for the drag on the system caused by neutron losses to fission products. Because of these two factors, it is necessary in practice to load reactors with more fuel than the theoretical minimum requirement. This extra fuel furnishes excess reactivity against which the system can draw to sustain the chain reaction as the operation of the reactor proceeds.

For stable operation, there must be a means of compensating for the excess reactivity that is present in the reactor core. In other words, there must be a way of controlling the rate at which the excess fuel is consumed. Usually this is done by introducing a balancing amount of negative reactivity in the form of substances that are highly efficient neutron absorbers. (They can be thought of as neutron blotters.) By moving these substances into and out of the reactor core with adjustable control rods, the neutron population of the core can be decreased or increased, thereby slowing down or speeding up the chain reaction. In effect, they serve to control the rate at which neutrons are fed to the fuel.

In all reactors, neutron sensing instruments are used to monitor the neutron population of the reactor core. On signals from these instruments, reactivity is added to or subtracted from the system by control rod adjustments or other means. In reactors designed for central-station service, several independent neutron monitoring circuits are employed and are wired into safety mechanisms that stop the reactor automatically if the neutron readings exceed predetermined limits. This is done by rapidly inserting control rods into the reactor core.

Similarly, other instruments monitor other aspects of reactor operation, such as the level of coolant in the reactor vessel, the temperature of the coolant leaving the reactor vessel, and the pressure of the primary reactor system. These instruments are also wired into safety mechanisms that stop the reactor automatically if an abnormal condition develops.

It should be added that every effort is made to design the safety mechanisms to operate in a fail-safe manner, meaning that if a component were to fail, the mechanism would automatically be triggered into operation. For example, control rods are held in standby position by power operated electrical or mechanical devices. Should a power failure occur, the rods would automatically be released and enter the reactor core, thereby shutting down the reactor.

Accidental criticality refers to the possibility of a fission chain reaction starting by accident. The answer to accidental criticality is safe geometry, which means ensuring that a critical mass cannot be assembled under any circumstances. The safeguards include designing shipping containers so that it is physically impossible to load an unsafe number of fuel elements into them, and equipping fuel storage vaults with spacer devices so that safe geometry is assured.

As we have seen, there are multiple physical barriers in a central-station atomic power plant against the escape of radioactive substances into the environment. There is, first of all, the ability of the fuel material to retain fission products. Then there is the fuel element cladding, through which fission products must pass in order to get into the reactor coolant. Then there are the walls of the reactor vessel and of other massively constructed equipment which must be breached before radioactive substances can get out of the reactor system proper. And, finally, in most of the plants being built today, there is what reactor designers call the vapor containment system, which encloses the reactor installation and, in the event of a major accident, serves to limit the escape of radioactive substances from the plant to the environment.

A containment system is designed by imagining what is usually referred to as the maximum credible accident—that is, the most serious reactor accident that could be expected to happen if major design safeguards failed. This involves hypothesizing not one but a combination of several highly improbable things going wrong simultaneously. In the design and construction of the vapor containment system, the rule of conservatism applies. (3)

18 ◊ *Why Nuclear Power?*

THE availability of energy (or power) for the operation of machinery has been a decisive factor in the improvement of man's standard of living. With the steady increase in population and in the use of mechanical devices to increase the productivity of labor, the world's power requirements are growing rapidly. In the past, the chief sources of energy were the fossil fuels—coal, oil, and natural gas—and, to a smaller extent (4%) water power. The increasing demand has spurred worldwide exploration for fuels, especially for oil, and so far the newly discovered reserves have kept pace with consumption.

Although this appears to be a satisfactory situation, there are limitations to be borne in mind. First, the time must come, possibly by the end of the present century, when the reserves begin to dwindle and a shortage of fossil fuels becomes a definite prospect. And, second, the distribution of oil and coal reserves is such that many industrialized countries are already compelled to import their fuel supplies at considerable cost. A new source of energy, especially one for which the basic fuel is cheap and available to all, would thus be a great boon to mankind. (7)

When estimates of U.S. reserves of fossil fuel are examined in the light of the present pattern and projected growth of the national energy demand, two points stand out. First, our present pattern of fossil fuel consumption is decidedly out of balance with our resources. Coal, which is estimated to account for more than three-quarters of recoverable fossil-fuel re-

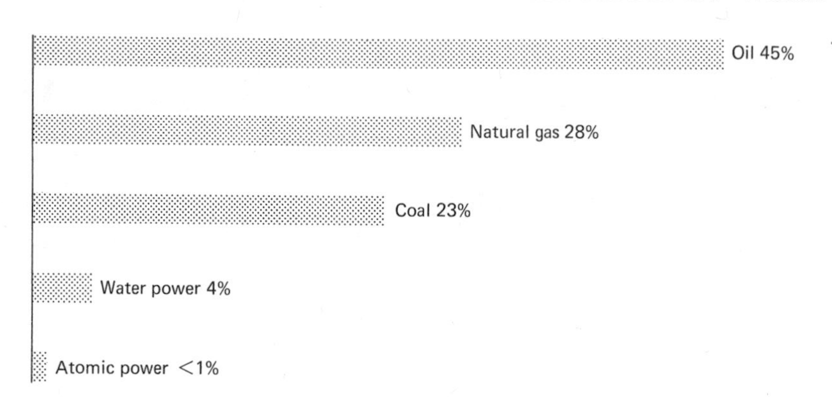

Energy patterns today (35)

serves, today fills less than one-quarter of the energy demand. Conversely, oil and natural gas, which are estimated to account for less than one-quarter of the recoverable reserves, today fill more than three-quarters of the demand. We are thus depleting our stocks of oil and gas at a much higher rate than our stocks of coal. To be sure, when necessary we can make synthetic oil and gas from coal, but not without increasing energy costs.

Unfortunately, increased use of atomic energy will not correct this imbalance in our present use of fossil fuel. The reason is that atomic fuel is likely to be used chiefly in electric power generation, and in this field coal now supplies more energy than oil and gas combined. However, to the extent that electric utilities elect to use atomic fuel instead of burning oil and gas, atomic energy will help postpone the depletion of these valuable resources.

The second point that stands out is that while we face no early fossil-fuel shortage, our reserves of these fuels are not to be classed as inexhaustible. A report of the National Fuels and Study group estimates that, at today's rate of fuel consumption, our total recoverable reserves of fossil fuel (coal, oil, and gas combined) would last some 800 years. But when projected increases in the rate of consumption are taken into account, the estimate of 800 years shrinks to 200 years or less. And, if low-grade sources such as lignite and oil shale are left out of the calculation, the estimate shrinks to 100 years or less. These numbers are by no means to be taken as definitive, since at present we can only roughly infer the extent of our fossil-fuel reserves and since there is also much uncertainty in projecting future energy demand.

If we turn to the long-range significance of atomic power, the first question that arises is: How large are our reserves of atomic fuel? The answer is that they are very large indeed. Our reserves of uranium potentially represent ten to fifty times or more the energy equivalent of our reserves of fossil fuel. (This estimate is derived from AEC data using figures given for uranium reserves recoverable from ore at costs up to twelve times present levels. There are almost unlimited reserves of uranium in lower-grade deposits.) And we have additional reserves of atomic fuel in the form of thorium. (2)

The only natural material directly suitable for nuclear fission is an isotope of uranium, uranium 235. This comprises only about 0.7% of natural, or normal uranium—7 parts in 1000. Another isotope, uranium 238, makes up almost all the remaining 99.3%, and this is not fissionable in the same sense as is ^{235}U. [^{238}U is fissionable only by fast neutrons.] Uranium 238, however, can be converted in a nuclear reactor into a useful fissionable material—a plutonium isotope, plutonium 239.

Thorium is the only other natural nuclear fuel source of

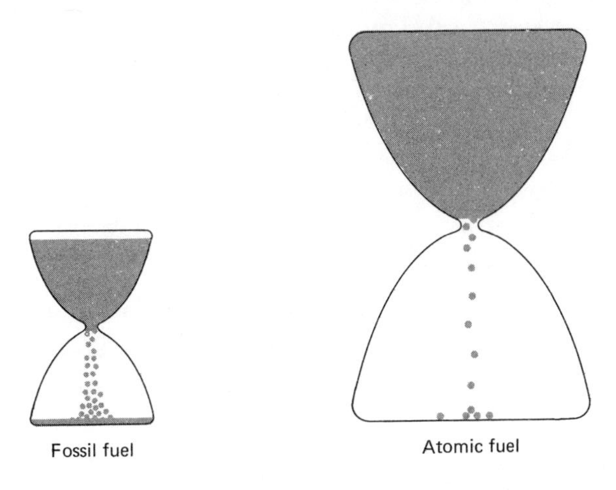

A way of looking at the present fuels situation

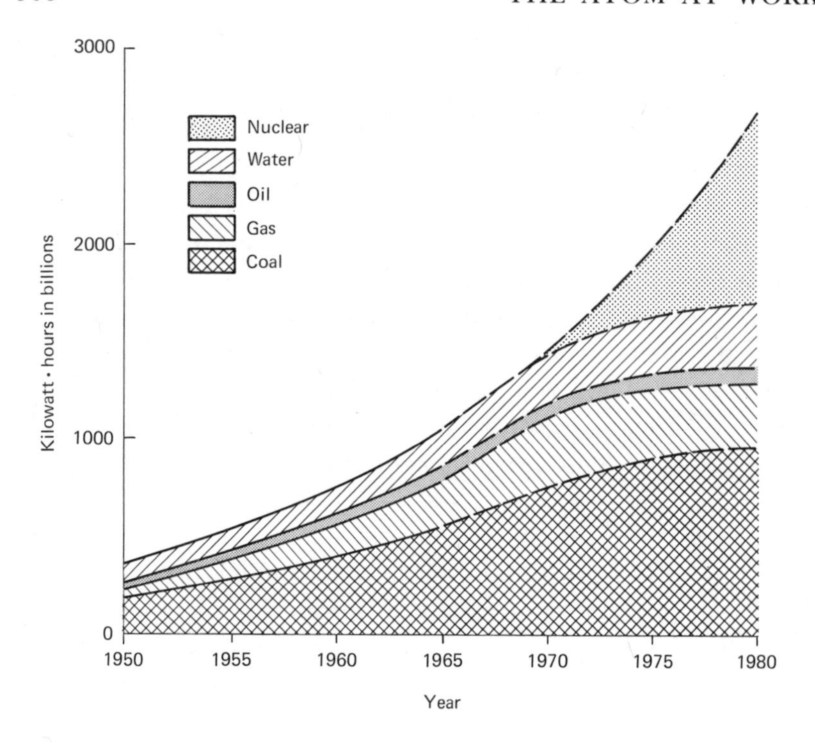

Annual energy requirements for generation of electricity (35)

consequence. Thorium is not fissionable, as uranium 235 is, but is fertile, like uranium 238, so it also can be converted in nuclear reactors into a fissionable isotope—in this case uranium 233 (which is not found in nature). (35)

Thorium is discussed later.

The economic benefits of nuclear power are twofold. First, it is helping to reduce the cost of generating electricity in areas of the country dependent in the past on high-cost fossil fuels. Secondly, the ability to draw on atomic fuel resources greatly strengthens the country's long-range energy position.

In the first connection, a point worth mentioning is that, because of the compactness of atomic fuel and consequent elimination of fuel transportation cost differentials, nuclear power will in time act to standardize electricity costs across the nation. How compact is atomic fuel? The fissioning of 1 gram

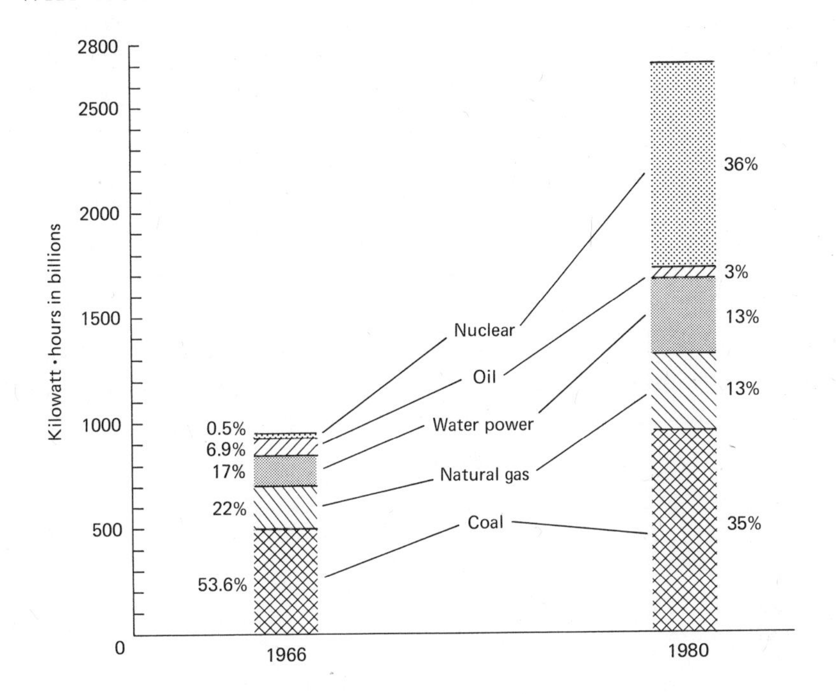

Power sources for electrical energy (35)
(Percent of total for each source)

of fissionable material releases 23,000 kilowatt hours of heat. This means that 1 ton of uranium has roughly the same potential fuel value as 3 million tons of coal or 12 million barrels of oil. In practice only a small fraction of the potential energy value of atomic fuel is extracted during a single cycle of reactor operation. Even so, a ton of reactor fuel still substitutes for many fully loaded freight trains of conventional fuel.

In the second connection, it is a remarkable fact that, with barely more than 5% of the world's population, the United States produces and consumes more than one-third of the world's electricity. An equally remarkable statistic is that, taking all forms of energy into account, the United States will probably use as much energy from fuel in the next 20 years as in all its history. And the rate of fuel consumption is expected to double in the 20 years thereafter. If this trend continues, our reserves of fossil fuels, vast as they are, will progressively be depleted. Opinion varies on this point, but even allowing for the discovery of new deposits, the chances are that if fossil fuels continue to carry as large a share of our energy burden as they do now, we will begin to experience some depletion effects as early as the turn of the century. Our resources of nuclear fuels are large and if we successfully develop technology for breeding (producing more fissionable fuel from fertile material than is consumed in the operation of a reactor), they will be almost limitless. Ultimately, though, we may have to look to still other energy sources, and that may be where *thermo*nuclear power comes in (see chapter 19). (23)

When a self-sustaining nuclear fission reaction was first achieved in a converted squash court at the University of Chicago in 1942, would-be prophets predicted that coal and oil would soon be obsolete and that the wheels of the world would turn on nuclear power. More than 30 years have passed, and the road to nuclear power has been a rocky one. Not only have nuclear power plant costs been higher than anticipated, but coal and oil plants have been made more efficient. The economic battle is still going on. Only now are large nuclear power stations becoming economically competitive. (27)

At the start of the 1960s the U.S. roster of civilian nuclear power projects was limited to a few demonstration units with capacities in the neighborhood of 200,000 kilowatts and a number of small experimental or prototype units with capacities in the range of 3000 to 90,000 kilowatts. By the end of 1963 several large-scale (400,000–600,000 kilowatt) projects had been undertaken. In mid-1965 utility announcements of commercial nuclear power projects began to gather momentum. By January 1968 nearly 50 million kilowatts of nuclear power capacity had been scheduled for construction. By the end of 1968 the total had risen to more than 72 million kilowatts.

One way of providing perspective on this figure is to say that it represents a financial commitment, for plants alone, of somewhere in the neighborhood of $10 billion. Another and even more impressive way is to point out that the installed capacity of the U.S. electrical industry at the start of World War II totaled only 42 million kilowatts.

As the dimension of the nuclear construction program grew, the unit size of the plants increased. At the end of 1968, the largest nuclear units on order were in the 1–1.2-million-kilowatt range, and the average unit size was above 800,000 kilowatts. These figures become all the more impressive when one realizes that up to the mid 1950s the largest conventional steam-electric unit in service in the U.S. had a capacity of only 250,000 kilowatts.

The expectation is that by 1975 some 70 million kilowatts of nuclear capacity will be in commercial service, and that by 1980 the total will have increased to about 150 million kilowatts. If the latter expectation is borne out, nuclear power would then account for approximately 25% of the country's total electrical capacity and, because of operational economies, for an even higher percentage of the total electrical output. Looking still further into the future, many believe that by the end of this century nuclear power will account for half of the total electrical output and for essentially all power generating capacity built thereafter. (23)

The next subject is nuclear fuels. Uranium has been discussed in connection with Becquerel's discovery of radioactivity in 1896. The element itself was discovered a century earlier, in 1789 to be precise. That year heralded two revolutions, the quieter one in science and technology.

This unsung incident occurred in the relatively obscure laboratory of a German scientist named Martin Heinrich Klaproth (1743–1817), who died without ever realizing its true significance. It was to him an accidental and passing discovery.

Klaproth was working on projects involving pitchblende ores. In the course of his experiments he isolated from the ores a black powdery material with chemical properties strikingly different from those of any elements he knew.

He showed the substance to other scientists. No one could identify it or see a use for it. But it was new and it needed a name, and in honor of the planet Uranus, which had been discovered a short time earlier, Klaproth called his puzzling material uranium.

For more than a century, this strange substance remained little more than a laboratory curiosity. Until relatively recently, American stocks of uranium were extremely low because there was little industrial demand for it. The United States, which had led the way in the application of nuclear fission, was dependent on foreign sources of supply. From mines deep in the Belgian Congo came much of the uranium needed for the development and production of the first atomic bombs. Every ounce of Congolese ore was hauled out through 1200 miles of rocky canyons and steaming jungles, then shipped across submarine-infested seas to the United States.

But the supply was just not enough and a widespread search for domestic sources was launched in the late 1940s.

Dreams of riches spurred the searchers on. The scene often was reminiscent in many respects of the gold rush days in California or Alaska, without the lawlessness, but with all the glitter and romance of overnight success, sudden wealth, and of the triumph, by a few lucky men, over wilderness, loneliness, and bad luck.

Grizzled old-time prospectors and trained geologists were among the searchers, but so were clerks, ranchers, teachers, students, businessmen, off-duty soldiers, and even housewives.

The exploration carried them into remote regions where panoramic vistas seldom seen by men stretched before their eyes. Most of the adventurers found little or no telltale radioactivity with their Geiger counters. Yet a surprising number did locate important ore deposits, not infrequently in the abandoned workings of old radium mines, and later sold their interests to established mining companies.

A very few struck it rich and disposed of their rights for thousands or even millions of dollars. When the rush at last subsided, about a score of mines, many of them by this time owned by large metal or chemical corporations, accounted for about 85% of the known ore reserves in the Colorado Plateau, where most of the great search was concentrated.

Uranium is a metal. The ore is not in a pure state, but in the form of an oxide. Industry is able to produce pure metal from the oxide form, however, and in its freshly milled and polished state, metallic uranium has a silvery luster. However, like many metals, it oxidizes rapidly and soon becomes coated with a black layer of oxide by the action of air and moisture on the metal surface.

Uranium is heavy—one of the heaviest of all metals. It weighs about 65% more than lead. A piece as large as a soft-drink can weighs about 17 pounds.

Uranium is also radioactive, as we know, and this characteristic helps prospectors locate it in the earth. Using Geiger counters and other devices for detecting radioactivity, serachers can locate deposits of uranium ore that are not apparent by other means.

Early in the intensive period of seeking uranium, only a few really good sources were known. These were the known pitchblende ore deposits at Shinkolobwe in the Belgian Congo, Great Bear Lake in the Northwest Territories of Canada, and the Joachimsthal area in Czechoslovakia. (Our word dollar is derived from the Austrian coin, the thaler, made of Joachimsthaler silver, that is, silver from Joachimsthal.) Prospectors knew about, but had little concept of the extent or richness of deposits in the Colorado Plateau in the American West.

But as uranium became a prime target for prospectors, new sources turned up. In South Africa, for example, tailings remaining from treatment of the high-grade gold ores were reprocessed for uranium. In 1949, uranium ore was found in the Todilto limestone near Grants, New Mexico, and the Happy Jack copper deposits in Utah were first mined for uranium, although uranium had been known there since 1920. And in 1952, geologists discovered the now-famous Jackpile Mine in

New Mexico. This mine has since produced millions of tons of primary black ore, rich in usable uranium.

Within 3 more years, many additional ore deposits had been discovered—in the Black Hills of South Dakota, in the Gas Hills of Wyoming, and in Utah, where the famous Mi Vida and Delta Mines made millionaires out of prospectors. Canada, France, Australia, and Argentina also reported discoveries.

Uranium makes up about 2 parts per million of the earth's crust, and traces of it are found almost everywhere. Its abundance is greater than that of gold, silver, or mercury, about the same as tin, and slightly less than cobalt, lead, or molybdenum. There is an average of 1 pound of uranium in every 500,000 pounds of the earth's rocks.

A prospector or miner following a uranium discovery in the western United States often finds visible indications of the metal's ancient past. A petrified log 200 million years old may be encrusted with uranium. Ripples worn by long-vanished inland seas may be seen in sandstone beds that have been raised thousands of feet above sea level in one of the most arid portions of North America. The fossil remains of dinosaurs, prehistoric crustaceans, and ancient trees are common in this vast wasteland.

About 95% of the uranium ore mined in the United States is found in pitchblende and coffinite deposits. These were buried many millions of years ago in sandstones formed by rushing streams in what was once a gigantic semitropical savannah teeming with turtles and crocodiles. At some point in geologic history, ground waters carried dissolved uranium into these ancient stream channels where the chemicals produced by decaying plants could precipitate it. The chemical encounter between the water-soluble uranium and the organic products resulted in the formation of literally thousands of ore bodies containing pitchblende, coffinite, carnotite, and many other uranium minerals. Today, geologists look upon these deposits as having been formed in petrified rivers since much of the original plant material buried in them has been changed to mineral or stone.

How these deposits were formed has long been a puzzle to geologists seeking to unravel the secrets of the ore's whereabouts. Many feel that tiny microbes, with special food requirements, created chemical sinks containing hydrogen sulfide that precipitated the uranium and other metals as they were introduced into the sands by ground waters. They surmise that the ore deposits are not static, that they will not always remain where they are, for the ore is still subject to the chemical reactions of ground waters that continually leach away and redeposit the uranium. This may happen again and again, until all of the river-laid organic material is destroyed and nothing is left in the sandstone to precipitate the uranium. Fortunately for us, it would take millions of years to destroy the existing ore deposits. However, geologists must continually appraise new prospecting areas in terms of the ravages wrought by these destructive processes during past geologic time. Many rocks that might contain uranium, but do not, are known to have been chemically altered in this manner. (35)

Only 1 uranium atom out of every 140 found in nature is ^{235}U. It isn't easy to separate these readily fissionable atoms from other atoms or to enrich natural fuel by increasing the percentage of ^{235}U that it contains. Is there a practical substitute for ^{235}U that could help stretch our atomic resources?

As was mentioned earlier, there are luckily at least two such supplements—both man-made. One is plutonium, which is produced in a two-step process when a neutron is absorbed by the most common variety of uranium—^{238}U. The second is uranium 233, an isotope that appears as a result of radioactive decay after neutrons have been absorbed in still another material—thorium 232.

Uranium 238 and ^{232}Th are both called fertile materials because they provide a base from which nuclear engineers can grow new nuclear fuel, in a breeder reactor.

So far, no commercial power reactor has used fuel elements manufactured from ^{233}U. They may begin to appear before long, however, and they can be expected to play an important role in conserving our natural resources of fissionable material. Power plants are already operating in various parts of the world to convert thorium into ^{233}U; and in some ways this third fuel is the best of all.

Baron Jöns Jakob Berzelius (1779–1848) could not have chosen a more appropriate name for the element he isolated (from minerals on a little island off the coast of Norway) in 1828. (Berzelius, a Swedish chemist, discovered several other elements and was the inventor of the symbolic language of chemistry, that is, the use of one or two letters to signify an element.) Thor was an enormously powerful Scandinavian god, who often gave humans a helping hand. Thor controlled thunder and lightning. He was a source of power to be reckoned with and so is thorium.

Thorium is neither rare nor commonplace. It is the thirty-fifth most abundant chemical element in the earth's crust, ranking just behind lead and molybdenum. Thorium is several times more abundant than uranium, but statistics are inexact because even today there is no thorium mining industry. Most thorium is produced as a by-product from the processing of the rare earths or uranium ore.

The first practical use of thorium developed more than 75 years ago, when an Austrian baron named Carl Auer von Welsbach noticed that burning gas gives off a bright, steady light when surrounded by a fabric cylinder covered with thorium oxide. The brilliant glow of incandescent gas mantles was thorium's major contribution to the energy industry until the discovery of nuclear power. As late as 1952 more than 65% of the country's thorium production went into this original application. By 1964, this figure was down to about 30%.

The coming of electricity didn't completely eliminate the value of thorium in illumination, because it is also useful in light-bulb filaments. The addition of about 1% thorium to tungsten makes the material easier to draw into filaments, and it also keeps the hairlike wires from becoming brittle and cracking. The principal use of thorium today, however, is in metal alloys.

Thorium itself isn't a nuclear fuel. The nucleus of a thorium atom generally doesn't fission when struck by a neutron. Instead, the neutron is absorbed; and the most common thorium nucleus, which consists of 90 protons and 142 neutrons, suddenly includes an extra neutron, making it much less stable than it was before.

Within 23 minutes half the ^{233}Th atoms in any given mass will decay, a neutron being converted into a proton, with the emission of an electron (and a neutrino). But as soon as the number of protons in a nucleus changes, the atom itself is transformed into a completely different element. An atom with 91 protons in its nucleus isn't thorium at all; it's protactinium.

Protactinium 233 is only slightly more stable, having a half-life of 27 days. It goes through a second decay process, losing another electron; and the result this time is an atom containing 92 protons and only 141 neutrons. This is ^{233}U, which can fission when struck by a neutron—transforming part of its mass directly into energy. Uranium 233 can support a chain reaction in a nuclear reactor. It is the third fuel of the atomic age.

Thorium doesn't exactly compete with uranium; it can never replace it completely. Thorium may supplement ^{238}U as a fertile naterial, but you can't fuel a reactor with thorium alone. It must be used in conjunction with one of the fissionable materials—^{235}U, plutonium, or ^{233}U.

Our voracious appetite for electricity would really threaten to exhaust both our fossil fuel and our supply of reasonably priced uranium if it weren't for the promise of breeder reactors, which can produce new fuel at a faster rate than the original fuel is used up.

If we kept using only ^{235}U—even if it were practical to pay as much as $100 a pound for unenriched uranium oxide (that's about sixteen times what it costs now)—we would reach the bottom of our natural fuel reservoir early in the next century. Luckily, however, breeders can postpone that day almost indefinitely by transforming ^{238}U into fissionable ^{239}Pu.

As wonderful as this is, there are certain dangers in becoming too complacent about the plutonium breeders that are expected to come into their own in the 1980s. For instance, we must not forget that the breeders have a doubling time of their own. This is the minimum time a reactor takes to produce enough new fissionable material (allowing for the amount that burns in place) to fuel another reactor of the same size and power output. For the early plutonium breeders this is likely to be about 20 years, although by the end of this century it might be reduced to as little as 6 years. If the demand for electricity keeps rising, however, we probably won't be able to stop mining uranium even then. Breeders just don't breed fast enough. (38)

19 ◊ Controlled Nuclear Fusion

FUEL requirements are skyrocketing. The role of nuclear fission, the splitting of the heaviest nuclei, has been discussed. The other nuclear process that can yield energy is fusion, the combining of some of the lightest elements.

Actually, there are many other nuclear processes which are accompanied by a liberation of energy. But only with nuclear fission and fusion is there a possibility of producing more energy than is consumed in causing the reaction to occur. In other words, there is some prospect that the process, once started, can be self-sustaining—like a fire.

Fission is not the complete solution to the energy problem. Although it is true that the world resources of the basic materials, namely uranium and thorium minerals, are fairly abundant, there are many countries that either do not possess these minerals or do not have the means for producing the best nuclear fuels from them.

It is such considerations that make nuclear fusion of exceptional interest as a possible source of power. The essential fuel material is a form (isotope) of hydrogen, called heavy hydrogen or deuterium, that is present in all water; for every 6500 or so atoms of ordinary (or light) hydrogen in water, there is 1 atom of deuterium. Calculations show that the energy that could, in theory, be produced by the fusion of the deuterium nuclei present in a gallon of water is equal to that obtainable from the combustion of 300 gallons of gasoline. The enormous amounts of water available on earth thus represent a virtually inexhaustible potential source of energy.

The cost of obtaining the deuterium fuel from water is not large. At the present, it costs about 4 cents to extract all the deuterium in a gallon of water. If the fusion process were operative, even at a low efficiency, the fuel costs would thus be insignificant. Here then is apparently the ideal source of energy—cheap, abundant, and available to all. But, unfortunately, this is not the whole story. For one thing, as in a nuclear fission system, the price of the fuel would represent only a small portion of the cost of the electric power produced. For another, there are tremendously difficult problems to be solved before fusion power can be a reality.

Before describing the conditions that must be realized if fusion energy is to be released in a practical manner, it is of interest to review the evidence showing that nuclear fusion is indeed possible. In the first place, it is reasonably certain that fusion is the energy source of the sun and other stars. The sun's fuel material, however, is not deuterium but ordinary hydrogen; in a series of nuclear reactions, four hydrogen nuclei are fused together to form a helium nucleus. It is known, however, that the processes take place much too slowly to be useful on earth. In the sun, the ability to generate energy at a high rate depends on the enormous quantity of hydrogen present.

Further evidence that nuclear fusion can be achieved is obtained from laboratory experiments. Deuterons, or deuterium nuclei, can be accelerated to high velocity (and kinetic energy) in a charged-particle accelerator, such as a cyclotron or similar machine. If these deuterons are allowed to strike a solid target containing deuterium, fusion reactions take place. However, of the collisions between the accelerated deuterons and those in the target, only a very small proportion leads to fusion. In the great majority, the impinging deuteron is merely deflected (or scattered) and at the same time is deprived of some of its energy; it then becomes essentially incapable of fusing with another deuteron. In effect, most of the energy of the accelerated deuterons is dissipated as heat in the target. Much more energy is consequently spent in accelerating the deuterons than is produced by the small number of fusion reactions that occur. Although the acceleration procedure is not a basis for the practical liberation of energy, it does show that fusion between two deuterium nuclei is possible.

Finally, nuclear fusion is the source of the large amount of energy released with great rapidity in exploding the so-called H-bomb. An attempt might perhaps be made to scale down or slow down an H-bomb and thus release fusion energy in a controlled manner. But this is not possible. The H-bomb requires the use of a fission bomb as a trigger, and this cannot be scaled down. It has been suggested that an H-bomb might be exploded underground where the resulting energy would be stored as heat; this could be drawn upon gradually by injecting water to produce steam as required. Such energy might be a valuable by-product of the utilization of H-bomb explosions for peaceful purposes.

In order to see how controlled nuclear fusion might be realized, let us examine the essential requirements. For fusion to occur, the two light nuclei must first come close enough to permit interaction to take place. Since each nucleus carries a positive electrical charge, the two nuclei repel one another more and more strongly as they come closer together. Consequently, for the nuclei to interact they must have enough energy to overcome the force of electrostatic repulsion which tends to keep them apart. The magnitude of the repelling force increases with the electrical charges on the two nuclei. To keep this force small, therefore, the interacting nuclei should have the lowest possible charge (or atomic number).

The element with the smallest atomic number is hydrogen, since its nuclei (and those of its isotopes) carry but a single charge. The obvious choice for fusion reactions on earth is thus some form of hydrogen. The fact that it happens to be cheap and abundant is, of course, advantageous. Three isotopes of hydrogen are known. The lightest, with a mass number of 1, is ordinary hydrogen, H; the nucleus of an atom of this isotope is a proton. Ordinary hydrogen is the isotope that undergoes fusion in the sun.

The next isotope is deuterium, mass number 2, represented by ^2H or, more commonly, by the symbol D; its nucleus, the deuteron, is also indicated by D, although D^+ is used when it is necessary to distinguish between the neutral atom and the (positively charged) nucleus. It occurs in all natural water, as mentioned earlier, and can be extracted without too much difficulty. Finally, there is tritium, mass number 3, represented by ^3H or T; the nucleus is called a triton. This isotope is radioactive and is very rare in nature. It can be made by the interaction of neutrons with lithium 6 nuclei and is fairly expensive. (7)

(Proton, deuteron, and triton come from the Greek words for one, two, and three.)

Fusion processes with deuterons and tritons take place fast enough to make them reasonably possible sources for the release of energy at a useful rate. These isotopes are, therefore, the most practical fusion fuels. Because of the low cost and availability of deuterium, it would be preferable to utilize this isotope alone; the fusion process would then involve only deuterons. Two such reactions, occurring with roughly equal probabilities, are known; they are

$$D + D \rightarrow {}^3He + n + 3.2 \text{ MeV}$$

and

$$D + D \rightarrow T + H + 4.0 \text{ MeV}$$

where n represents a neutron. In accordance with the universal convention, the energy release is expressed in units of MeV, that is, millions of electron volts. In the first of these two reactions, the products are a helium 3 nucleus and a neutron, and in the second they are a triton and a hydrogen nucleus (proton). The triton formed in this manner can then react fairly rapidly with another deuteron, that is,

$$D + T \rightarrow {}^4He + n + 17.6 \text{ MeV}$$

leading to the formation of a helium 4 (ordinary helium) nucleus and a neutron, plus the large energy release of 17.6 MeV.

Since the two deuteron-deuteron, or D-D, reactions take place to about the same extent, and one of them is immediately followed by the D-T reaction, the net energy release for the fusion of five deuterons is $3.2 + 4.0 + 17.6 = 24.8$ MeV. From this result it can be calculated that complete fusion of the deuterons in 1 gram of deuterium will yield 5.6×10^{10} calories. One gallon of ordinary water contains ⅛ gram of deuterium and so is equivalent to 7×10^9 calories. This is approximately the combustion energy of 300 gallons of gasoline. The total amount of deuterium in the ocean is estimated to be 4.5×10^{19} grams; its fusion energy content is thus 2.5×10^{30} calories or roughly 3×10^{20} kilowatt-years. At the world's current rate of power consumption of 5×10^9 kilowatts, the reserves should last billions of years.

The fusion of deuterium would be the preferable reaction for the release of energy, but it is possible that it may be too slow to be practical. It would then be necessary to resort to the fusion of deuterium and tritium nuclei, in accordance with the D-T reaction given above. To do this, the tritium would have to be made by the reaction of the lithium 6 isotope with neutrons, that is,

$${}^6Li + n \rightarrow T + {}^4He$$

Initially, the neutrons would have to be obtained from a nuclear fission reactor. But once the fusion process was under way, the neutrons released in the D-T reaction could be utilized to produce the tritium.

The three fusion processes of interest are shown in the diagram, which indicates the rearrangements among the constituent neutrons (n) and protons (p). A deuteron consists of one proton and one neutron, and a triton of one proton and two neutrons. The energy released appears as kinetic energy and is divided between the two products in inverse proportion

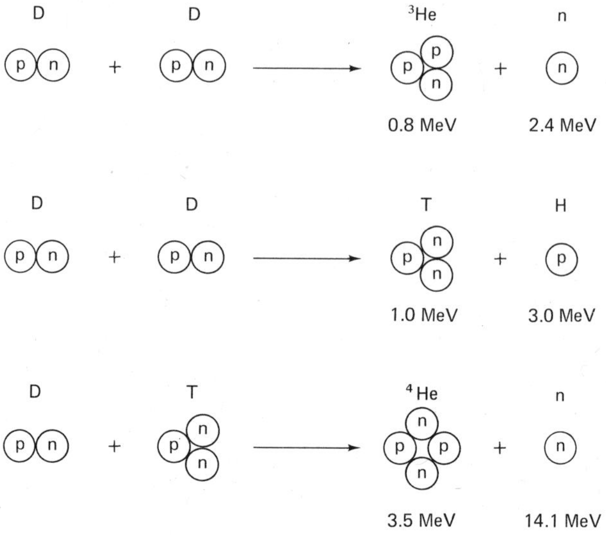

D + D ⟶ ^{3}He + n
0.8 MeV, 2.4 MeV

D + D ⟶ T + H
1.0 MeV, 3.0 MeV

D + T ⟶ ^{4}He + n
3.5 MeV, 14.1 MeV

Fusion reactions with deuterium (D) and tritium (T) nuclei. The number of MeV in each case indicates how the fusion energy is divided between the products.

to their masses. The amount of energy carried by each of the products is indicated. For the type of fusion system envisaged at present, the uncharged particles, that is, the neutrons, would escape with their energy. Only the kinetic energy carried by the charged particles—the H, ^{3}He, T, and ^{4}He nuclei—would remain to make the fusion process self-sustaining.

So far, it has been established that the nuclear reactions of interest for the controlled release of fusion energy involve either deuterium alone or deuterium and tritium. The forces of repulsion between these light nuclei are the minimum possible, and the resulting reactions could be expected to take place at a reasonable rate. The next point to consider is how to confer sufficient energy on the nuclei to permit them to overcome the repelling forces. One obvious way is to make use of an accelerator to provide the necessary energy. But this procedure is much too wasteful of energy to be of practical value. It has been used extensively, however, in the laboratory to study the probabilities of the D-D and D-T reactions. In fact, much of what is known about these reactions was determined in this manner.

Another way of supplying energy to the nuclei is to raise the temperature. The kinetic energy of an atom (or nucleus) is proportional to the absolute temperature. (7)

The centigrade or Celsius temperature scale designates the freezing point of water as 0° and the boiling point as 100° (both under specified standard conditions). The lowest Celsius temperature attainable is −273° C. The laws of physics can be stated more simply using the Kelvin or absolute temperature scale, on which the lowest attainable temperature is 0° (freezing, 273°; boiling, 373°).

An energy of 1 electron volt corresponds to an absolute temperature of 1.16×10^4 degrees Kelvin. Hence, it is only a matter of attaining a sufficiently high temperature to permit fusion reactions to occur. This is precisely what happens in the sun. At first sight, it might appear that the situation is somewhat similar to that in which particles are accelerated. But, if the nuclei are confined in some manner, so that they cannot escape, the consequences are very different. Although many of the nuclear collisions in a high-temperature system result in scattering instead of fusion, the effect is a redistribution of energy rather than a loss of energy. Temperature and average energy are unchanged. In a confined space, the nuclei moving in random directions collide repeatedly until fusion reactions take place.

Fusion reactions brought about by means of high temperatures are referred to as thermonuclear reactions.

The thermonuclear approach, through the use of very high temperatures, appears to be the most promising for controlled fusion. A great advantage is that the process could be made self-sustaining. The fuel gas, either deuterium alone or a mixture of deuterium and tritium, is first heated in some manner to a temperature at which nuclear fusion occurs at an appreciable rate. Part of the energy released then serves to raise the temperature of the incoming gas, the remainder being utilized for power production. Continuous operation of a thermonuclear fusion reactor might thus be possible.

At these very high temperatures, all the hydrogen atoms will be stripped of their electrons. Such a gas, consisting of positively charged nuclei (or ions) and free negative electrons, is said to be ionized. An ionized gas is commonly called a plasma. It should be remembered that, although a plasma contains free positive ions and free negative electrons, the numbers of positive and negative electrical charges balance exactly, and the plasma as a whole is neutral. Because of the presence of the electrically charged particles, plasmas have a number of interesting properties; some of these can be utilized to advantage in controlled fusion research, but others are a decided drawback. Their unusual behavior has led to the revival in recent years of the expression the "fourth state of matter," first used by the English scientist William Crookes (1832–1919), to describe plasmas.

Suppose for the moment that it is known how to produce a deuterium (or deuterium-tritium) plasma at a temperature of some 100 million degrees or more. How is such a plasma to be confined? The difficulty does not lie in the very high temperature because, at the low gas densities in a fusion reactor, the total energy content of the plasma would be insufficient to cause any significant damage to a containing vessel.

The problem arises from the loss of energy by the nuclei as a result of striking the walls. At a temperature of 100 million degrees, the nuclei (and electrons) in a plasma are moving randomly in all directions at average speeds of several thousand miles per second. Consequently, within less than a millionth of a second, all the particles will have hit the walls of the containing vessel; as a result, they would lose essentially all their kinetic energy. In other words, the plasma is rapidly cooled. Even if the high-temperature plasma could be generated, it would not last long enough to allow a significant amount of fusion to take place.

A method must therefore be found to prevent the plasma

particles from striking the walls of the containing vessel. For this, the electrical charges carried by the nuclei and electrons can be turned to advantage. It is difficult for charged particles to cross the lines of force of a magnetic field. A plasma can therefore be confined by a magnetic field of suitable form.

In effect, a confining magnetic field exerts an inward pressure which, in the ideal situation, just balances the outward pressure of the plasma particles. This concept is often referred to as a magnetic bottle for confining a plasma. In a sense, the magnetic field serves to support the plasma pressure and to insulate it, so that it does not lose its energy by the particles' striking the walls of the containing vessel. (7)

(See chapter 12 for a description of how the earth's magnetic field confines particles in space.)

The rate at which energy would be generated per unit volume of a fusion reactor increases with the gas density. [The more nuclei there are to fuse, the more the fusion that takes place.] For energy to be released at a useful rate, therefore, the density must not be too low. An important aspect of this problem is that of the confinement time. Even under the best conditions, high-energy particles will soon escape the confinement of a magnetic field. In a practical fusion reactor, the confinement time must be sufficiently long to permit a reasonable amount of fusion to occur. The minimum time is inversely proportional to the particle density. Hence, a moderately high density will make it possible to operate the reactor even if the confinement time is short, as it may well be.

On the other hand, if the excess energy, over and above that required to heat the incoming gas to the thermonuclear temperature, is not removed as fast as it is generated, the temperature (and pressure) may become excessively high. The magnetic field may then not be able to confine the plasma, and it will escape and be cooled. The operating density selected for the fusion reactor thus represents a balance between magnetic field strength and rate of energy removal, which set an upper limit, and the rate of energy generation and confinement time, which determine a lower limit.

As far as can be estimated at present, the plasma in a controlled fusion system will probably have a density of 10^{15}–10^{17} particles per cubic centimeter. This may be compared with 3×10^{19} particles (molecules) in a gas at normal temperature and pressure. At the temperatures which must be attained in a deuterium fusion reactor, a density of 10^{15} particles per cubic centimeter would represent a pressure of about 1500 pounds per square inch. It is this pressure which must be contained by the magnetic bottle. (See photograph section.)

Several different lines of endeavor are being followed in the researches on controlled fusion. It may well be asked therefore: Why dissipate the effort in so many directions? Wouldn't it be better to concentrate on one promising system and develop it into an operating device for producing useful power? The answers to these questions lie in the great complexity of the behavior of hot plasmas in magnetic fields. The problems are so great that the only hope of solving them is by a simultaneous attack from several directions.

Although the approaches differ from each other in certain respects, they are, in fact, interrelated. Information obtained from one project helps to advance the others. As a result, overall progress is greater. Moreover, it is clear that each method of plasma confinement studied so far has both advantages and drawbacks. It is probable, therefore, that the solution to the problems will not be found in any single system, but in a composite of two or more that combines their advantages while eliminating or minimizing their drawbacks.

Temperatures in excess of 20 million degrees K, hotter than the interior of the sun, have been attained in plasmas with densities of nearly 10^{17} ions per cubic centimeter. These are close to thermonuclear fusion conditions. But the confinement time of such plasmas has been no more than a few millionths of a second, compared with a tenth of a second or so thought to be the minimum for the given density in a practical fusion system.

On the other hand, at very much lower plasma densities, confinement times approaching a minute in duration have been realized; the problem here is to build up the density. In general, the limitations so far have been the instabilities of the plasma in a magnetic field. As a result of extensive theoretical and experimental studies, a better understanding is being obtained of the underlying phenomena. With this understanding, there are hopes that methods will be found for controlling or avoiding the instabilities, so that a high-temperature, high-density plasma may be confined for an appreciable time. A self-sustaining fusion reactor could then become a reality.

As far as we can see at present, there appear to be no fundamental obstacles to success; how long it will take to achieve, however, is quite impossible to predict. There are problems of enormous difficulty to be solved before controlled nuclear fusion can be utilized as a source of power. But the rewards of this achievement would be so great that it represents one of the great scientific challenges of our century. (7)

20 ◊ Some Power Applications

Power from Radioisotopes

ON January 16, 1959, a device that turned heat from radio activity into electricity was demonstrated publicly for the first time on the desk of the President of the United States. The device was the size of a grapefruit. It weighed 4 pounds and was capable of delivering 11,600 watt-hours of electricity for about 280 days. This is equivalent to the energy produced by nickel-cadmium batteries weighing nearly 700 pounds. It was called SNAP 3.

The U.S. Atomic Energy Commission had begun developing a series of these compact devices in 1956 to supply power for several space and terrestrial uses. The devices were all described by the general title Systems for Nuclear Auxiliary Power. The initials form the word SNAP.

The kinetic energy possessed by radioisotope decay particles is transformed into thermal or radiant energy that may be used to produce electrical power through a number of energy conversion techniques.

One early objective was a 500-watt 60-day generator (SNAP 1) to power instruments in space satellites. Cerium 144, a beta emitter with a 290-day half-life, was selected as the heat source. A small turboelectric generator with high-speed rotating components was developed to convert the heat into electricity. This project was abandoned in favor of a thermoelectric conversion system. Thermoelectric devices, with no moving parts, could greatly extend the usefulness of the generator beyond its expected 2-month life.

A milestone was reached in the U.S. space program with the orbiting in 1961 of a satellite which carried a radioisotope generator as a supplementary source of electricity for its radio transmitters. This marked the first use of atomic power in space.

To do a job effectively over a long period of time, a satellite needs a dependable, long-lived electrical supply. And because pounds are precious in payloads rocketing into space, the electricity source must be light in weight. It must be rugged to withstand the rigors of a rocket ride. It must be so safe that in the event of accident there will be no serious consequences from radioactive contamination. (26)

Other SNAP units were developed for use in manned and unmanned moon landings.

SNAP 27 was used to power the Apollo lunar surface experiment package left by astronauts on the surface of the moon. These packages automatically transmit measurements to earth for 1 year.

Isotopic power units are only one type of space power supply. Solar cells, fuel cells, and batteries are also used in space. Each type has its advantages and limitations.

Space missions where radioisotope generators have significant advantages over other power supplies include the following:

Satellite orbits passing through radiation belts around the earth, where solar cells deteriorate.

Operations on the moon, where long periods of darkness require heavy batteries to supply power while the solar cells are deprived of sunlight.

Space-probe missions into opaque atmospheres such as that of the planet Venus, where solar cells would be useless for lack of light.

For long-lived, high-powered missions where fuel cells or batteries are too heavy and where solar-cell arrays are too large.

There are many out-of-the-way places on earth where electrical power is needed for weather stations, navigational beacons, and other special installations. In some isolated spots men are needed primarily to operate the machinery that provides the electricity. Others, unmanned, have relatively short-lived batteries to supply power. It is costly to provide men with fuel and the necessities of life in remote locations, and to replace batteries at distant unmanned stations. We must forego the benefits of more stations because of the high expense of maintaining them.

The development of cheap, reliable, long-lived radioisotope generators may help solve this problem. Generators with no moving parts have an inherent reliability and freedom from maintenance. The slow decay of isotopes, such as strontium 90 and cesium 137, can provide years of operating life.

The world's first atomic powered weather station was placed on bleak, uninhabited Axel Heiberg Island in Canada, only 700 miles from the North Pole, in 1961. The unmanned station, designed to collect and relay data on temperature, wind, and barometric pressure, was located to fill a gap in the weather network between two manned outposts at Resolute Bay and Eureka. It was part of a joint project of the U.S. and Canadian weather bureaus, and its data were made available to all nations. In a completely successful 2-year test, the station proved one important advantage of atomic power by remaining in operation longer than has any unattended battery-operated station. Waste heat from the generator keeps the temperature of the buried equipment nearly constant. This prevents freezing, a common failing of battery-operated stations in polar regions. (26)

The frigid and desolate reaches of the Arctic and Antarctic are typical remote areas where small, packaged nuclear power plants are potentially superior to those burning oil and coal.

Scientific and military bases in these inhospitable regions are hundreds of miles from sources of conventional fuel. Ships and planes cannot carry in supplies during much of the year. Still the men and equipment at McMurdo Sound in Antarctica must have some source of energy to ward off the sixty-below temperatures of the polar night. Heat from the fissioned uranium nucleus has provided this energy in place of flown-in diesel fuel. (27)

SNAP units have been used to provide power for flashing-light navigational buoys and for lighthouses. In a lighthouse near the entrance to Baltimore harbor a SNAP unit replaced a lighthouse crew of three men. (One cannot help feeling sad at the passing of one of the last outposts of "getting away from it all," which is the way lighthouse life is usually portrayed in the movies.)

Another SNAP unit has been used in a floating weather station.

Such stations measure and transmit air temperature, barometric pressure, and wind velocity and direction. Storm detectors trigger special hourly transmissions during severe weather conditions.

Safety considerations play an important part in the choice of fuel for satellite power generators. Plutonium 238 emits alpha radiation, plus very low levels of gamma and neutron radiation. We have noted that alpha radiation is the least penetrating of all. Even without shielding, plutonium 238 radiation presents no hazard to man unless the plutonium source is inhaled or ingested.

All forms of plutonium, however, are extremely poisonous to living organisms, and rigid safety requirements for plutonium generators have been established. For example, fuel capsules are designed to survive accidents on a rocket vehicle launch pad, including fire and explosion. They must withstand impact against the earth if the rocket fails in flight. The generators must be designed so that when they reenter from orbit, they either remain intact or burn up in the atmosphere at altitudes greater than 100,000 feet. Realistic tests are conducted to prove that these requirements are met.

Strontium 90, the fuel for many generators used on earth, presents a special biological problem. It is a bone-seeking radioisotope that can be harmful if absorbed in quantity by a man or animal. It tends to enter and remain in bone structure. Nevertheless, it is an excellent heat source. What is done to make this radioisotope safe?

To begin with, the possibility of absorption is eliminated by using the insoluble compound strontium titanate. Strontium 90 is locked chemically in this compound, which remains stable even beyond its melting point of 1910° C. The compound's solubility in seawater is extremely low, and its solubility in fresh water is so low that it has never been measured.

In addition, fuel pellets for generators using strontium 90 are encased in an alloy called Hastelloy C. This alloy forms strong, ruptureproof fuel capsules and will not corrode even if immersed in the ocean. Further protection for generators is provided by additional layers of Hastelloy C. Generators are designed so sturdily that they can survive a direct hit from a crashing airplane without releasing their fuel.

Beta particles from strontium 90 are stopped in the material surrounding the fuel and do not constitute a radiation hazard. As the beta particles are stopped, however, secondary radiations similar to X rays are produced. These radiations, called bremsstrahlung, are dangerous to man and may require protective shielding. This is provided by casings of lead more than 4 inches thick, depleted uranium (containing less of the fissionable isotope uranium 235 than the 0.7% found in natural uranium) nearly 3 inches thick, or cast iron 8 inches thick.

Such shielding adds weight to generating systems, but this disadvantage is less important in an earth-based generator than in one rocketed into space. Moreover, the bulk provided by the shielding material is often useful. For example, it serves to protect deep-sea generators from tremendous water pressures on the ocean bottom. (26)

Energy for Space Travel

THE secret of space travel is energy—immense amounts of energy. The first stage of the Saturn V moon rocket generates as much energy each second as a million automobile engines.

Our quest is for a means of uniting the almost limitless energy of the atomic nucleus and the rocket's unique ability to thrust through the vacuum of space, the wedding of nuclear fission (age, about 30 years) to the rocket (age, about 1000 years).

In the words of Glenn T. Seaborg: "What we are attempting to make is a flyable compact reactor, not much bigger than an office desk, that will produce the power of Hoover Dam from a cold start in a matter of minutes." (22)

Some day, perhaps 15 years hence, a rocket will thrust a manned spacecraft from its parking orbit around the earth and inject it into an elliptical transfer orbit intended to intercept the planet Mars 7 months later. The men in this interplanetary craft will require electrical power for several purposes, for, according to an old rule of thumb, a man can live for only 40 days without food, 4 days without water, and 4 minutes without air. Enough food can and will be carried along on that first Mars journey, but there will not be room enough in the adventurous craft for all the water and air that will be required, unless it is possible for small amounts of these vital fluids to be used over and over again. The purification and regeneration of water and air will require electricity. So will the craft's instruments and radios. Still more power will be needed to keep the cabin at a livable temperature.

For some long space voyages requiring large power supplies, chemical forms of energy—rocket fuels, battery fluids, and hydrogen—do not have enough energy per unit mass (kilowatt-hours per pound); they weigh too much for long-life space missions (although they are best for missions involving less power or shorter duration). Similarly, solar power has limitations for some missions. The sun's contribution of energy, which is 1400 watts of power per square meter, or 150 watts per square foot, on the earth's surface, will steadily decrease as the spacecraft swings outward toward Mars. Mars is about 1.5 times as far from the sun as the earth is, so the solar energy density is reduced by a factor equal to the square of 1.5, or 2.25. Huge arrays of mirrors or solar cells would therefore be needed to capture enough solar energy for a spacecraft operating near Mars.

In a situation where large amounts of power are needed over long periods of time, the best source of electricity is a nuclear reactor, which uses energy contained in fissionable uranium. Uranium 235 contains 100,000 times as much energy per unit mass as the best chemical fuels. (34)

How a Nuclear Rocket Works

THE rocket concept was grasped by the first caveman when he pushed off from a lakeshore on a raft; every action has an equal and opposite reaction according to Newton's third law of motion. The caveman's action was pushing the shore away with his foot; the reaction was the surge of the raft onto the lake.

Actually, it is not necessary to have something solid to push against. The caveman could have propelled himself out on the lake by hurling rocks shoreward; as each rock left his hand, he and the raft would have moved a bit farther. It is the same way in airless space; propulsion in a given direction means throwing something away in the opposite direction. Instead of rocks, the ordinary chemical rocket expels a roaring jet of hot combustion gases. But the effect is the same. To make a nuclear rocket, the uranium nucleus must be fissioned in such a way that something is expelled from the spaceship.

Chemical rocket engines, jet engines, automobile engines—in fact, most of mankind's engines—extract heat from a fuel and turn it into motion through the expansion of hot gases. The nuclear rocket sprouts from the same family tree; it also creates hot high-pressure gas and turns it into thrust. The nuclear rocket is a direct descendent of the aeolipile, a steam-spouting reaction engine reputedly built a century or two before Christ by the famous Alexandrian engineer Hero.

It is relatively easy to see how a hot gas expands against a piston in an internal-combustion engine to do useful work. The piston presents the gas with something solid to push against. And the rocket engine operates in much the same way, except that the piston is replaced by the rocket nozzle. The hot gases created by chemical combustion or nuclear heaters issue from the throat of the rocket nozzle and expand against its flared sides, pushing the nozzle (and the whole rocket) upward. The pressure against the nozzle walls is the reaction to the expulsion or pushing away of the hot exhaust gases.

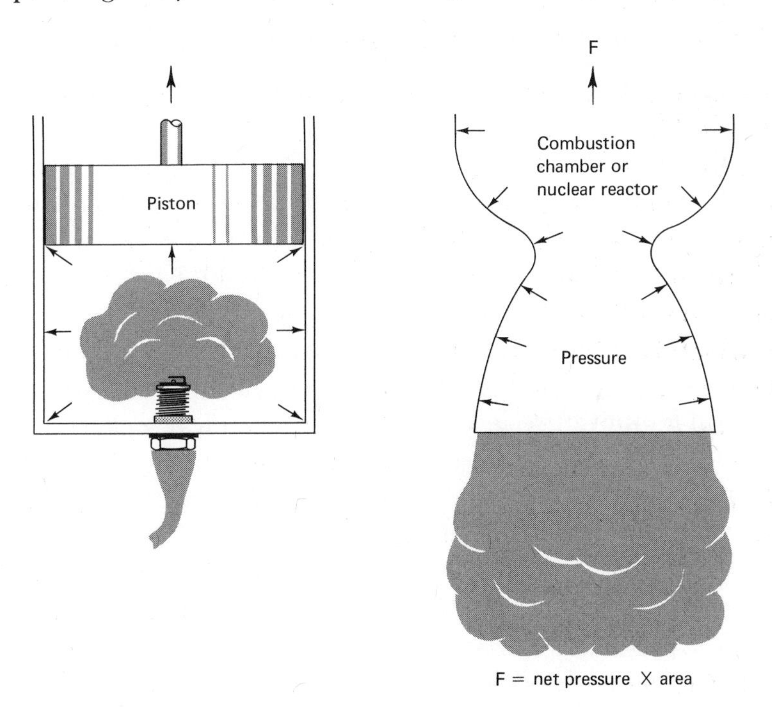

At left, hot gases expand against a piston, pushing it upward. At right, hot gases expand against a rocket nozzle, pushing it upward. (The reaction thrust F can be computed from the total net pressure against the nozzle and the combustion chamber area.)

The higher the propellant velocity, the more thrust we get from each kilogram of gas that roars out the nozzle each second. We want to have a high exhaust velocity for good rocket performance because we can thereby accomplish a space mission with less propellant. We will show later that the nuclear rocket produces about twice the exhaust velocity of the best chemical rocket. High exhaust velocity is the greatest advantage of the nuclear rocket.

The question then is: How does a nuclear rocket generate high exhaust velocities? The exhaust velocity v of any rocket is proportional to $\sqrt{T/M}$, where

T = the temperature of the hot gases just before they enter the nozzle throat
M = the average molecular weight of the exhaust gases

It is obvious that for a high v value we wish to maximize the quantity $\sqrt{T/M}$.

Knowing that T and M control v, let us first try to manipulate T. Since T is under the square root sign, it will have to be quadrupled to double v, the exhaust velocity. Chemical rockets already operate at temperatures close to 3000° K; if the nuclear fuel is to stay in solid form, it is apparent it cannot go to 12,000° K, which is hotter than the sun's surface. In fact, today's nuclear rocket reactor fuels barely survive at 3000° K, the same general temperature level achieved in chemical rockets.

To avoid this impasse, something must be done to reduce M, the molecular weight, rather than to increase T. In a chemical (combustion) rocket, M does not drop much below 18, because the most convenient oxidizers have quite heavy atoms. For example, the advanced Centaur engine burns hydrogen with oxygen to form water (H_2O) for which $M = 18$. Chemical rocket exhaust velocities are thus limited mainly by the high molecular weights of the combustion products. As long as chemical fuels must be burned with oxygen or fluorine, chemical rocket exhaust velocities cannot be greatly improved.

In a nuclear rocket, however, combustion is not required:

The nucleus fissions without chemical stimulation, and the propellant is not an engine fuel but a separate substance that is heated by the fissioning nuclei in a nuclear reactor. A nuclear rocket designer can heat up any propellant he wishes as long as it does not chemically attack his reactor fuel elements. Therein lies the secret of the nuclear rocket's high exhaust velocity—it can make use of a propellant with a low molecular weight.

Hydrogen won out over all competition, mainly because the dangerous handling problems were licked and because the nuclear rocket needed the higher exhaust velocity promised by pure hydrogen if it was to achieve its full potential.

The first operational nuclear rockets will spew hot molecular hydrogen out of their nozzles. With $M = 2$ instead of 18, as in hydrogen-oxygen chemical engines, the nuclear rocket exhaust velocity still will be more than double that of the best chemical rocket for the same temperature.

The great master of science fiction Jules Verne, in an amazing 1865 novel, *De la Terre à la Lune,* wrote that the Baltimore Gun Club fired a manned projectile to the moon from a huge cannon emplaced near Cape Canaveral, Florida. (22)

Nuclear Power and Merchant Shipping

EARLY in 1819 a small sailing ship, the *Savannah,* made the first crossing of the Atlantic with the assistance of a steam engine. Hers was a daring pioneering accomplishment sponsored by American merchants. But the shipping industries of the world were not ready for her, and the SS *Savannah* (the letters SS stand for steamship) was a commercial failure, ending her days as a simple sailing ship. Change came slowly, and not until 20 years later did the first vessel, the British ship *Sirius,* cross the ocean propelled entirely by steam. This venture pointed the way for the development of the great British steam merchant ship fleet.

Steamships cost more to build and more to operate than

sailing ships. Nevertheless in the early days of steam, merchants began to ship the best and most profitable of their cargoes in steamships. Why? Because steam offered, at a price, the reliability demanded by the booming commerce of the nineteenth century. Today a new technological pioneer, the merchant ship NS *Savannah* (the letters NS stand for nuclear ship), is demonstrating nuclear power at sea. Flying the American flag, she is an experiment: a merchant ship with nuclear engines, whose job is to demonstrate to the world's shipping industries what nuclear power can do.

Why take nuclear power from snug stationary power plants ashore and send it to sea? What advantages are expected which justify the strenuous, expensive engineering effort and the investment of public and private funds? They are four: freedom from frequent refueling, low fuel costs, higher speeds, and greater cargo capacity.

The owner of a nuclear ship can expect its fuel costs to be less than the costs of oil for a conventional ship. Since all the fuel needed for several years can be bought at one time, the owner need not consider costs of fuel in remote ports or worry about fluctuating prices.

In view of the independence of fuel supply and lower fuel costs, it is probable that a shipowner can afford to order nuclear ships with more powerful engines and sail them at higher speeds over longer routes than would be practicable for conventional ships. Some marine architects and economists believe that higher-speed express service can make new types of cargo economically feasible for movement by sea. These promised advantages could change the pattern of maritime trade by encouraging new designs for merchant ships and improved cargo handling techniques that would cut costs and increase the delivered value of perishable items.

Because the fuel in a nuclear ship is compact and long-lasting, much of the capacity normally preempted by fuel can be converted to revenue-producing cargo.

This listing of the advantages expected of nuclear energy

in the merchant fleet is frankly optimistic. Not everyone shares this optimism. Some naval architects, shipbuilders, and ship-owners are not yet convinced; others actively doubt the reality or degree of these advantages. Still others believe the costs and troubles of building and operating nuclear merchantmen over-balance possible benefits. Today shipbuilders and shipowners know precisely the costs, advantages, and disadvantages of con-ventional ships; the costs, advantages, and disadvantages of nuclear shipping are still conjectural and will remain so until enough nuclear ships are at sea to provide adequate informa-tion.

On land a nuclear power station faces few dangers from nature. Putting a nuclear power plant into a ship and sending it to sea exposes it to a vastly different, sometimes hostile environ-ment. Even in calm seas a nuclear ship faces the ancient mari-time hazards of grounding, collision, and sinking. Additionally, a nuclear ship must be so designed and built that passengers and crew are protected from uncontrolled release of radiation or radioactive materials if an accident should occur.

Since a nuclear power plant generates radiation and radio-active materials, a nuclear ship's passengers and crew and the people living and working near harbors and along nearby coasts also must be protected.

With the NS *Savannah* a reality, shipbuilders and ship-owners in many countries are giving serious thought to nu-clear power. To date, however, the only nations to build nu-clear powered ships for civilian purposes are the United States and the Soviet Union.

Much of the Soviet Union lies in or near Arctic regions. The economic development of the Soviet Union's northern regions makes the use of powerful icebreakers most desirable to clear the way for convoys along a northern sea route and to extend this route deeper into the Arctic. Ice breaking is heavy work. Conventional icebreakers need frequent refueling and therefore cannot navigate the whole Arctic Basin. Fuel limita-tions restrict them to a narrow coastal strip where ice conditions are generally severe. To overcome these adversities, the Soviet Government designed and built a nuclear-powered icebreaker, the *Lenin,* which joined the Soviet Arctic fleet in December 1959. By the end of 1963 the *Lenin* had sailed about 60,000 miles, 40,000 in ice. If atomic icebreakers can open the Arctic Ocean for shipping throughout the year, convoys of ships then could sail between Murmansk and Vladivostok across the top of Siberia, a distance of 5805 miles. By way of the Suez Canal, it is almost 13,000 miles.

The Arctic Ocean remains one of the last unknown areas. Even a few years ago an underwater trade route through the Arctic would have been considered only a subject of science fiction. Today it is a distinct possibility. The navy's nuclear submarines *Nautilus, Skate,* and *Seadragon* have explored that ocean and have sounded out its passages. On historic cruises beneath the ice, they found that an underwater Northwest Passage does exist. Although the ice above may bar the routine passage of surface ships, beneath this barrier there is safe passage for nuclear submarines. (19)

Our atomic powered submarines are capable of cruising several times around the world on a single fuel loading. Even Jules Verne, who had the imagination to write *20,000 Leagues Under the Sea* a century ago, did not foresee so concentrated an energy source. (This novel deals with the voyage of the *Nautilus,* a futuristic craft powered by electrochemical means. The atomic-powered USS *Nautilus* is its namesake.) (2)

Commercial submarines traveling Arctic sea-lanes would greatly reduce the trading distances of the world. The present route from Tokyo to London is 11,200 nautical miles; by polar passage it shrinks to 6500 miles. Likewise, a ship bound for Oslo, Norway, from Seattle via the Panama Canal sails 9300 miles, while a submarine navigating the Arctic shortcut would travel only 6100 miles.

No one can say how soon commercial nuclear submarines will sail from Alaska across the North Pole. But that day doubtless will come. Until then, more exploration is needed to see how ice conditions change with the seasons. When the first commercial cargo submarine does make its first voyage across the top of the world, it will depend on nuclear energy to run its engines. (19)

21 ◊ Plowshare

THE peaceful uses of chemical high explosives are many. They helped dig the Panama, Erie, and Corinth canals. Without them modern mining would be impossible. Explosives start reluctant oil wells flowing, blast menacing rocks from ship channels, remove tree stumps, and dig ditches. They make possible many of the feats of modern construction. They have found their way into production processes—such as the explosive forming of metals—and are employed as space-age laboratory tools—such as shock tubes. Highly specialized explosions power our cars and aircraft. Small charges are used in rockets and flares that help rescuers find lost seafarers. Others give us holiday fireworks. Still others reveal secrets of the earth's inner structure, or effect the stage separations of spacecraft.

Although it took nearly 400 years for the chemical black powder explosives to be adapted from military purposes to mining, the imaginative mind of man is already defining—with the atomic age roughly three decades old—an array of peaceful wonders and benefits from nuclear explosives. Uses are as varied as those of chemical high explosives. Even as nuclear energy has brought a new dimension to the term explosion, so man has begun to think in terms of heretofore impossible things he can now do with explosions. Not only is he thinking, he is doing—a nuclear explosion technology for peaceful applications is being developed.

Geographical engineering describes the use of nuclear explosives to change the geography of our planet—digging sea-level canals between oceans, stripping overburdens (the waste rock lying between the surface and a mineral deposit) from deep mineral deposits, cutting highway and railway passes through mountains, creating harbors and lakes where none existed before, and altering watersheds for better distribution of water resources.

Nor do proposals for peaceful uses of nuclear explosives stop with large-scale earth-moving. Also envisioned are constructing underground reservoirs, increasing gas-well productivity, and controlling subterranean water movement. Eventually, the energy from nuclear explosives may even be used for underground desalting of seawater, for producing steam, and for creating basic industrial chemicals directly from mineral deposits.

In mining, nuclear explosives might be used to break up ore bodies to obtain minerals whose recovery is not now economically feasible. The shattering effect and heat of nuclear explosions may one day enable recovery of vast oil reserves from sand and shale formations that are now uneconomical to exploit.

As a tool of research, a nuclear explosion is many things—the most intense source of high-energy neutrons available, a made-to-order seismic signal, a package of extreme pressures and temperatures, and a means of producing transplutonium elements. Nuclear explosion research possibilities range from studies of the inner structure of the earth to studies of the basic structure of matter.

Possible peaceful uses for nuclear explosives are many and varied. Some have been appraised as feasible, both economically and technically, for use in the immediate future; others have been set aside for consideration until the distant future. Almost all require further evaluation and testing, and a program to provide the needed studies and experiments is under way. Plowshare is the name given to this endeavor.

One of the first proposals for using nuclear explosives for peaceful purposes came from the famous Hungarian-

American mathematician John von Neumann (1903–57) in the late 1940s. (25)

(For a small country, Hungary has produced an astonishing number of great scientists. Besides von Neumann, who is credited among other things with the development of much of the logic of modern computers, Hungarian-born scientists whose names appear in these pages or in the newspapers include Georg von Hevesy, Eugene Wigner, Leo Szilard, and Edward Teller.)

Earth could be excavated for a few cents per cubic yard in some projects in which conventional methods would cost from 20 cents to $5 per cubic yard.

In terms of economics, if the return does not justify the investment, nuclear explosives will find few users. The economy of nuclear explosives for large excavation projects has been demonstrated, and has been predicted by calculations for a number of other purposes. A planned series of cratering and cavity-making explosions in different rock and earth media is designed to enable engineers to construct—with precision and safety—harbors, dams, underground reservoirs, and mountain passes.

A sea-level canal across the Central American isthmus has been a dream since Balboa first saw the Pacific Ocean. In fact the present Panama Canal was originally begun by a French Company as a sea-level canal, but the overwhelming amount of rock to be excavated forced the company to redesign it as a lock canal.

A 1960 Panama Canal Company report indicated that digging a sea-level canal with nuclear explosives would be feasible and safe. Furthermore, to dig such a canal by nuclear excavation methods would cost only a fraction of the expense of conventional excavation.

Petroleum recovery is another activity that has been subjected to considerable study, and mining of mineral deposits may be made easier by using nuclear explosives.

Gas fields exist from which little or no gas can be produced due to the low permeability of the host rock. Preliminary studies indicate that increased production can be achieved in reluctant oil and gas fields by using nuclear explosives as fracturing tools.

Hydroelectric power development in the desert of North Africa awaits only the introduction of water from the Mediterranean Sea, no more than 35 miles away, into two below-sea-level depressions. One is the 8000-square-mile Qattara Depression in Egypt's western desert, which is as much as 400 feet below sea level. The other is the 50,000-square-mile Chotts Depression, starting just 20 miles from Tunisia's coast. Studies have been made of the possibility of connecting these depressions to the sea by canals so that large hydroelectric plants could be powered by the flow of salt water into the depressions to form shallow new inland seas. It is predicted that natural evaporation from the new seas would reduce their level rapidly enough to assure a continuous inflow from the sea for many years. It is also believed that canals into these depressions might open up vast, now unusable, areas to commerce and induce human migration to the vicinity. Nuclear explosives might make building these canals feasible.

New harbors, particularly in such areas as the west coasts of Africa, Australia, and South America, would greatly assist economic development of these regions. These coasts adjoin areas of extensive mineral resources and some of the world's most fertile fishing grounds. Well-placed harbors can open these regions to development, but in some cases only nuclear explosives are powerful enough to do the required work.

In the control and conservation of water supplies, nuclear explosives have been suggested to alter watersheds, interconnect aquifers (water-carrying underground rock formations), create or eliminate connections between surface and underground water supplies, and—where evaporation loss is high—create underground reservoirs. One of the most promising suggestions is the use of nuclear explosives to connect the surface with existing potential aquifers. This would be especially important in the arid regions where infrequent torrential downpours punctuate long, dry periods. In such areas, unless there is a way to impound rainwater quickly—preferably underground to minimize loss through evaporation—it is lost.

Other proposals to develop natural resources include the use of explosives to bring down canyon walls to form dams, or to aid in releasing natural geothermal heat to produce steam for desalting seawater or for electric power. Synthesis of chemicals in the ground also has been proposed. For example, calcium carbide might be produced from an explosion in a formation of coal and limestone; then by adding water, acetylene gas could be made.

For scientists, a nuclear explosion provides an intense source of many things needed in research: high pressure, high temperature, fundamental particles such as neutrons and neutrinos, and most forms of electromagnetic radiation, such as gamma rays.

New elements have been created in nuclear explosions that do not occur naturally on earth. Einsteinium, element 99, and fermium, element 100, were first identified in the products of a thermonuclear explosion. Since then, several nuclear explosions designed specifically to produce heavy elements have been conducted. Analysis of the data shows that many transplutonium isotopes were created.

This method of making transplutonium elements involves exposing heavy-element target atoms to the intense neutron flux produced by the explosion. The resulting instantaneous capture of many neutrons in each of the nuclei of the target atoms creates unstable neutron-rich isotopes of the target material. These isotopes then undergo beta decay, in which electrons are ejected from the nuclei. Isotopes with higher atomic numbers and masses greater than that of target element result.

Under the extremely high pressures achieved in a nuclear explosion, the electron shells of the atomic structure are deformed, and matter acts in strange ways. For example, at 10 million times our atmospheric pressure carbon can be compressed into a state denser than diamonds, and iodine, usually a nonconductor, becomes an electrical conductor.

As a source of neutrons, a nuclear explosion makes it possible to improve measurements of neutron capture and fission excitation values for numerous elements, and offers a possible means of measuring these values for highly radioactive isotopes that cannot be measured in the laboratory. (25)

The fact that nuclear power has been abused by the major political powers, which use it as a threat, is no reason for not using it for peaceful purposes.

The imagination and effort devoted to the Plowshare program must be great and relentless. For at stake is a source of tremendous energy, capable of doing great good for mankind. Surely as man discovered means to free nuclear energy, he is capable of finding ways to use it for his benefit. (25)

22 ◊ *Nuclear Energy for Desalting*

A tremendous volume of water—324 million cubic miles of it—covers three-quarters of the globe. As drinking water, it's useless to us in its present state because it contains 3½% salt, and man can only tolerate 0.2%. Beyond that, the salt burden is more than human kidneys can secrete, and the body becomes dehydrated in its efforts to rid itself of the excess. The purer the water the better, although our taste buds do not agree: absolutely pure water is tasteless, and most people prefer to drink water with some slight mineral content.

Could the burgeoning technologies of nuclear power and water desalting be joined to help solve civilization's immediate and projected needs for potable water? The idea was not entirely new.

In 1959 plans for a small seawater distillation plant utilizing a nuclear heat source were made. The plan called for the use of an experimental process heat reactor coupled to a million-gallon-per-day demonstration seawater distillation facility.

Unfortunately, it was difficult to find the proper site for the reactor portion of the plant. However, a highly successful distillation plant was built in San Diego, California. This is the same plant that in 1964 was dismantled and reassembled at Guantanamo Bay, Cuba.

The uprooted San Diego unit was one of the plants built by the Office of Saline Water to demonstrate the different promising processes for water conversion. The plant was producing 1.4 million gallons of water per day at approximately $1 per thousand gallons, which is about three times the cost usually regarded as economic. Demonstration plants, however, are not expected to achieve the economy and efficiency possible in second- and third-generation operating facilities. The San Diego plant was successful since it did acceptably demonstrate a promising technique, and is still doing so in Cuba.

With the improvement of nuclear reactors as sources for the production of electrical energy, and their increasing acceptance by utilities and the public, interest was again stimulated in the possibility of dual-purpose nuclear-power and water-production plants.

All processes for making potable water from saline water consume energy, and it doesn't matter whether that energy comes from a fossil-fuel-burning plant or one that uses nuclear fuel. The choice of a heat source can therefore be based upon economic considerations, and it has been found that nuclear power is the more economical alternative in large sizes.

Currently the most promising process for large-scale production of fresh water—50 million gallons per day or over—is distillation. This system requires quantities of low-temperature steam, and one of the places that steam of this temperature can be found is in the turbines of modern thermoelectric power plants.

In these plants high-temperature steam is produced to turn the large turbogenerators that generate electrical power. As the steam passes through the turbine its temperature decreases until a point is reached where it is low enough so that it might be used for saline water distillation. Thus, the steam used to generate electricity could also be used to provide energy to the water plant.

Plant economics are enhanced by using the higher temperature energy for power generation and the lower temperature exhaust steam for desalting. By combining the two processes (power generation and desalting) in one plant, larger heat sources can be used, requiring lower unit capital investments. In dual-purpose plants important items of equipment and facilities (such as water intake and discharge lines, control rooms, maintenance shops, etc.) could be shared to provide significant cost savings.

The world's gathering water-shortage crisis has many causes. In only a few specific places can natural aridity be blamed. In areas well endowed with water resources man has allowed his rivers and lakes to become polluted with industrial and municipal wastes. He has not prudently guarded his water fortune but has spent it with lavish recklessness, and now finds himself impoverished.

Desalting of water is not a panacea for all mankind's water needs. Every sound approach, every technologically feasible means of increasing the supply of potable water will be needed to meet the staggering demands of the future. Nuclear powered desalting units will be prominent among the methods by which we will obtain large amounts of additional fresh water. Fortunately for the world's thirst, nuclear power for desalting will be available soon, rather than late. (18)

23 ◈ *Food Preservation by Irradiation*

ON February 8, 1963, the Food and Drug Administration of the U.S. Department of Health, Education, and Welfare ruled that bacon preserved by radiation is safe and fit for unlimited human consumption. With that ruling a new chapter may have begun in man's agelong battle to preserve his food from spoilage.

The FDA has also approved the use of radiation for sprout inhibition of white potatoes (see photograph section) and disinfestation of wheat and wheat flour in this country.

Losses from insect damage to wheat run to millions of dollars annually. Chemical pesticides can control adult insects, but their eggs remain in the wheat and hatch when the temperature rises. Exporters believe radiation is a most effective way of destroying the eggs without harming the grain.

Interest in radiation processing of food is worldwide. Up to 30% of food harvests are lost in some parts of the world because of animal pests and microorganisms.

Before man knew what caused his food to spoil, he was busy trying to prevent it. Over the centuries he has devised many plans for keeping edible food on hand—drying, salting and smoking, dry and cool storage, fermentation and pickling, canning, refrigeration and freezing, making sugar concentrates (like jam and jelly), and using chemical preservatives.

Even cooking and use of spices are short-term ways of

preserving food. Many of our basic foodstuffs originated in attempts to keep food. Butter and cheese, for instance, are ways of extending the life of milk.

The use of radiation is the most recent step in this activity.

Drying is an ancient method (still widely used) of preserving food. The sun, which sustains man in so many other ways, dries cereal grains and some vegetables before they are harvested. This natural process is so efficient that no help from man is needed. Even prehistoric farmers knew, however, that rain or cloudy periods would rot these staples in the fields; so they brought some of their crops into their caves to dry.

As the world's population grew, man could not depend solely upon sun-dried food for all his needs. Nonetheless, until the late eighteenth century attempts to assist nature were relatively simple and consisted of placing foods near the fire, spreading them more thinly to dry in the sunlight, and sheltering them from rain. About 1795, however, two Frenchmen, Masson and Challet, made an important improvement. They built a dehydrator in which air at about 105° F was blown over thinly sliced vegetables. Since that time more efficient dehydrators with huge capacities have been developed.

Chemical and biological forces are at work in all stages of food growth. After foods are harvested, these processes tend to cause deterioration rather than produce maturation. The chemical forces can be controlled by chemical additives, and biological activity can be stilled by blanching (quick exposure to moist heat) or drying.

Although most food preservation systems aim at destroying microorganisms, or at least inhibiting their growth, a few foods are formed by action of the tiny organisms. One of these processes is winemaking, which is probably as ancient as the caveman and certainly has been familiar in all known civilizations.

A yeast, *Saccharomyces ellipsoideus,* is the agent that creates alcohol from sugar. Usually this is present on grape skins; when yeast cells have access to sugar in the juice, through bruises or breaks in the skin, fermentation takes place. The process of fermentation is one of decomposition of carbohydrates; it is easily distinguished from putrefaction, which results from the action of microorganisms on protein materials. In modern winemaking man crushes the grapes purposely, heats the juice to kill contaminating organisms, and reinoculates it with pure wine yeast to control the fermentation process.

Fermentation also is involved in preparation of beer, cottage cheese, buttermilk, cheese, sauerkraut, bread, and many other foods.

The French made another major contribution to food preservation. All eighteenth-century soldiers were hampered by diets of putrid meat and other inferior foods. So in the 1790s Napoleon offered a prize of 12,000 francs for the invention of a method of food preservation for his fighting forces. A confectioner named Nicolas Appert observed that food cooked in sealed bottles remained unspoiled as long as the container didn't break. He didn't know why this happened (nor did any scientist of the day), but the process worked. Appert won the prize in 1809, and canning came into use. (11)

By 1820 the first commercial canning plants were operating in the United States and by 1840 canning was common throughout the country. But the mechanism of food spoilage and the reasons why canning prevented it were still mysteries. In 1864 Louis Pasteur reported to the French Academy of Sciences that the cause of a disease then ravaging the beer and wine industries of France was a microscopic vegetation and that, when the wine contained no living microorganisms, it remained unspoiled. Wine or beer heated to 135° F and sealed in jars did not sour. Here was the explanation of Appert's success of 55 years earlier. Unlike dehydration, canning is not an improvement on a natural process, but represents man's first attempt to control Nature for the preservation of food.

Freezing, a more recent technique in food preservation, had scattered users and advocates long before its commercial exploitation. Patents for freezing fish, for instance, were issued to H. Benjamin in England in 1842. Cooling with natural ice or in caves had been known for centuries. Mechanical refrigeration, developed in the late nineteenth century, allows foods to be stored at temperatures of about 40° F and remains an important way to preserve raw foods for a limited time.

Because of their high water content, most foods freeze solidly at temperatures between 25° and 32° F. If freezing is rapid, many small ice crystals are formed, and the changes in the food tissues are reversible. Slow freezing, according to a theory of ice crystal damage, permits large, uneven crystals to build up. These puncture the food cells as they grow, and, when the food is thawed, it will be mushy.

Although the factors in successfully freezing food are well under control, the complex physical, chemical, and biological changes involved are still not completely understood. In freezing, as in canning, Nature is controlled, even though man is not always quite sure why. (11)

Recently, interest has been expressed in freezing people—for defrosting at some future time, when cures for their illnesses have been found.

For maximum nutrition, food plants and animals should be eaten as near to the harvest or slaughter as possible. Once a plant or animal is dead, decomposition begins and nutrients are lost.

Food spoils because of physical, chemical, or biological deterioration and by the activities of microorganisms and insects. Microorganisms specialize in the kind of organic matter they decompose; only a few consume more than one kind of food. The result is, however, pretty much the same: All living tissues ultimately decay to the minerals, water, carbon dioxide, and ammonia from which their complex organic molecules are made.

Although people around the world do not agree on what constitutes spoiled food—consider the variety of tastes in cheese, for instance—food that is truly spoiled makes anyone who eats it sick. Most cases of food poisoning come from intestinal infections caused by bacteria.

The most common type of bacterial food poisoning is caused by *Staphylococcus aureus,* a common bacterium found on the skin and in the nose and throat. This kind of illness typically begins suddenly. Victims usually recover after several days' illness. It can be prevented by cleanliness, proper cooking, and refrigeration of foods.

Another kind of food-borne illness, trichinosis, can come from the parasite *Trichinella spiralis,* which is often found in pork products. Trichinae, or larvae of the parasite, are destroyed by cooking pork well or keeping it below freezing temperature for 20 days. This infestation is common; often it is not serious enough to cause sickness, but it can be dangerous.

A most serious but quite rare form of food poisoning, fatal to 70% of its victims, is botulism. It comes from a poison formed in food by the bacterium *Clostridium botulinum.* This organism lives in the soil and is not itself harmful to man, but, when it multiplies in food under anaerobic conditions (without air), it produces a potent toxin. If that toxin is consumed, even in minute amounts, botulism will result. The only known treatment for botulism is an antitoxin, but frequently the disease is not diagnosed soon enough for antitoxin to be obtained, and the patient dies.

Botulism is more common in home-canned foods than in commercial products. Prevention of botulinum toxin formation is important in all food preservation.

Except for canning, radiation processing is the only original—that is, nonnatural—method of preserving food developed since the dawn of history. Quickly, economically, and safely, without raising internal temperatures more than a few degrees, ionizing radiation can preserve foods by inhibiting or destroying bacteria and other microorganisms. Radiation speeding through food ionizes some atoms in its path and causes an alteration of vital macromolecules in bacteria and other microorganisms and these are destroyed.

If food atoms are ionized, they suffer no harmful effects. The food does not become radioactive, and with low doses of radiation there is less loss of vitamins than in canning, freezing, or drying. Some vitamins are lost if higher radiation doses are applied, but they can be replaced, as they sometimes are in other processed foods.

Radiation preservation of food is accomplished in two ways: pasteurization, which is accomplished with low dosages, and sterilization, which requires higher levels. Food can be irradiated by bombarding it with beta particles or with gamma rays.

The amount of radiation to be delivered depends upon the food itself and the result desired. If the goal is prolongation of shelf life, or storage time, a pasteurization dose, generally 200,000–500,000 rads, is sufficient. If the aim is to sterilize food for long-term storage without refrigeration, the required dose is 2 million–4.5 million rads.

At even lower doses, radiation can perform effective preservation chores. Applied to potatoes or onions it is highly effective as a sprout inhibitor. Grains and cereals can be disinfested of insects. It is also possible to sterilize the larvae of insects that lodge inside fruits.

Since the same radiation that will kill a bacterial cell also will destroy some of the cells in the tissues of the food product, extensive testing has been undertaken to learn the effects of radiation on food wholesomeness.

Organoleptic or acceptability problems—alterations in color, taste, or odor—that plagued early radiation processing attempts, have been solved for the most part by more recent research. Meats, in particular, had suffered alteration in odor, flavor, texture, and color. These changes have been eliminated or substantially reduced by irradiation at very low temperatures, application of adsorbents as odor scavengers, skillful use of spices and condiments, and appropriate cooking practices.

It has been found that if irradiation is done at ultralow temperatures, from −32° to −78° F, off flavors are markedly reduced. Chicken, precooked and then irradiated at −78° F, has been evaluated by a trained taste panel as having little if any detectable irradiation flavor.

Sterilized cooked hamburger held at room temperature for over a year does not appear to differ from fresh-cooked hamburger.

Among foods which could benefit most from low-dose radiation, fish, a highly perishable product, is one of the most important commercially.

The marketing of irradiated fish fillets would bring about a radical departure from existing fresh fish distribution practices. At present most of the fresh fish landed in Boston, for example, is sold in the fresh state within a 200-mile radius of that city. Although some fresh fish is sold in all cities, most inland consumers buy frozen fish because it offers more variety, better quality, and often lower price. A high percentage of fresh fish spoils during shipment.

Even in coastal cities, where the demand now is fairly stable, fresh fish prices vary sharply with the supply: If fish are not available, prices are high; if supplies are plentiful, prices plummet because of the highly perishable nature of fish.

The effect of radiation pasteurization on the demand-supply picture would be dramatic. Irradiators might be placed on a mother ship to provide a light dose to whole fish soon after the catch and reduce bacterial populations; another pasteurizing dose could be applied later, after processing and packaging had been completed at shore plants. Fishermen could operate at greater distances from home ports and stay out for longer periods.

The longer keeping time of 21–30 days would be sufficient so that gluts and shortages would nearly cancel out. Movement of supplies could be planned to handle peak demands, such as those on special holidays. Surpluses could still be frozen, and untreated fresh fillets could be distributed locally.

Radiation pasteurization, by providing better quality food at a favorable price, could open a new market for fresh fish in the high-population areas of the Midwest.

Fruit in its raw state is more widely desired and enjoyed

than any other food, yet it has been estimated that one-fifth of the fruits and vegetables grown in the United States is never consumed because of postharvest deterioration.

Postharvest deterioration stems from a variety of causes, including disease organisms, overripeness, chilling injury, sprouting, and undesirable dehydration. Chemical fungicides and cool storage are used to control these hazards. Radiation pasteurization promises to keep these commodities edible and acceptable longer.

Research into the utilization of radiation for food preservation has had several beneficial side effects.

Of great importance has been the discovery that radiation is an effective tool in the control of *Salmonella* foodborne illnesses, high on the list of public-health problems both here and abroad.

Radiation may also permit elimination of some chemical sprays now in use, such as biphenyl, a preservative used on oranges, which has some undesirable characteristics.

Another by-product of food irradiation research is new general knowledge of nutrition. In evaluating studies of long-term animal feeding experiments, scientists have profoundly increased our knowledge of the interaction of vitamins, the destruction of vitamins and the other trace elements in food, the destruction of proteins, and the multitude of other changes which take place in food through cooking, canning, and freezing. (11)

Although we do not normally associate France with milk, we are aware that pasteurization is a French invention, named after Pasteur, and that it originated with wine, which we *do* associate with France. That canning is French comes as a shock, canning not usually being associated with *haute cuisine.* Those gourmets who view irradiation with a botulated eye should be heartened, therefore; if Americans cannot manage to retain the good taste of irradiated food, surely we can depend upon the French to learn how.

24 ◊ *Nondestructive Testing*

DROP a coin in the slot, pull the plunger, and reach down for your candy bar, cigarettes, or bottled drink.

Unless you're in the business of making vending machines—or slugs—you probably think little at such a time about the machine or the coin you drop into it. You may be surprised to learn that the machine subjects your coin to a great many tests, to assure its genuineness, before the product you bought is delivered.

The coin is tested for size, shape, and magnetic properties (and, in some machines, for its weight and elastic properties), all in the few seconds between the time you insert it and the time your purchase pops out.

These tests must be made quickly and in such a manner that the coin is still useful when the tests are finished. In this respect, the tests have much in common with many nondestructive tests made in industry.

Many people might think of X radiography—the inspection of an object by obtaining a visible image of the X radiation transmitted through the object—as a typical nondestructive test. For most cases this is certainly a valid assumption: X radiography is a very widely used method of nondestructive testing. However, if we were to inspect a box of photographic film by this method, the test probably would not be nondestructive.

The purpose of the examination may be to detect internal or external flaws, to measure thickness, to determine material structure or composition, or to measure or detect any of the object's properties. The test method may be a simple visual one, or it may involve some form of electromagnetic energy other than visible light, such as X rays, infrared rays, or microwaves.

The purposes behind these tests are as varied as the tests themselves. Aircraft maintenance workers perform periodic nondestructive tests to find out if components of an aircraft are still capable of functioning without the likelihood of failure in flight. A tomato-squeezing shopper makes her test to help her decide if the tomato is ready for use. She may be compared with the inspector who drives a just-finished automobile off the end of the assembly line to determine whether or not the car is ready for the consumer.

Actually we find nondestructive testing entering the production process well ahead of this last step and continuing even after the product leaves the factory. It is used to detect faulty material before it is formed or machined into component parts, to detect faulty components before assembly into the product, to measure the thickness of metal or other material, to determine the level of liquid or solid contents in opaque containers (from huge gasoline storage drums to beer cans), to identify and sort materials, and to discover defects that may have developed while the material was being processed or used. In addition, manufacturers use nondestructive testing to improve and control manufacturing processes.

X radiation was discovered in 1895, and medical use of X radiation was initiated in many parts of the world soon afterward. However, industrial use of nondestructive testing lagged behind. Although this seems surprising on first consideration, there were valid reasons for the delay. Nondestructive testing was not urgent because of the large safety factors which typically were engineered into almost every product. Service failures did take place, of course; and railroad axles and crankshafts for rotating equipment, for example, did break in service despite large safety factors. However, the role of material imperfections in such failures was not then fully recognized, and therefore little concentrated effort was made to find them.

During the 1940s, the significance of imperfections on the useful life of a product came to be more fully appreciated. In aircraft design, in nuclear technology, and, more recently, in space exploration, the factors of high hazards and costs have made maximum reliability vital.

There are five basic elements in any nondestructive test.

First, a source is needed which provides some probing medium—a medium that can be used to inspect the item under test.

Second, this probing medium must change as a result of discontinuities or variations within the object being tested.

Third, a detector capable of detecting the changes in the probing medium is required.

Fourth, a means of indicating or recording the signals from the detector is necessary.

Fifth, some method of interpreting these indications must be provided.

Let us now see how these apply in the case of X radiography. The source of the probing medium is the X-ray generator from which the radiation is emitted. The rays are modified as they are transmitted through the test specimen. The detector in this process normally is X-ray-sensitive photographic film. Darkening of the film provides an indication and a record of the test, and the interpretation usually is provided by a human observer.

X-ray machines also come in a variety of sizes and shapes. The smaller machines, such as the one your dentist uses, usually operate at less than 100,000 volts. In industry, machines of this voltage are used to inspect the smaller thicknesses of light materials such as aluminum. To inspect heavier materials, such as steel, larger machines operating at higher voltages and producing X rays of shorter wavelength are necessary. Typical industrial machines operate at up to 300,000 volts and are used to inspect steel up to about 3 inches thick. (Even larger machines are available and are gradually becoming more common.) Accelerators and betatrons which can accelerate electrons to energies of 15–20 million volts are X-radiographic sources which are useful for inspecting steel many inches thick or other heavy materials.

The method typically used to make an X radiograph is illustrated here. Notice how the X rays diverge from the target of the X-ray tube and cast a shadow of anything in their path. The flaw in the specimen illustrated might be a blowhole in a casting or a light metal impurity in some heavy material.

The radiation entering the specimen is absorbed less by the flaw than by the rest of the material, and therefore more radiation passes through to strike the film beyond the flaw than beyond the rest of the specimen. This would show up on the developed film as a dark area, revealing the presence, location, and general shape of the flaw. Just as we can see a flaw by a darkening of the X-ray film, we can also see changes in thickness or material. On the radiographs the dentist makes of your teeth, for example, the gums appear much darker than the teeth because the lighter materials in the soft tissue are less absorbing for the X-ray beam. Cavities appear dark within the shadow of the tooth because there is less material to absorb the radiation. Fillings, which contain heavy materials such as silver,

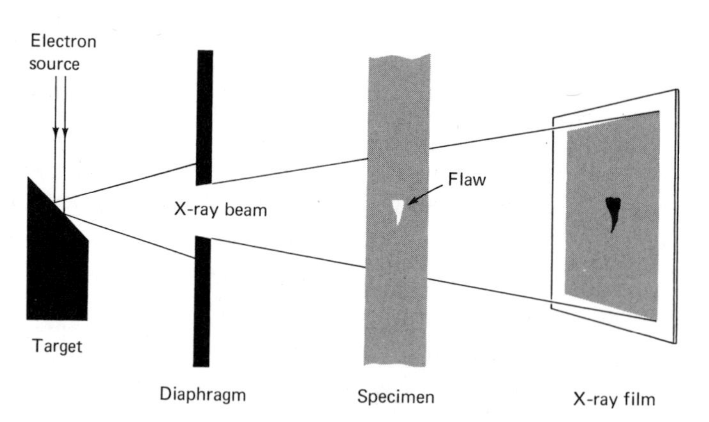

Fundamentals of an X-ray exposure. The target, or anode, of the X-ray tube is bombarded by the electron beam in the tube and is the source of the X radiation. The X-ray intensity is decreased as the radiation passes through the thickness of the specimen.

are very absorbing and appear as light areas (see photograph section).

Another imaging technique is fluoroscopy. In this technique the arrangement is similar to that for film radiography except that the film is replaced by a phosphor screen. The phosphor, when stimulated by X rays, emits visible radiation (light) of an intensity proportional to the X-ray intensity. Therefore the picture is like that obtained with film methods, except that the dark and the light areas are reversed.

Two related advantages of fluoroscopy are that there is no waiting for a film to be developed and that the object, therefore, can be moving while it is being inspected. A fluoroscope can be used to inspect the contents of a suitcase. (16)

A practical matter, these hijacking days.

Neutron Activation Analysis

A criminal goes to jail because he carried a minute speck of evidence away from the scene of a crime. A meteorite is analyzed to billionths of a gram to reveal traces of elements it brings from space. Testing of contraband opium reveals its geographical source.

What do these dissimilar events have in common? All involve identification of materials by activation analysis, a sensitive, versatile analytical tool employing nuclear energy.

In activation analysis a sample of an unknown material is first irradiated (activated) with nuclear particles. In practice these nuclear particles are almost always neutrons. The irradiated atoms are made radioactive by the neutrons. They disintegrate with the emission of gamma rays. The gamma rays are next counted (analyzed), a process which reveals the half-lives of the radioactive nuclei and also their gamma-ray energies. These nuclear fingerprints can then be located in specially prepared tables of data to identify the artificially created nuclei and by inference the elements in the original nonradioactive material as well (see diagram, page 366).

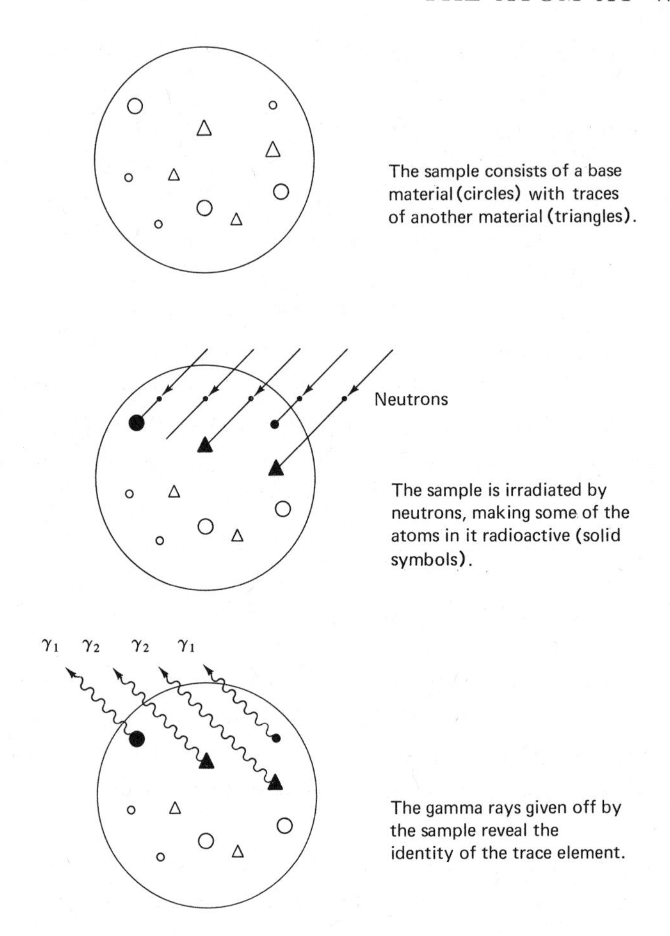

How activation analysis works. Traces of various elements can be identified and measured by analyzing the gamma rays they give off after being irradiated with neutrons or other nuclear particles.

The value of activation analysis as a research tool was recognized almost immediately upon the discovery of artificial radioactivity by Frédéric and Irène Joliot-Curie, in 1933. The first activation analysis experiment was carried out in 1936 by Georg von Hevesy (one of the first users of radioactive tracers) and Hilde Levi in Copenhagen when they bombarded impure yttrium with neutrons to activate and measure the contaminant, a small quantity of dysprosium. From this point, activation analysis caught on rather slowly, awaiting developments in the two basic components in any activation analysis facility, namely, the radiation source and the gamma-counting equipment.

Later, nuclear reactors with their copious neutrons became available. In addition, substantial strides have been made recently in the manufacture of other effective neutron sources, costing far less than reactors. Coupled to these advances has been the wizardry of modern electronics which has given us gamma detection and measuring equipment with a sensitivity unheard of at the time of the Joliot-Curies.

Activation analysis is now usefully engaged in such diverse areas as:

PETROLEUM	Analyzing oil refinery feeds for vanadium
AGRICULTURE	Detecting pesticide residues on crops
ELECTRONICS	Measuring impurities in silicon semiconductors
ASTRONAUTICS	Possible determination of the compositions of planetary surfaces
METALLURGY	Measuring trace impurities
CRIMINOLOGY	Identifying gunpowder residues
GEOLOGY	Analyzing minerals
MEDICINE	Tracing metals in metabolism

Its major disadvantage is the fact that activation analysis costs may be a factor of two or more higher than those of more conventional techniques if high-flux reactors have to be used. In addition, activation analysis cannot discern combinations of atoms (chemical compounds), which various methods of chemical analysis are able to do.

The uncanny ability of activation analysis to detect and identify very tiny amounts of certain elements has come to the aid of law-enforcement officers in a variety of ways.

Human hair, for example, contains small traces of metallic elements like sodium, gold, and copper. Activation analysis has shown that the quantities of these elements present in each hair are relatively constant for an individual but vary from person to person. A recent murder conviction in Canada was based partly on the fact that a hair found in the hand of the murder victim matched the suspect's hair when the nuclear fingerprints were compared. The French courts also have admitted activation analysis data as evidence in criminal cases.

The fact that a person died from poisoning can sometimes be determined through activation analysis. Even nonlethal doses of arsenic, for example, produce arsenic-rich regions in the subject's hair that gradually move from root to tip as the hair grows. Hair even hundreds of years old can be analyzed successfully for arsenic and other residues.

English scientists recently found an unusual amount of arsenic in a relic of hair from Napoleon's head. The suspicion now is that he was slowly poisoned to death.

The case of King Eric XIV of Sweden is similar. A murder legend has persisted in Sweden for four centuries. When the king's body was exhumed recently, activation analysis showed that his body contained traces of arsenic.

Activation analysis goes beyond these ghoulish activities in helping to solve crimes. Besides being able to match human hair, it can also compare infinitesimal grease spots, specks of dirt too small to be seen with the naked eye, and tiny flakes of paint from automobiles in accidents. It can either match bits of material left behind at the scene of a crime to a suspect, or it can identify minute traces of substances that he has carried away from the scene. What's more, identification can be made without damaging the specimens. Thus, they can later be admitted as evidence in court.

Activation analysis of wipings taken from a suspect's hands will reveal not only whether he has fired a gun recently but also the type of ammunition used, the number of bullets fired, and the hand in which the gun was held. This is possible because when a gun is discharged gunpowder residues spread over a wide area, including the holder of the gun. These residues contain small amounts of various metals that can be measured easily by activation analysis.

The publication of Rachel Carson's *Silent Spring* (Houghton Mifflin, 1962) focused attention on the dangers of pesticides. Activation analysis permits agricultural scientists to detect pesticides on crops and in prepared foods by monitoring traces of bromine and chlorine that are left on the foods, even after processing.

Activation analysis not only identifies dangerous chemical culprits in biology, but it can also exonerate elements under suspicion. For example, it was thought for some time that trace amounts of selenium in peas caused a muscular paralysis called lathyrism in underfed parts of the world where peas are a main source of food. It was shown through activation analysis that selenium is not guilty. The cause of lathyrism has recently been shown to be a copper deficiency induced by beta-aminopropionitrile (BAPN) present in two varieties of peas.

Another biological application, this time with underworld overtones, is the use of activation analysis to identify the geographical source of opium. This can be done by identifying the trace elements the poppy plants absorb from the different soils in which they are grown around the world.

A basic problem in cosmology is to estimate the relative abundances of the chemical elements throughout the solar system and the rest of the universe. An astronomer can tell what the incandescent stars are made of by using his spectroscope, but the composition of colder bodies in the reaches of outer space is practically unknown. Careful analysis of those extraterrestrial messengers, the meteorites, is the best source of firsthand data from outside our atmosphere.

Using activation analysis, scientists have studied the relative abundances of the rare-earth elements in a number of meteorites. The rare earths all have similar chemical properties and are supposed to have evolved as a unit throughout the history of the cosmos. Any differences observed between terrestrial and meteoric rare-earth abundances will call for a special explanation.

Only a few of the manifold possible uses of activation analysis have been mentioned. Other, perhaps less exciting, examples, such as macroanalysis, impurity control in semiconductor manufacturing, and the measurements of contaminants in chemical plants, could also have been mentioned. Moreover, this new method has just begun to prove its usefulness to science and industry, so that new applications can be expected over the next several years. (15)

25 ◊ *Radioisotopes in Industry*

ACCORDING to Lord Snow, the noted British physicist and writer, the Scientific Revolution began about 30 years ago, when atomic particles were first used in industry. Man-made radioactive elements have been serving industry on a routine basis half that long. Radioisotopes are used in industry primarily for measuring, testing, and processing. One property of radioisotopes that makes them valuable is that they are detectable in extremely small amounts. Their detectability recalls an unconfirmed story about Georg von Hevesy. Hevesy ate in a boarding house, and he began to wonder whether his landlady was making stew out of food left over from the day before. To find out, he put a small amount of a radioisotope on his uneaten food. When he examined the stew the next day with a counting instrument, he found that some of its ingredients had, indeed, come from scraps on his own plate.

Today isotopes are so common that they are familiar materials in many industries. Once in a while, however, they still appear in surprising and unexpected places. For example, a radioactive form of the noble or inert gas, krypton, can help detect impurities in the air. In one type of smog alarm, krypton is caged in a crystalline chemical. When sulfur dioxide, a notorious air pollutant, enters the alarm device, krypton is released as a result of chemical action. The gas is scanned by a Geiger counter, and the number of radioactive disintegrations recorded gives a fair idea of how much sulfur dioxide displaced the krypton. Information from a series of these devices placed strategically above a city can alert industry and government officials to order cutbacks on smoky operations, leaf burning, and other smog-causing activities.

The chances are that the paper in this book—like any paper product you pick up nowadays—was atom-inspected. For several years large paper manufacturers all over the world have been using isotope gauges to "feel" the thickness of paper while it is rolled out by machines operating at runaway speed.

Manufacturers take these pains in measurement so that their products will be uniform. Uniformity is necessary in all mass production. It is vital in precision manufacturing to assure the proper performance of such equipment as space exploration missiles, in which as many as 300,000 parts, made at hundreds of different places throughout the country, must be incorporated. With that many interdependent parts, a missile that has components even 99.9999% reliable is not good enough; statistical analysis tells us such a missile would have only a 50–50 chance of making a successful flight.

MEASUREMENTS THROUGH THE AGES

Modified from *Precision, A Measure of Progress,* General Motors Corporation, 1952.

UNIT OR METHOD	ERA	ACCURACY
Yard (king's arm)	3000 B.C.	1–2 inches
Barleycorn (length of grain—⅓ inch)	14th century	$^{1}/_{10}$ inch
Micrometer	17th century	10^{-3} inch
Johansson gauge blocks (steel standards)	late 19th century	10^{-5} inch
Interferometry (beams of light)	20th century	10^{-6} inch
Radioisotopes (radiation)	20th century	10^{-6} inch

How do you check 300,000 parts for uniformity as they are being made?

One type of isotope gauge, the thickness gauge, though as busy as a workhorse, is so inconspicuous that you could miss it in visiting a factory where it is used. It usually consists of a box containing a radioisotope (such as cesium 137), a radiation detector, and an indicator. The material being measured passes in a sheet between the radioisotope source and the detector. The action of this gauge may be compared to the action of sunshine on your eyes: The amount of light reaching your eyes depends partly on whether or not you are wearing sunglasses and, if you are, how dark the lenses are. Similarly, the amount of radiation reaching the thickness-guage detector is proportional to the thickness and density of the sheet of material that separates it from the source of radiation (see diagram). The measurements are recorded by a meter on a chart. Instruments now in commercial use detect and record variations in the thickness of paper or other sheet material rushing by at hundreds of feet a minute.

In a refinement of this instrument, an electronic signal from the detector may be fed back by a servomechanism to control the thickness of the sheet remotely by changing a roller setting. Thus the isotopic device can not only measure and monitor the thickness of material continuously but can also control it.

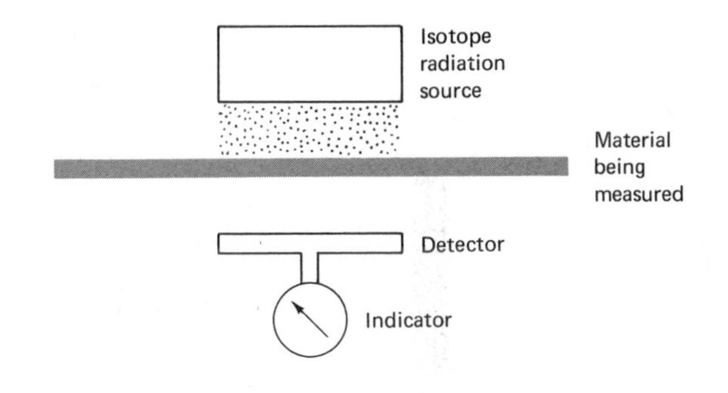

Principle of the thickness gauge

Another type of problem is the measurement of a thin layer of one substance adhering to a backing material of another. Knowing how much tin is used to coat steel sheet, for example, is vital in a tin-can factory. The manufacturer can measure the thickness of the tin with isotopes by backscatter gauging or with X-ray fluorescence.

The principle of the backscatter gauge is similar to that of the thickness gauge. Radiation is scattered when it strikes any material. The thicker the material, the more radiation is scattered, up to a point. In a backscatter gauge the detector is shielded from all but the scattered radiation; therefore the more radiation striking the detector, the thicker the material is (see diagram). As with a thickness gauge, the indicator and recorder give a running account of small thickness variations. This gauge is also used to measure materials like steel pipe wall and other products for which a two-sided thickness gauge is unsuitable.

In X-ray fluorescence the radiation excites both the coating and the backing to produce X rays that are characteristic of the elements of which these materials are formed. As the coating material is excited, the intensity of the induced radiation increases with the thickness of the coating. When the backing is excited, the induced X rays are absorbed in the coating before reaching the detector, and so the intensity of X rays decreases with thickness. In both measurements a detector picks up intensity changes, and these are recorded as thickness values.

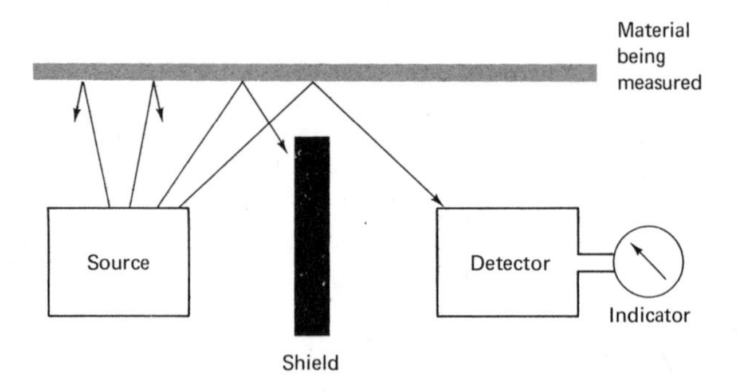

Principle of the backscatter gauge

The chief advantage of backscatter gauging is that the thickness of an object can be measured if access is possible from one side only. An advantage of X-ray fluorescence is the extreme precision of its measurements.

Another type of gauge is the level indicator. This instrument "sees" the level of a substance in a container. It is used in the beverage industry, for example, to sort out at high speed partly filled cans. These gauges are also used in chemical plants to keep inventories of solutions of corrosive materials in tanks, in the coal industry to report which ore cars in the belly of a mine are full, and in many other industries.

Level gauges (see diagram) work on the go–no-go principle, like a light switch. When there is a substance between the

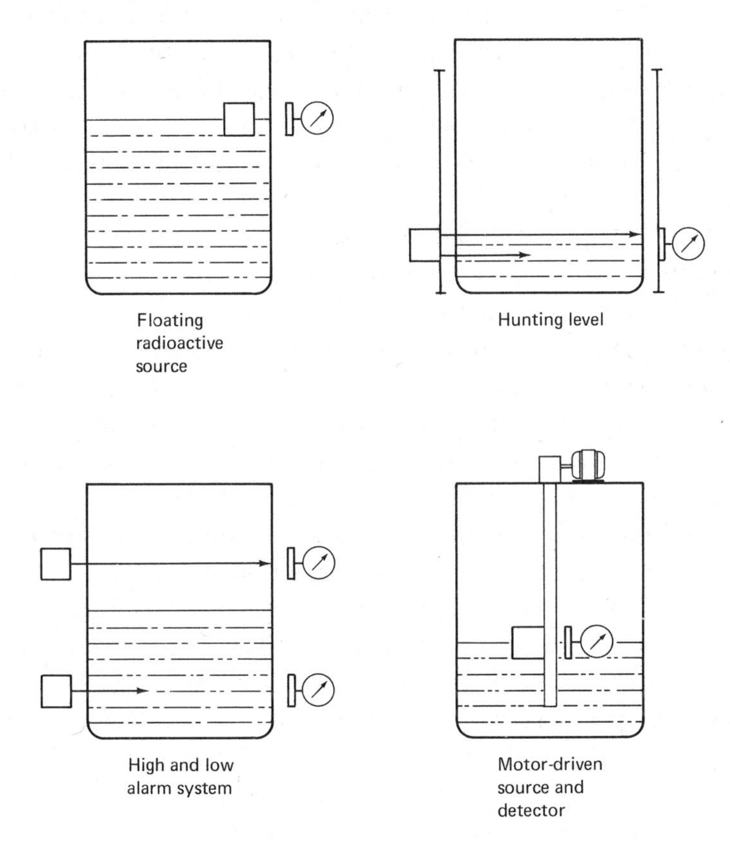

The principle of various types of level gauges

source and detector, the indicator is off; when the material drops below the level of the source, it is on. This gauge also is often used to control the rate of processes: By opening a remote valve when the indicator is off, the gauge will change the inflow of materials to a container in which a processing operation occurs.

The simplest type of level gauge has a radioactive source floating on top of the liquid or in a well surrounded by the liquid. A detector is moved up and down on the outside of the container until its maximum reading shows the level of the source.

When a source cannot be placed inside a container, source and detector can be mounted on opposite sides of the container and moved up and down together to hunt the level of the liquid. A sharp rise in radiation indicates that the rays are no longer passing through the liquid and that the surface level has been passed. Another type, the high-and-low-level alarm system, uses fixed detectors at the top and at the bottom of a tank.

More complicated devices use motor-driven servomechanisms that move the source and detector up and down the tank at the same level as the liquid.

Radioisotopes are important tools used in research to extend the life of our automobiles. The point of highest wear in any engine is where the piston ring rubs on the cylinder wall. Small iron particles are worn off the piston rings and are carried away by the lubricating oil. The amount of this wear can be measured either by weighing the piston ring before and after running the engine or by measuring the iron that accumulates in the oil. Before the isotopic method was developed, an engineer had to run an engine for about 10 days to get a detectable weight change, even with the most sensitive balance. With radioactive tracers similar measurements can be made in a matter of minutes with results that are more reliable and at least 50 times as sensitive as those from weight tests.

Isotopic tracers were first used for the measurement of piston wear in 1947. The method makes use of the fact that extremely small amounts of radioactive isotopes in the particles abraded from the pistons can be detected and measured. There are several variations of the method. In one technique whole piston rings are put into a nuclear reactor and bombarded with neutrons to convert some of the iron atoms into radioactive iron. These rings are installed in an engine. The engine is run, and the oil is pumped through a detecting instrument to count the radiations from the radioactive iron particles that have worn off and accumulated in the oil.

Many industries use radioisotopes in a few operations, and a few use them in many operations. In the petroleum industry, for instance, isotopes are found everywhere—from the oil well to the gasoline storage tank.

In petroleum exploration one method of finding oil-bearing geologic formations is to lower radioactive needles into the ground or into test wells and measure the changes in their scattered radiation as they pass through various types of rock. A geologist makes a record, or log, of the amount of radiation at several places and draws a profile of the underlying layers of earth. This profile tells him whether there is justification for drilling in the area.

At the refinery radioisotopes are used to follow the flow of material through various processes.

Finally, oil is commonly shipped long distances through pipelines. Several companies often use the same pipeline, just as cattlemen use the same range for their herds. Cattle can be distinguished by their brands, but different oil batches ordinarily look alike. However, oil can also be branded by tagging the leading edge of each batch in the pipeline with a very small amount of radioisotope. Some of the isotope precedes company A's oil, for example, as it moves along, and another tracer is carried along ahead of company B's shipment. When the oil reaches the corraling area (in this case, a series of tanks at a distribution point), a man at a valve uses a radiation monitor to detect the radioisotope and then routes A's oil into tank A and B's products into tank B. The routing also can be controlled by machine.

Many biologically important substances, such as vitamins, hormones, and enzymes, are very complex organic chemical molecules or mixtures. Workers in the pharmaceutical industry have done extensive research to find out how much of a vitamin, for example, is found in various food substances. In the past, the only way to find out was to assess the effect of the vitamin on animals. For example, if an animal is fed a diet containing no vitamin B_{12}, it will develop anemia, a deficiency disease; if it is then fed a vitamin B_{12} preparation, it will get well. The amount of B_{12} in the preparation may be estimated by observing how rapidly the animal recovers. This assay method is obviously tedious, expensive, and not very precise.

With radioactive tracers, complex substances can be analyzed by the process known as isotope dilution. An analysis of the amount of vitamin B_{12} in a sample of mixed vitamins may be made in the following manner:

1. Prepare a pure sample of vitamin B_{12} containing radioactive cobalt as a tracer. Determine the amount of radioactivity in a given weight of the sample.
2. Add the radioactive tracer-tagged B_{12} to a known amount of the vitamin mixture that you want to analyze.
3. Mix tracer and vitamins thoroughly.
4. Take a sample of the mixture.
5. Separate a small amount of the vitamin B_{12} from the mixture and purify it.
6. Determine the weight of this new sample and its radioactivity. The activity of the new sample of vitamin B_{12} per unit of weight will not be as high as that of the original sample because it has been diluted with the vitamin B_{12} in the mixture. You can then calculate how much vitamin B_{12} must have been in the original mixture to cause the change in the amount of the radioactivity of the tracer.

This procedure is exceedingly simple compared with the one using experimental animals.

Companies that make soap or detergents are interested in how effectively their products will remove dirt from soiled clothes. Although it has been difficult to make radioactive dirt for testing that is as good as the real dirt normally found on clothes, some fairly good substitutes have been developed. Radioactive dirt is washed from clothes with new soap products, and the dirty water is examined with instruments to measure the amount of radioisotope removed.

The Esso Research and Engineering Company has installed a radiation source for making detergents that can be destroyed by bacteria so that waste suds will not clog sewers, waterworks, and rivers. Chemists are studying ways of producing other chemicals with radiation, including hydrazine, a rocket fuel.

Methods of modifying the properties of textiles by irradiation are being developed. Wool, a very complex molecule, is improved by small amounts of radiation. When cotton, another very complex molecule, is joined to a simple organic chemical by radiation, an improvement of the fibers results.

Polyethylene, a good plastic for packaging, is often wrapped around food and shrunk into place with heat. Ordinary polyethylene is too weak to be handled this way. If it is treated with just the right amount of radiation, however, some of the molecules break and combine with others to make the plastic strong. A spoon made from irradiated plastic does not become misshapen in hot water (see photograph section).

Wood may be made harder and therefore more useful by soaking it in a simple chemical and irradiating the soaked piece so that the chemical molecules combine between the wood fibers to make a plastic. The beauty of the wood is unchanged.

Complete sterilization is needed for hospital supplies and companies in the United States, England, and Australia sterilize hypodermic syringes, pharmaceuticals, and surgical sutures by irradiation. The radiation does not destroy the protective wrappers on the sutures as moist-heat sterilization sometimes does.

In an Australian plant, goat hair is irradiated to destroy the harmful anthrax bacterium before the hair is sold for making cloth or rugs.

At the time that Rutherford and Hevesy were using natural radioisotopes or even when man-made isotopes became available from the first cyclotron, radioisotopes couldn't turn a dollar because they were scarce and expensive to make. All the naturally radioactive radium ever refined amounts to only about 3 pounds. By contrast, a large nuclear reactor now produces more radiation each year than would be emitted by 100 tons of radium.

An exotic use for radioisotopic tracers has been proposed to help determine whether there are living organisms on Mars. Project Gulliver space scientists propose an unmanned device that will land on Mars, shoot out a string covered with silicone grease, and then pull it back. The retrieved string, covered with Martian soil, will be soaked in a culture broth containing radioactive carbon. If bacterial life is present in the soil, radioactive carbon dioxide gas will be given off and automatically measured by a radiation counter. From the counting rate sent to earth by radio, scientists will be able to tell whether there is life on Mars and, if so, something about the kind of life. (30)

26 ◊ *Nuclear Clocks*

How old is the earth? The records of every civilization disclose attempts to delve into the past beyond the memory of the oldest man; beyond recorded history, beyond earliest legend.

Curiosity about the remote past may be very ancient, but the only reliable method of measurng very long intervals of time is new. The possibility of doing so became apparent only after the discovery of radioactivity in 1896. If something gradually transforms itself into something else, if this transformation goes on at a known pace, and if all the products of the activity are preserved in some kind of a closed system, then it is theoretically possible to calculate the time that has elapsed since the process started. The theory was clear for years; the only problem was how to satisfy all those ifs.

By 1910 it was well established that the earth must be extremely ancient. Analyses of some minerals containing uranium showed them to be hundreds of millions of years old, even though the uranium came from rocks that were known to be relatively young. Measurements still were inaccurate, however, and only a few rare and unusually rich radioactive minerals contained enough of the products of radioactive decay to allow analysis of their age by the crude methods then available.

Not much progress was made for about 30 years until A. O. C. Nier perfected the mass spectrometer. In 1946 Arthur Holmes in England and F. G. Houtermans in Germany realized that Nier's mass spectrometer analyses of lead made it possible, for the first time, to make rational calculations about the age of the earth. The two scientists independently calculated that age at about 2 billion to 3 billion years, using the handful of data available to them from Nier's measurements. It is interesting that today, thousands of analyses later, our planet's age usually is given as 4.5 billion years. The early estimates were not far off.

In a large number of radioactive nuclei of a given kind, a certain fraction will decay in a specific length of time. Let's take this fraction as one-half and measure the time it takes for half the nuclei to decay, the half-life of that particular nucleus. During the interval of 1 half-life, one-half of the nuclei will decay, during the next half-life half of what's left will decay, and so on. We may tabulate it like this:

ELAPSED TIME (NO. OF HALF-LIVES)	AMOUNT LEFT OF WHAT WAS ORIGINALLY PRESENT
1	1/2
2	1/4
3	1/8
4	1/16
5	1/32
6	1/64
7	1/128
. . .	. . .

In other words, after 7 half-lives, less than 1% of the original amount of material will be radioactive and the remaining 99% + of its atoms will have been converted to atoms of another nuclide. This kind of process can be made the basis of a clock. It works, in effect, like the upper chamber of an hourglass. Mathematically it is written:

$$N = N_0 e^{-\lambda t}$$

where N = the number of radioactive atoms present in the system now

N_0 = the number that was present when $t = 0$, (in other words, at the time the clock started)

e = the base of natural logarithms (the numerical value of $e = 2.718\ldots$). Natural logarithms are also called Napierian logarithms after their creator, John Napier, a Scottish mathematician (1550–1617) who also invented the decimal point

λ (lambda) = the decay rate of the radioactive material, expressed in atoms decaying per atom per unit of time

t = the time that has elapsed since the origin of the system, expressed in the same units

Obviously, in ordinary computations that would not be enough information to calculate the time, because there still are two unknowns, N_0 and t. In a closed system, however, the atoms that have decayed do not disappear into thin air. They merely change into other atoms, called daughter atoms, and remain in the system.

And at any point in time, there will be both parent and daughter atoms mixed together in the material. The older the material, the more daughters and the fewer parents. Some daughters are also radioactive, but this does not change the basic situation. Thus it follows that

$$N_0 = N + D$$

where D = the number of daughter (decayed) atoms. We may then substitute into the first equation

$$N = (N + D)\, e^{-\lambda t}$$

and solve

$$t = 1/\lambda \cdot \ln\ (1 + D/N)$$

where ln = the natural logarithm, the logarithm to base e.

This kind of system can be represented crudely by an old-fashioned hourglass, which has the parameters of these equations marked. (Keep in mind, however, that this is only a gross analogy. Nuclear clocks run at logarithmically decreasing rates, but the speed of a good hourglass is roughly constant.)

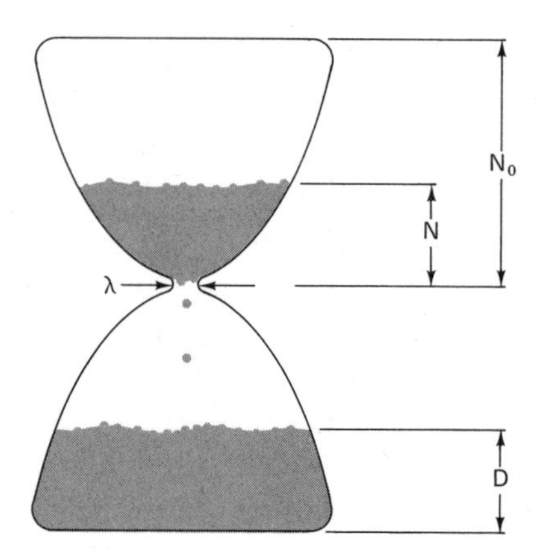

An hourglass illustrates an ideal closed system. Nothing is added and nothing is removed—the sand just runs from the top bulb to the bottom.

Remember that the decaying nucleus does not disappear. It changes into another nucleus, and this new nucleus forms an atom that may be captured and held fixed by natural processes. The decayed nuclei are thus collected, so that here we have the bottom chamber of the hourglass.

But sometimes we need only the top chamber of an hourglass.

The Carbon 14 Clock

CARBON 14 decay is the best example of a top-only hourglass. Carbon 14 is constantly being produced in the upper atmosphere from atoms of nitrogen 14 being struck by neutrons that had their origin in cosmic rays. The reaction is written:

$$^{14}\text{N} + \text{neutron} \rightarrow {}^{14}\text{C} + \text{proton}$$

Radioactive decay then follows, with a half-life of 5800 years for the ^{14}C.

$$^{14}\text{C} \rightarrow {}^{14}\text{N} + \text{electron}$$

The radiocarbon emits an electron and changes back into nitrogen.

As far as anyone can tell, ^{14}C was produced at a constant rate above the earth for at least 50,000 years before the first atomic bomb was exploded. In other words, the ^{14}C cycle is like an hourglass in which the sand in the upper part is replenished as fast as it runs out through the hole in the waist. A process of this sort, where production equals decay, is called a secular equilibrium.

The newly produced ^{14}C soon is evenly mixed with the carbon dioxide in the air, is taken up by all living plants, and then finds its way into all living animals. In effect, all carbon in living organisms contains a constant proportion of ^{14}C. If any of this carbon is taken out of circulation—when a tree branch is broken off, for instance, or when a shellfish dies in the ocean—no more new ^{14}C is added to that particular system, but the old ^{14}C continues to run out. In effect it now starts measuring time as an hourglass should.

When we find a piece of charcoal in a cave or a piece of wood in some ancient structure, for example, we can measure the amount of carbon in it, determine how much of it is ^{14}C,

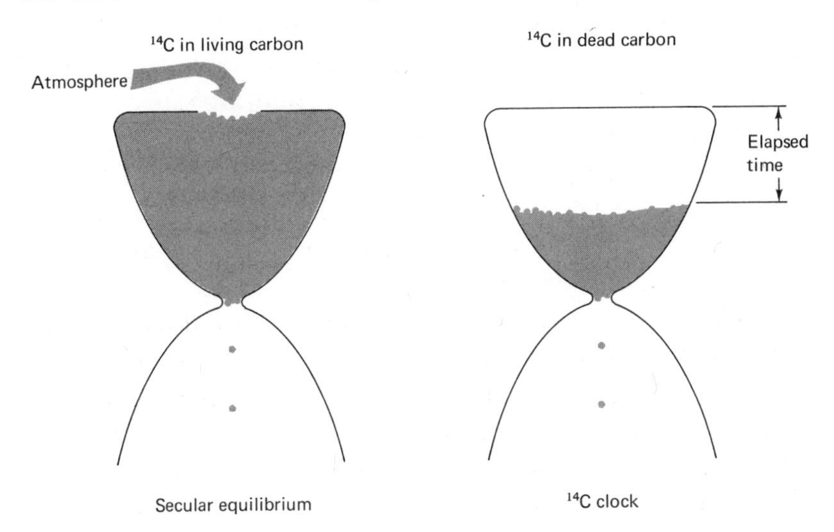

To illustrate secular equilibrium one must imagine an hourglass in which the sand in the top bulb is continuously replenished as fast as it runs out through the hole in the waist and disappears.

and then calculate back to the time when the radioactivity from the ^{14}C was the same as we now find in living wood. In other words, if we assume that we know from the observed secular equilibrium how much ^{14}C originally was present in living material, then we can calculate the time of death of any similar but ancient material. That is the basis of the ^{14}C method of age determination.

For example, a bit of a rafter from a prehistoric cliff dwelling or a remnant of charcoal from an ancient fire may be analyzed for its remaining ^{14}C content, and its age determined accurately within the margin of a few hundred years. This fixes the time at which the wood for the rafter or the firewood was broken or cut from the living tree, and hence the period in which the men lived who used the wood.

Carbon 14 measurements are made by taking a known amount of carbon, reducing it to a gas, and then counting the ^{14}C disintegrations in the gas. This may sound simple, but in reality the measurement process is a formidable undertaking, because the amount of the ^{14}C isotope in the carbon is so extremely small. (The remainder of the carbon, of course, consists of other isotopes—^{12}C or ^{13}C, which are stable.) The age of the sample is calculated from the sample counting rate minus the background; the lower the counting rate the higher the age. The upper limit of the age that can be measured is determined by the instrument accuracy in the net count. In very old samples this error may be great enough so that the calculated age of the sample may have little or no meaning.

Carbon 14 is by far the most widely used method of measuring geologic time. It has become the mainstay of archaeology and geology for studies of events of the past 50,000 years or so, and also has wide applications in climatology, ecology, and geography. One important contribution has been in study of the early inhabitants of North America. With the aid of ^{14}C it has been possible to date human living sites from many points in the western United States. The first appearance of these sites, about 11,500 years ago, apparently coincided with the time when a land bridge was open from Asia to America over what is now the Bering Strait. An ice-free passage extended from this bridge through present-day Alaska and western Canada to the United States. This may have been the route taken by the first immigrants to America—a population of mammoth-hunters, who made the characteristic flint Clovis arrow and spear points.

By about 11,000 years ago, these Clovis people had spread across the area of the United States and into Mexico. It may have been they who killed off the mammoths and then gradually assumed the characteristics of the Folsom culture. The Folsom people were bison-hunters, and long were thought to have been the first population in America. It was with the use of ^{14}C that it finally was possible to place these two cultures in proper sequence—the Clovis first—and to correlate them with major natural changes, especially the advance and retreat of glaciers across the continent.

The Long-lived Clocks

ALL other practical age-determination schemes are based on a few long-lived isotopes, with half-lives relatively near the age of the earth (4.5 aeons). They are listed on the table on page 388.

Some of the nuclides that are theoretically available are useless on a practical basis, because they are so rare in nature. Many others cannot be used for reasons that are fundamental to the whole process of nuclear age determination by whole hourglass (that is, parent-daughter) methods. Let's look at these reasons.

These methods are based on closed systems in which the daughter products of the radioactive decay are locked with the parent material from the beginning of the system, and nothing is added or removed thereafter. To state it in terms of our analogy, the hourglass must be in perfect working order—no leaks or cracks permitted.

There is another fundamental requirement: At the

LONG-LIVED ISOTOPES

ISOTOPE	EMITS	DECAYS TO	HALF-LIFE (AEONS)
Uranium 238	8 alpha particles *	lead 206	4.51
Uranium 238	spontaneous fission	2 fragments	10 million †
Uranium 235	7 alpha particles	lead 207	0.713
Thorium 232	6 alpha particles	lead 208	14.1
Rubidium 87	beta particle	strontium 87	4.7
Potassium 40	electron capture	argon 40	1.3
Rhenium 187 ‡	beta particle	osmium 187	40

* This means that uranium decays through successive steps in which the entire series emits eight alpha particles.

† Remember, this enormous period of time is a measure of the rate of spontaneous fission, not of the age of ^{238}U.

‡ The rhenium-osmium scheme is shown below the dotted line because the method is still in an early experimental stage and its general utility is not yet established.

beginning, the bottom part of the hourglass must be empty. If some sand were already in the bottom at the start, we would mistakenly be led to conclude that the time elapsed was longer than it actually was. That necessity places a severe limitation on the type of system we can use.

Consider, for example, the decay of potassium 40 into calcium 40. Measuring this process is perfectly suitable from the point of view of half-life, but the daughter product is identical with the most common isotope of ordinary calcium. And calcium is present everywhere in nature. Even the purest mineral of potassium, sylvite (the salt, potassium chloride), contains so much calcium impurity that the radiogenic daughter calcium, produced by the decay of potassium in geologic time, is negligible in comparison. We can say that the bottom of this potassium 40 hourglass has been stuffed with so much sand from the very beginning that the few grains that fall through the waist are lost in the overall mass. This demonstrates that schemes involving the decay of a relatively rare nuclide into a relatively common one are not usable.

The decay of rubidium 87 into strontium 87 is perhaps the most useful scheme for geologic age determination. The same problem shows up here, but at least there is a way out of the wilderness. It is not exactly simple, but a consideration of it is fundamental to understanding the process of nuclear dating. The diagram shows patterns from mass spectrometer charts; each peak represents an isotope of strontium, and the height of every peak is proportional to the relative abundance of that isotope. The mass-spectrum of a rock or mineral containing common strontium (which is a mixture of several isotopes) is shown in chart A. The peak of ^{87}Sr is small compared to the others. Chart B shows the mass-spectrum of strontium from an old rubidium-rich mineral crystal, drawn to the same scale, as far as the nonradiogenic isotopes, ^{84}Sr, ^{86}Sr, and ^{88}Sr, are concerned. The ^{87}Sr peak in this spectrum is obviously larger than in the common strontium in A. This is because this isotope is radiogenic and has been accumulating from the decay of rubidium since this crystal was formed.

The question we must answer is: How much of this ^{87}Sr was formed from ^{87}Rb decay and how much originally was

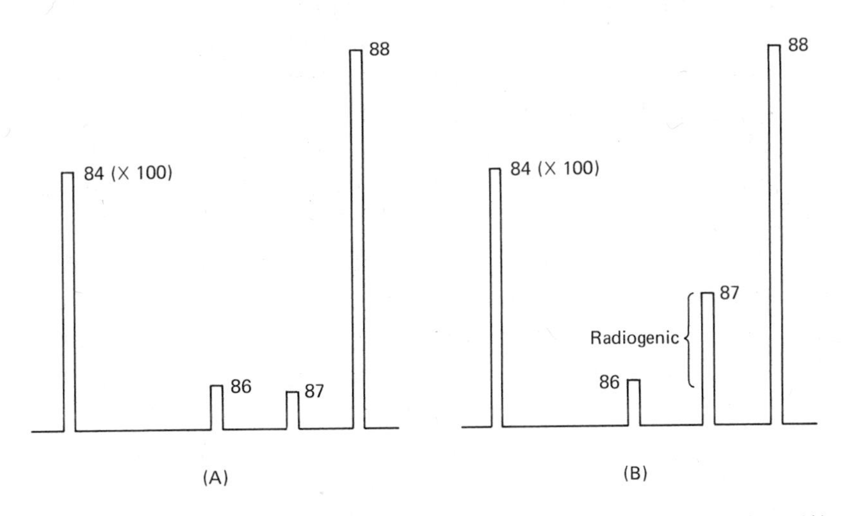

Mass-spectrograph charts showing the isotopic spectra of two kinds of strontium: (A) common strontium and (B) strontium from an old mineral rich in rubidium. (See page 24 for the mass spectrograph.)

present in the crystal as an impurity? If the amount of this original strontium is not too large, the problem can be solved by simple arithmetic.

First, we must find a good sample of common strontium—that is, ordinary strontium, the kind shown at A in the diagram. We cannot require that this strontium be entirely uncontaminated by radiogenic strontium, because all strontium is more or less contaminated. What we need is strontium contaminated to just the same extent as the strontium that was taken as an impurity into the closed system when it first formed. In geological specimens such a material is usually available.

Let us take as our closed system a mica crystal in a mass of granite. Mica contains a fair amount of rubidium, and it retains its radiogenic strontium very well. Furthermore, mica crystals are often associated or even intergrown with the slender, rod-shaped crystals of a mineral called apatite—a phosphate of calcium. It is justifiable, on the basis of geological knowledge, to say that the mica and the apatite grew at roughly the same time and thus presumably from the same liquid medium that became granite when it later solidified. Now strontium is geochemically similar to calcium, and some strontium will have gone into the apatite crystal in place of calcium. Apatite contains no alkalis—hence apatite will have virtually no rubidium (which is an alkali) in it to contaminate the ^{87}Sr. Consequently, when we find apatite in an old granite, we know the apatite will still contain the kind of common strontium that was taken into the mica crystal when it grew originally.

We can separate the apatite from the granite by standard mineralogical techniques, extract the strontium from the apatite chemically, and analyze it on a mass spectrometer to obtain the isotopic spectrum—the relative amount of each isotope that is present. We can then perform the same isotopic analysis on the strontium extracted from the mica, and subtract the original (apatite) strontium from the total (mica) strontium, to obtain the radiogenic component or daughter product.

The Uranium Fission Clock

WHEN a neutron strikes the nucleus of uranium 235 or plutonium 239, it may cause the nucleus to split into two roughly equal fragments in the well-known process of neutron-induced fission. The most common uranium isotope, ^{238}U, also breaks up by fission, but does so all by itself, without the need for any external neutrons. That process is spontaneous fission and it goes on at random, very much like radioactive decay. It is a relatively rare process and the fission half-life is long—about 10 million aeons (10^{16} years). That means that only about 1 spontaneous fission occurs in uranium 238 for every 2 million alpha decays. That is enough to make a useful clock, however, because ^{238}U is present almost everywhere.

Imagine an atom of ^{238}U in some mineral. When the atom suddenly fissions, the two fission fragments rip like cannon balls through the surrounding crystalline structure in opposite directions, creating havoc along the way. They travel a distance something like 4 millionths of an inch before they are finally slowed down and stopped by all their collisions with other atoms. Each fragment's path remains behind as an intensely damaged tube through the crystal.

The process was known for a long time before anyone was able to find these fission tracks (the damaged tubes) in the crystals. Finally, about 1960, physicists fell upon the idea of etching freshly broken surfaces of crystals with acid. They reasoned that a region so intensely disturbed by the passage of a fission fragment should be etched more easily and deeply than the undisturbed surrounding crystal. Fission tracks have now been found in almost every common mineral (since almost all minerals contain small amounts of uranium).

The fission clock method works this way: A cleavage face or a polished surface of a crystal or glass fragment is etched with a suitable solvent. The etching brings out the fission tracks so they can be seen (usually as little conical pits) and counted under a microscope.

After this, the sample is exposed to a known amount of slow neutrons in a nuclear reactor. New fissions are produced, but this time only in ^{235}U (which is present in all natural uranium in the proportion of 1 atom of ^{235}U to 137.7 atoms of ^{238}U), because slow neutrons do not produce fissions in ^{238}U. After the neutron irradiation, the same surface is etched again, and the new tracks counted. The old tracks, having been etched twice, now appear larger and thus can be distinguished from the new ones that were caused by ^{235}U fission. (17)

Knowing the number of induced and spontaneous fission tracks, the number of slow neutrons to which the sample was exposed, the likelihood that a neutron will induce ^{235}U to fission, the spontaneous fission decay rate λ_f of ^{238}U, and the alpha decay rate λ_α of ^{238}U, the age of the crystal or glass can be calculated.

Plumbology

THE most complicated and therefore probably the most interesting decay scheme of all is the decay of uranium to lead, discovered well over half a century ago and still intensively studied. There are several reasons for the interest.

First, uranium and lead are geochemically separated to a high degree, not only on the small scale of an ore deposit but also on the scale of the earth as a whole. Second, natural uranium has two isotopes with half-lives that are neither too long nor too short to be useful (the greater half-life almost exactly equaling the age of the earth), and these half-lives differ from each other by a factor of about 6.3. That leads to very important consequences, as we shall see. Third, uranium and lead are both common, and techniques are available for extracting them in measurable quantities from almost any natural material.

The greatest achievement of the plumbologists has been the calculation of the age of the earth. This was first proposed by Houtermans, who jokingly called the method plumbology, from *plumbum,* the Latin word for lead, which also provides its symbol, Pb. The calculation was also made independently by Arthur Holmes in 1946 and was finally perfected by C. C. Patterson in 1958. It is actually a rather simple calculation, although the way to discovering it was far from easy. Before we look at it in detail, however, let us consider some basic assumptions and explain what is meant by the age of the earth.

From studying the mechanics of the solar system, scientists have become reasonably certain that the earth and the other planets and their satellites all were formed in a common process in a relatively short period of time, geologically speaking. Perhaps it took a dozen million years or so, but compared to the time that has elapsed since, that is a twinkling. At some time soon afterward, the earth became molten, or at any rate fluid enough to allow much of its iron to settle toward the center to form the earth's core. Similar cores presumably formed in other planets. As the iron went down, it took some lead with it, and as the silica went up, uranium followed it toward the surface, because of the chemical affinity between these kinds of elements. In the present earth, we have found, almost all the uranium is concentrated in the top layer, or crust, which is only about 25 miles thick under the continents and even thinner under the oceans.

The time of this early and relatively rapid separation of uranium and lead on a worldwide scale is the event that plumbologists can determine, and the period since then is what they mean by the age of the earth.

How is this done? We have said that one of the isotopes of uranium, ^{235}U, decays faster—about 6.3 times faster—than the other, ^{238}U. They decay into two different isotopes of lead. Therefore, if we can determine the isotopic composition of average ordinary lead in the earth's crust today, and if we can somehow obtain a sample of the kind of lead that is locked in the earth's core, we can calculate how long it took to change the primordial lead (like that in the core) into present-day lead in the crust by the gradual addition of radiogenic lead—lead that has resulted from the decay of uranium. Now, someone might logically ask, "Isn't it necessary to know also the actual amount of uranium involved in the process, and isn't this difficult to determine?" It turns out to be a remarkable aspect of the Holmes-Houtermans calculation that the uranium concentration terms cancel out in the equations and only the ratio of the isotopes and their decay constants need be considered. These are all known accurately.

Next, we must decide just what is average present-day lead. It isn't enough to go to a lead mine and get a sample, because, unfortunately, leads from different mines have widely varied isotopic composition—that is, a different mixture of four natural isotopes, ^{204}Pb, ^{206}Pb, ^{207}Pb, and ^{208}Pb—as a result of their geologic histories. However, geologists have been able to separate lead from recent marine sediments, obtained from the ocean bottom, far from land. These are of uniform composition, and are good samples of what the world's rivers bring into the ocean. Other useful samples can be found in plateau basalts, which are enormous bodies of dark volcanic rock that make up the bedrock in many parts of the world. The lead from these basalts is isotopically very much like the lead in the oceans.

Very well, but how about the lead from the core? Where can we hope to find a sample of it? It turns out to be easier than you might think. Astronomers believe it highly probable that most meteorites are fragments of a former planet that broke up for reasons that are not entirely clear. It is pretty definite, however, that this protoplanet (or these protoplanets, for there may have been more than one) had an iron core, and this core (or these cores) is the source of the iron meteorites sailing around in space. A large meteorite hit the earth not too long ago (geologically speaking) and caused the Meteor Crater near Canyon Diablo in Arizona.

Many fragments of the meteorite iron have been found around the crater, and it is reasonable to assume that this is the kind of iron we would expect to find in the core of the earth. Like the core iron, it is mixed with a little lead, which can be isolated and analyzed in a mass spectrometer for its isotopic composition. This lead is found to be much less contaminated with radiogenic lead, and hence is much more primitive than the oldest leads found on earth. Thus, meteorites presumably are as close as we can get to true primordial lead—the lead of the time when the earth (and the protoplanet) first formed.

Once these measurements were available, it was easy (as is shown below) to write the Houtermans equation for present-day and primordial leads in this way:

$$\frac{\left(\frac{^{206}Pb}{^{204}Pb}\right)\text{present} - \left(\frac{^{206}Pb}{^{204}Pb}\right)\text{primordial}}{\left(\frac{^{207}Pb}{^{204}Pb}\right)\text{present} - \left(\frac{^{207}Pb}{^{204}Pb}\right)\text{primordial}} = 137.7\,\frac{(e^{\lambda(238)t} - 1)}{(e^{\lambda(235)t} - 1)}$$

where 137.7 = the present ratio of ^{238}U to ^{235}U
e = the base of natural logarithms
$\lambda(235)$ = the decay constant of ^{235}U
$\lambda(238)$ = the decay constant of ^{238}U
t = age of the earth

Substituting the best experimental lead isotope ratios into the equation and solving for *t,* Patterson was able to calculate that the earth is 4.55 billion years old. Subsequent calculations based on other procedures generally have confirmed that result. (17)

To arrive at the equation, note that the distribution of the isotopes of lead, 204, 205, and 207, in the center of the earth, where there is no uranium, does not vary with time. On the other hand, the distribution of these isotopes of lead at the surface of the earth is changing, two of the isotopes being continuously augmented through the decay of uranium. In particular, in a long series of alpha and beta decays, ^{235}U generates ^{207}Pb while ^{238}U generates ^{206}Pb. In both series the first (alpha) decay takes much longer than the subsequent decays; therefore, the first decay rate determines the decay rate for arriving at ^{207}Pb and ^{206}Pb. There is no generation of ^{204}Pb. We have, therefore:

$$^{207}\text{Pb (present)} = {}^{207}\text{Pb (primordial)} + \text{no. of } {}^{235}\text{U decays} \quad (1)$$
$$^{206}\text{Pb (present)} = {}^{206}\text{Pb (primordial)} + \text{no. of } {}^{238}\text{U decays} \quad (2)$$
$$^{204}\text{Pb (present)} = {}^{204}\text{Pb (primordial)} \quad (3)$$

To determine the number of ^{235}U decays from primordial time to the present, we note that:

$$^{235}U(t) = {}^{235}U(0)e^{-\lambda(235)t}$$

or:

$$^{235}U(0) - {}^{235}U(t) = {}^{235}U(t)(e^{\lambda(235)t} - 1)$$

that is:

$$\text{no. of } {}^{235}\text{U decays} = {}^{235}\text{U (present)} (e^{\lambda(235)t} - 1) \quad (4)$$

Similarly, we have:

$$\text{no. of } {}^{238}\text{U decays} = {}^{238}\text{U (present)} (e^{\lambda(238)t} - 1) \quad (5)$$

Equations (1)–(5) give the result shown on page 395.

Minerals That Can Be Dated

MEASURING age by one of the long-lived radioisotopes requires a closed system. Usually this is some kind of crystal formed in a period of time that is short, compared to the time that has elapsed since, and that has remained unchanged since it formed. Specifically, neither the parent isotopes can have been added nor the daughter isotopes removed by any process other than radioactive decay.

The earth is a dynamic system, however. Things are always changing and moving—not very rapidly, perhaps, but fast enough, in geologic time, to raise mountains and shift oceans. Solutions are moving around, dissolving something here and depositing it again somewhere else. Temperatures are changing as one place is denuded by erosion and another area buried under layers of sediment. Under such conditions, few systems remain closed. It is perhaps surprising that we find any closed systems at all. Let us look at a few that are known to be reliable.

In the early 1950s, when the potassium-argon (parent-daughter) method was being developed, it was thought that the potash-bearing variety of the mineral feldspar would be an ideal closed system, because it was usually optically clear and free of flaws. This widely shared, logical, and perfectly scientific deduction soon turned out to be quite wrong. The scientific workers discovered that when feldspar and mica from the same rock (and thus of the same age) were analyzed side by side, the mica always came out older. Investigation showed that feldspar leaked argon (lost some of its radiogenic argon) even at room temperature, but the mica retained all or nearly all of the argon that had been generated in it.

With the development of the rubidium-strontium (parent-daughter) method came the realization that mica was also very useful for this analysis, for it usually contains ample rubidium and not much original strontium that would mask the presence of the radiogenic strontium. As a result, mica, especially black mica (the mineral biotite), has enjoyed great popularity as a good and easy-to-find closed system.

Everything has its limits, and mica is no exception: Even mica tends to leak argon at elevated, but still relatively low (geologically speaking), temperatures. These effects also depend on pressure and other factors, not all of which are well known.

That means that we cannot always rely on mica to give the date of the original crystallization of a rock—the time when it cooled from a molten state. Instead, mica will tell us when the rock last cooled from, say, several hundred degrees centigrade, regardless of what may have happened to the rock before that.

In spite of early disappointments with potash feldspar for argon dating, some of it is useful for rubidium-strontium procedures. Feldspar is an excellent closed system for rubidium and strontium; it remains closed even at temperatures high enough to melt many other minerals. It is not affected at all by the same degree of heating that will drive argon out of biotite. The rubidium-strontium age of feldspar usually comes close to the time of original crystallization of the rock.

Obviously, here is a geologically important tool. If we find feldspar and biotite in one rock, and if feldspar tested by the rubidium-strontium method gives the same age as biotite tested by potassium-argon decay, then we can say with confidence that the rock has not been reheated since shortly after it crystallized. Conversely, if the biotite comes out much younger than the feldspar, we can be sure that something has happened to this rock long after it first crystallized. Such information is not only valuable to pure science—it can also be useful in locating areas favorable for ore prospecting and in other practical ways.

BASIC MEASUREMENT METHODS

METHOD	MATERIAL	TIME DATED	USEFUL TIME SPAN (YR.)
Carbon 14	wood, peat, charcoal	when plant died	1000–50,000
	bone, shell	slightly before animal died	2000–35,000
Potassium-argon	mica, some whole rocks	when rock last cooled to about 300°C	100,000 and up
	hornblende sanidine	when rock last cooled to about 500° C	10 million and up
Rubidium-strontium	mica	when rock last cooled to about 300° C	5 million and up
	potash, feldspar	when rock last cooled to about 500° C	50 million and up
	whole rock	time of separation of the rock as a closed unit	100 million and up
Uranium-lead	zircon	when crystals formed	200 million and up
Uranium 238 fission	many	when rock last cooled	100–1 billion (depending on material)

One of the most talked-about age measurements in recent years was the determination of the unexpectedly great age of fossil ancestors of man found by the late British anthropologist Dr. L. S. B. Leakey (1903–73) in Olduvai Gorge, Tanzania. With the potassium-argon method, the age came out a little less than 2 million years, about twice as old as it should be in the view of many scientists. Human remains of such great antiquity had never been found before, and much doubt was raised about the validity of the figures.

Time periods as short as 2 million years are not easy to measure by potassium-argon. The amount of argon produced in that time is extremely small, and contamination by argon from the air is a serious problem. Still, the measurements were repeated, the rocks were studied again, and the result did not change: the fossils were still about 2 million years old.

In cases like this, one tries to find some other method to check the results in an independent way. After many attempts it was discovered that the same rock strata dated by potassium-argon also contained some pumice—a porous volcanic glass—and that this glass was suitable for uranium fission-track dating. What was the result? Just about 2 million years.

Granted that the age of rocks in many parts of the world is now suddenly known—and that this was a total mystery some dozen years ago. Granted that enormous strides forward have been made. It's only a beginning.

Vast areas of the world are still geologically unexplored. The geologic time scale is till fragmentary and crude. Thousands of important geologic questions remain to be defined, explored, and answered by nuclear age determination. And—as is inevitable in science—many of them will lead to new questions. It is apparent that dating techniques have barely begun to be used and understood by geologists. But apart from geologic work, what else is in store? (17)

We now know some of what was in store only a few years ago. Rocks have been brought back from the moon, rocks that indicate that the moon is the same age as the earth: a giant step in the quest for understanding the origin of the universe.

Oxford University has a research laboratory for archaeology and the history of art. There, techniques in addition to nuclear decay, such as neutron activation, X-ray fluorescence, and mass spectrometry, lumped into the term archaeometry, have been developed and applied to the dating of ancient pottery and the detection of fakes.

Recently several major museums have been embarrassed to learn that prized pieces, assumed to be ancient Etruscan, are considerably more modern.

Besides the detection of forgeries, all sorts of information can be and has been inferred: information about ancient metallurgical techniques, trade routes, and even fluctuations in ancient economies. One fascinating finding is that when the Anglo-Saxons debased their gold coins by adding silver, the silver was always added in multiples of one-seventh, the same number seven that has had mystical significance from the Bible to modern crap shooting.

The atom has been examined in this book and traced through the cells of the human body to the moon and beyond. We have seen it used to preserve our "bread," and, not living by bread alone, to preserve our art as the genuine article. With apologies to the Scarlet Pimpernel:

> We find it here,
> We find it there,
> We find the atom everywhere. . . .

The atom is truly ubiquitous.

APPENDIX: UNITS AND PREFIXES
SOURCES
GLOSSARY
SELECTED BIBLIOGRAPHY
INDEX

Appendix: Units and Prefixes

UNIT	ABBREVIATION	MEASURED PROPERTY
angstrom	Å	wavelength
curie	c	radioactivity
electron volt	eV	energy
gram	g	mass
meter	m	length
rad	rad	radiation absorbed dose
rem	rem	radiation dose
roentgen	r	radiation dose
second	sec	time
watt	w	power

PREFIX	MEANING	
		multiply by
pico-	0.000000000001	(10^{-12})
nano-	0.000000001	(10^{-9})
micro-	0.000001	(10^{-6})
milli-	0.001	(10^{-3})
centi-	0.01	(10^{-2})
kilo-	1000	(10^{3})
mega-	1,000,000	(10^{6})
giga-	1,000,000,000	(10^{9})

Sources

BOOKLETS in the series Understanding the Atom, issued by the Atomic Energy Commission, arranged alphabetically by title; the numbers correspond to numbers in parentheses in the text

(1) Animals in Atomic Research. Edward R. Ricciuti
(2) Atomic Fuel. John F. Hogerton
(3) Atomic Power Safety. John F. Hogerton
(4) Atoms, Nature, and Man. Neal O. Hines
(5) Atoms in Agriculture. Thomas S. Osborne
(6) Chemistry of the Noble Gases, The. Cedric L. Chernick
(7) Controlled Nuclear Fusion. Samuel Glasstone
(8) Elusive Neutrino, The. Jeremy Bernstein
(9) Fallout from Nuclear Tests. Cyril L. Comar
(10) First Reactor, The
(11) Food Preservation by Irradiation. Grace M. Urrows
(12) Genetic Effects of Radiation. Isaac Asimov and Theodosius Dobzhansky
(13) Microstructure of Matter. Clifford E. Swartz
(14) Natural Radiation Environment, The. Jacob Kastner
(15) Neutron Activation Analysis. William R. Corliss
(16) Nondestructive Testing. Harold Berger
(17) Nuclear Clocks. Henry Faul
(18) Nuclear Energy for Desalting. Grace M. Urrows
(19) Nuclear Power and Merchant Shipping. Warren H. Donnelly
(20) Nuclear Power and the Environment
(21) Nuclear Power Plants. Ray L. Lyerly and Walter Mitchell, III
(22) Nuclear Propulsion for Space. William R. Corliss
(23) Nuclear Reactors. John F. Hogerton
(24) Our Atomic World. C. Jackson Craven
(25) Plowshare. Carl R. Gerber, Richard Hamburger, and E. W. Seabrook Hull
(26) Power from Radioisotopes. Robert L. Mead and William R. Corliss
(27) Power Reactors in Small Packages. William R. Corliss
(28) Radioactive Wastes. Charles H. Fox

(29) Radioisotopes and Life Processes. Walter E. Kisieleski and Renato Baserga
(30) Radioisotopes in Industry. Philip S. Baker, Domenic A. Fuccillo, Jr., Martha W. Gerrard, and Robert H. Lafferty, Jr.
(31) Radioisotopes in Medicine. Earl W. Phelan
(32) Rare Earths, the Fraternal Fifteen. Karl A. Gschneidner, Jr.
(33) Research Reactors. Frederick H. Martens and Norman H. Jacobson
(34) SNAP: Nuclear Space Reactors. William R. Corliss
(35) Sources of Nuclear Fuel. Arthur L. Singleton, Jr.
(36) Space Radiation. William R. Corliss
(37) Synthetic Transuranium Elements. Earl K. Hyde
(38) Thorium and the Third Fuel. Joseph M. Dukert
(39) Your Body and Radiation. Norman A. Frigerio

Glossary

(Adapted from *Nuclear Terms: A Brief Glossary,* 2d ed., U.S. Atomic Energy Commission)

absorbed dose When ionizing radiation passes through matter, some of its energy is imparted to the matter. The amount absorbed per unit mass of irradiated material is called the absorbed dose, and is measured in rems and rads.

absorber Any material that absorbs or diminishes the intensity of ionizing radiation. Neutron absorbers, like boron, hafnium, and cadmium, are used in control rods for reactors. Concrete and steel absorb gamma rays and neutrons in reactor shields. A thin sheet of paper or metal will absorb or attenuate alpha particles and all except the most energetic beta particles.

absorption The process by which the number of particles or photons entering a body of matter is reduced by interaction of the particles or radiation with the matter; similarly, the reduction of the energy of a particle while traversing a body of matter. This term is sometimes erroneously used for *capture* (see).

accelerator A device for increasing the velocity and energy of charged elementary particles, for example, electrons or protons, through application of electrical and/or magnetic forces. Accelerators have made particles move at velocities approaching the speed of light.

actinide series The series of elements beginning with actinium, element 89, and continuing through lawrencium, element 103, which together occupy one position in the periodic table. The series includes uranium, element 92, and all the man-made transuranic elements. The group is also referred to as the actinides.

activation analysis A method for identifying and measuring chemical elements in a sample of material. The sample is first made radioactive by bombardment with neutrons, charged particles, or gamma rays. The newly formed radioactive atoms in the sample then give off characteristic nuclear radiations (such as gamma rays) that tell what kinds of atoms are present and how many. Activation analysis is usually more sensitive than chemical analysis. It is used in research, industry, archaeology, and criminology.

AEC See *Atomic Energy Commission.*

aeon One billion (10^9) years.

alpha particle [Symbol α (alpha)] A positively charged particle emitted by certain radioactive materials. It is made up of two neutrons and two protons bound together, hence is identical with the nucleus of a helium atom. It is the least penetrating of the three common types of radiation (alpha, beta, gamma) emitted by radioactive material, being stopped by a sheet of paper. It is not dangerous to plants, animals, or man unless the alpha-emitting substance has entered the body.

angstrom [Symbol Å] A unit of length, used in measuring electromagnetic radiation, equal to 10^{-8} centimeter. Named for A. J. Ångstrom, Swedish spectroscopist.

annihilation See *antimatter.*

antimatter (antiparticles) Matter in which the ordinary nuclear particles (neutrons, protons, electrons, etc.) are conceived of as being replaced by their corresponding antiparticles (antineutrons, antiprotons, positrons, etc.). An antihydrogen atom, for example, would consist of a negatively charged antiproton with an orbital positron. Normal matter and antimatter would mutually annihilate each other upon contact, being converted totally into energy.

atom A particle of matter indivisible by chemical means. It is the fundamental building block of the chemical elements. The elements, such as iron, lead, and sulfur, differ from each other because they contain different kinds of atoms. There are about 6 sextillion (6 followed by 21 zeros, or 6×10^{21}) atoms in an ordinary drop of water. According to present-day theory, an atom contains a dense inner core (the nucleus) and a much less dense outer domain consisting of electrons in motion around the nucleus. Atoms are electrically neutral.

atomic bomb — A bomb whose energy comes from the fission of heavy elements, such as uranium or plutonium.

atomic clock — A device that uses the extremely fast vibrations of molecules or atomic nuclei to measure time. These vibrations remain constant with time, consequently short intervals can be measured with much higher precision than by mechanical or electrical clocks.

atomic energy — See *nuclear energy*.

Atomic Energy Commission — [Abbreviation AEC] The independent civilian agency of the federal government with statutory responsibility for atomic energy matters. Also the body of five persons, appointed by the president, to direct the agency.

atomic mass — See *atomic weight; mass*.

atomic mass unit — [Symbol amu] One-twelfth the mass of a neutral atom of the most abundant isotope of carbon, ^{12}C.

atomic number — [Symbol Z] The number of protons in the nucleus of an atom, and also its positive charge. Each chemical element has its characteristic atomic number, and the atomic numbers of the known elements form a complete series from 1 (hydrogen) to 103 (lawrencium).

atomic reactor — See *nuclear reactor*.

atomic weight — The mass of an atom relative to other atoms. The present-day basis of the scale of atomic weights is carbon; the commonest isotope of this element has arbitrarily been assigned an atomic weight of 12. The unit of the scale is $^1/_{12}$ the weight of the carbon 12 atom, or roughly the mass of one proton or one neutron. The atomic weight of any element is approximately equal to the total number of protons and neutrons in its nucleus.

atom smasher — See *accelerator*.

autoradiograph — A photographic record of radiation from radioactive material in an object, made by placing the object very close to a photographic film or emulsion. The process is called autoradiography. It is used, for instance, to locate radioactive atoms or tracers in metallic or biological samples. See *radiography*.

background radiation — The radiation in man's natural environment, including cosmic rays and radiation from the naturally radioactive elements, both outside and inside the bodies of men and animals. It is also called natural radiation. The term may also mean radiation that is unrelated to a specific experiment.

backscatter — When radiation of any kind strikes matter (gas, liquid, or solid), some of it may be reflected or scattered back in the general direction of the source. An understanding or exact measurement of the amount of backscatter is important when beta particles are being counted in an ionization chamber, in medical treatment with radiation, or in use of industrial radioisotopic thickness gauges.

baryon — One of a class of heavy elementary particles that includes hyperons, neutrons, and protons.

beam — A stream of particles or electromagnetic radiation, going in a single direction.

beta particle — [Symbol β (beta)] An elementary particle emitted from a nucleus during radioactive decay, with a single electrical charge and a mass equal to $^1/_{1837}$ that of a proton. A negatively charged beta particle is identical with an electron. A positively charged beta particle is called a positron. Beta radiation may cause skin burns, and beta emitters are harmful if they enter the body. Beta particles are easily stopped by a thin sheet of metal, however.

betatron — A doughnut-shaped accelerator in which electrons, traveling in an orbit of constant radius, are accelerated by a changing magnetic field. Energies as high as 340 MeV have been attained.

BeV — Symbol for billion (10^9) electron volts.

binding energy — The binding energy of a nucleus is the minimum energy required to dissociate it into its component neutrons and protons. Neutron or proton binding energies are those required to remove a neutron or a proton, respectively, from a nucleus. Electron binding energy is that required to remove an electron from an atom or a molecule.

biological dose — The radiation dose absorbed in biological material. Measured in rems.

bone seeker — A radioisotope that tends to accumulate in the bones when it is introduced into the body. An example is strontium 90, which behaves chemically like calcium.

breeder reactor — A reactor that produces fissionable fuel as well as consuming it, especially one that creates more than it consumes. The new fissionable material is created by capture in fertile materials of neutrons from fission. The process by which this occurs is known as breeding.

bremsstrahlung — Electromagnetic radiation emitted (as photons) when a fast-moving charged particle (usually an electron) loses energy upon being accelerated and deflected by the electric field surrounding a positively charged atomic nucleus. X rays produced in ordinary X-ray machines are bremsstrahlung. (German; the term means braking radiation.)

bubble chamber — A device used for detection and study of elementary particles and nuclear reactions. Charged particles from an accelerator are introduced into a superheated liquid, each forming a trail of bubbles along its path. The trails are photographed, and by studying the photograph scientists can identify the particles and analyze the nuclear events in which they originate.

by-product material Any radioactive material (except source material or fissionable material) obtained during the production or use of source material or fissionable material. It includes fission products and many other radioisotopes produced in nuclear reactors.

capture A process in which an atomic or nuclear system acquires an additional particle; for example, the capture of electrons by positive ions, or capture of electrons or neutrons by nuclei.

carrier A stable isotope, or a normal element, to which radioactive atoms of the same element can be added to obtain a quantity of radioactive mixture sufficient for handling, or to produce a radioactive mixture that will undergo the same chemical or biological reaction as the stable isotope. A substance in weighable amount which, when associated with a trace of another substance, will carry the trace through a chemical, physical, or biological process.

cathode rays A stream of electrons emitted by the cathode, or negative electrode, of a gas-discharge tube or by a hot filament in a vacuum tube, such as a television tube.

chain reaction A reaction that stimulates its own repetition. In a fission chain reaction a fissionable nucleus absorbs a neutron and fissions, releasing additional neutrons. These in turn can be absorbed by other fissionable nuclei, releasing still more neutrons. A fission chain reaction is self-sustaining when the number of neutrons released in a given time equals or exceeds the number of neutrons lost by absorption in nonfissioning material or by escape from the system.

charged particle An ion; an elementary particle that carries a positive or negative electric charge.

cladding The outer jacket of nuclear fuel elements. It prevents corrosion of the fuel and the release of fission products into the coolant. Aluminum or its alloys, stainless steel, and zirconium alloys are common cladding materials.

closed-cycle reactor system A reactor design in which the primary heat of fission is transferred outside the reactor core to do useful work by means of a coolant circulating in a completely closed system that includes a heat exchanger.

cloud chamber A device in which the tracks of charged atomic particles, such as cosmic rays or accelerator beams, are displayed. It consists of a glass-walled chamber filled with a supersaturated vapor, such as wet air. When charged particles pass through the chamber, they trigger a process of condensation, and so produce a track of tiny liquid droplets, much like the vapor trail of a jet plane. This track permits scientists to study the particles' motions and interactions.

collision A close approach of two or more particles, photons, atoms, or nuclei, during which such quantities as energy, momentum, and charge may be exchanged.

containment The provision of a gastight shell or other enclosure around a reactor to confine fission products that otherwise might be released to the atmosphere in the event of an accident.

control rod A rod, plate, or tube containing a material that readily absorbs neutrons (hafnium, boron, etc.), used to control the power of a nuclear reactor. By absorbing neutrons, a control rod prevents the neutrons from causing further fission.

controlled thermonuclear reaction Controlled fusion, that is, fusion produced under research conditions, or for production of useful power.

conversion ratio The ratio of the number of atoms of new fissionable material produced in a converter reactor to the original number of atoms of fissionable fuel consumed.

converter reactor A reactor that produces some fissionable material, but less than it consumes. In some usages, a reactor that produces a fissionable material different from the fuel burned, regardless of the ratio. In both usages the process is known as conversion.

coolant A substance circulated through a nuclear reactor to remove or transfer heat. Common coolants are water, air, carbon dioxide, liquid sodium, and sodium-potassium alloy (NaK).

core The central portion of a nuclear reactor containing the fuel elements and usually the moderator, but not the reflector.

cosmic rays Radiation of many sorts but mostly atomic nuclei (protons) with very high energies, originating outside the earth's atmosphere. Cosmic radiation is part of the natural background radiation. Some cosmic rays are more energetic than any man-made forms of radiation.

counter A general designation applied to radiation detection instruments that detect and measure radiation in terms of individual ionizations, displaying them either as the accumulated total or their rate of occurrence.

critical assembly An assembly of sufficient fissionable material and moderator to sustain a fission chain reaction at a very low power level. This permits study of the behavior of the components of the assembly for various fissionable materials in different geometrical arrangements.

criticality The state of nuclear reactor when it is sustaining a chain reaction.

critical mass The smallest mass of fissionable material that will support a self-sustaining chain reaction under stated conditions.

crystal A periodic or regularly repeating arrangement of atoms, formed from a single element or compound.

curie — [Symbol c] The basic unit to describe the intensity of radioactivity in a sample of material. The curie is equal to 37 billion disintegrations per second, which is approximately the rate of decay of 1 gram of radium. A curie is also a quantity of any nuclide having 1 curie of radioactivity. Named for Marie and Pierre Curie, who discovered radium in 1898.

cyclotron — A particle accelerator in which charged particles receive repeated synchronized accelerations by electrical fields as the particles spiral outward from their source. The particles are kept in the spiral by a powerful magnetic field.

daughter — A nuclide formed by the radioactive decay of another nuclide, which in this context is called the parent.

decay chain — See *radioactive series*.

decay constant — The number of atoms decaying per atom per unit of time (0.693/half-life).

decay, radioactive — The spontaneous transformation of one nuclide into a different nuclide or into a different energy state of the same nuclide. The process results in a decrease, with time, of the number of the original radioactive atoms in a sample. It involves the emission from the nucleus of alpha particles, beta particles (or electrons), or gamma rays; or the nuclear capture of ejection of orbital electrons; or fission. Also called radioactive disintegration.

depleted fuel — See *depleted uranium, spent fuel*.

depleted uranium — Uranium having a smaller percentage of uranium 235 than the 0.7% found in natural uranium. It is obtained from the spent (used) fuel elements or as by-product tails, or residues, of uranium isotope separation.

detector — Material or a device that is sensitive to radiation and can produce a response signal suitable for measurement or analysis.

deuterium [Symbol ^2H or D] An isotope of hydrogen whose nucleus contains one neutron and one proton and is therefore about twice as heavy as the nucleus of normal hydrogen, which is only a single proton. Deuterium is often referred to as heavy hydrogen; it occurs in nature as 1 atom to 6500 atoms of normal hydrogen. It is nonradioactive.

deuteron The nucleus of deuterium. It contains one proton and one neutron.

device, nuclear A nuclear explosive used for peaceful purposes, tests, or experiments. The term is used to distinguish these explosives from nuclear weapons.

disintegration, radioactive See *decay, radioactive*.

dose See *absorbed dose; biological dose; threshold dose*.

dose equivalent A term used to express the amount of effective radiation when modifying factors have been considered. It is expressed numerically in rems.

dose rate The radiation dose delivered per unit time. Measured, for instance, in rems per hour.

dosimeter A device that measures radiation dose, such as a film badge or ionization chamber.

doubling time The time required for a breeder reactor to produce as much fissionable material as the amount usually contained in its core plus the amount tied up in its fuel cycle (fabrication, reprocessing, etc.). It is estimated as 10 to 20 years in typical reactors.

dual-purpose reactor A reactor designed to achieve two purposes, for example, to produce both electricity and new fissionable material.

electromagnetic radiation Radiation consisting of associated and interacting electric and magnetic waves that travel at the speed of light.

Examples: light, radio waves, gamma rays, X rays. All can be transmitted through a vacuum.

electron — [Symbol e^-] An elementary particle with a unit negative electrical charge and a mass $^1/_{1837}$ that of the proton. Electrons surround the positively charged nucleus and determine the chemical properties of the atom. Positive electrons, or positrons, also exist.

electron volt — [Abbreviation eV] The amount of kinetic energy gained by an electron when it is accelerated through an electric potential difference of 1 volt. It is equivalent to 1.603×10^{-12} erg. It is a unit of energy, or work, not of voltage.

element — One of the 103 known chemical substances that cannot be divided into simpler substances by chemical means. A substance whose atoms all have the same atomic number. Examples: hydrogen, lead, uranium. (Not to be confused with fuel element.)

elementary particles — The simplest particles of matter and radiation. Most are short-lived and do not exist under normal conditions (exceptions are electrons, neutrons, protons, and neutrinos). Originally this term was applied to any particle that could not be subdivided, or to constituents of atoms; now it is applied to nucleons (protons and neutrons), electrons, mesons, muons, baryons, strange particles, and the antiparticles of each of these, and to photons, but not to alpha particles or deuterons. Also called fundamental particles.

end product — See *radioactive series*.

energy — The capability of doing work.

enriched material — Material in which the percentage of a given isotope present in a material has been artificially increased, so that it is higher than the percentage of that isotope naturally found in the material. Enriched uranium contains more of the fissionable isotope uranium 235 than the naturally occurring percentage (0.7%).

excited state The state of a molecule, atom, electron, or nucleus when it possesses more than its normal energy. Excess nuclear energy is often released as a gamma ray. Excess molecular energy may appear as fluorescence or heat. See *ground state*.

excursion A sudden, very rapid rise in the power level of a reactor caused by supercriticality. Excursions are usually quickly suppressed by the negative temperature coefficient of the reactor and/or by automatic control rods.

experimental reactor A reactor to test the design of new reactors.

fallout Airborne particles containing radioactive material which fall to the ground following a nuclear explosion. Local fallout from nuclear detonations falls to the earth's surface within 24 hours after the detonation. Tropospheric fallout consists of material injected into the troposphere but not into the higher altitudes of the stratosphere. It does not fall out locally, but usually is deposited in relatively narrow bands around the earth at about the latitude of injection. Stratospheric or worldwide fallout is injected into the stratosphere and then falls out relatively slowly over much of the earth's surface.

fast neutron A neutron with energy greater than approximately 100,000 electron volts.

fertile material A material, not itself fissionable by thermal neutrons, which can be converted into a fissile material by irradiation in a reactor. There are two basic fertile materials, uranium 238 and thorium 232. When these fertile materials capture neutrons, they are partly converted into fissile plutonium 239 and uranium 233, respectively.

film badge A light-tight package of photographic film worn like a badge by workers in nuclear industry or research, used to measure possible exposure to ionizing radiation. The absorbed dose can be calculated by the degree of film darkening caused by the irradiation.

fission The splitting of a heavy nucleus into two approximately equal parts (which are nuclei of lighter elements), accompanied by the release of a relatively large amount of energy and generally one or more neutrons. Fission can occur spontaneously, but usually is caused by nuclear absorption of gamma rays, neutrons, or other particles.

fission fragments The two nuclei which are formed by the fission of a nucleus. Also referred to as primary fission products. They are of medium atomic weight, and are radioactive.

fission products The nuclei (fission fragments) formed by the fission of heavy elements, plus the nuclides formed by the fission fragments' radioactive decay.

fluorescence Many substances can absorb energy (as from X rays, ultraviolet light, or radioactive particles), and immediately emit this energy as an electromagnetic photon, often of visible light. This emission is fluorescence. The emitting substances are said to be fluorescent.

fluoroscope An instrument with a fluorescent screen suitably mounted with respect to an X-ray tube, used for immediate indirect viewing of internal organs of the body, internal structures in apparatus, or masses of metals, by means of X rays. A fluorescent image, really a kind of X-ray shadow picture, is produced.

flux (neutron) A measure of the intensity of neutron radiation. It is the number of neutrons passing through 1 square centimeter of a given target in 1 second. Expressed as nv, where n = the number of neutrons per cubic centimeter and v = their velocity in centimeters per second.

food chain The pathways by which any material (such as radioactive material from fallout) passes from the first absorbing organism through plants and animals to man.

fuel Fissionable material used or usable to produce energy in a reactor. Also applied to a mixture, such as natural uranium, in which only part of the atoms are readily fissionable, if the mixture can be made to sustain a chain reaction.

fuel cycle The series of steps involved in supplying fuel for nuclear power reactors. It includes mining, refining, the original fabrication of fuel elements, their use in a reactor, chemical processing to recover the fissionable material remaining in the spent fuel, reenrichment of the fuel material, and refabrication into new fuel elements.

fuel element A rod, tube, plate, or other mechanical shape or form into which nuclear fuel is fabricated for use in a reactor. (Not to be confused with element.)

fuel reprocessing The processing of reactor fuel to recover the unused fissionable material.

fundamental particles See *elementary particles.*

fusion The formation of a heavier nucleus from two lighter ones (such as hydrogen isotopes), with the attendant release of energy.

gamma rays [Symbol γ (gamma)] High-energy, short-wavelength electromagnetic radiation. Gamma radiation frequently accompanies alpha and beta emissions and always accompanies fission. Gamma rays are very penetrating and are best stopped or shielded against by dense materials, such as lead or depleted uranium. Gamma rays are essentially similar to X rays, but are usually more energetic, and are nuclear in origin.

Geiger-Müller counter (tube) A radiation detection and measuring instrument. It consists of a gas-filled (Geiger-Müller) tube containing electrodes, between which there is an electrical voltage but no current flowing. When ionizing radiation passes through the tube, a short, intense pulse of current passes from the negative electrode to the positive electrode and is measured or counted. The number of pulses per second measures the intensity of radiation. It is also often known as Geiger counter; it was named for Hans Geiger and W. Müller who invented it in the 1920s.

genetic effects of radiation — Radiation effects that can be transferred from parent to offspring. Any radiation-caused changes in the genetic material of sex cells.

graphite — A very pure form of carbon used as a moderator in nuclear reactors.

ground state — The state of a nucleus, atom or molecule at its lowest (normal) energy level.

half-life — The time in which half the atoms of a particular radioactive substance disintegrate to another nuclear form. Measured half-lives vary from millionths of a second to billions of years.

health physics — The science concerned with recognition, evaluation, and control of health hazards from ionizing radiation.

heavy hydrogen — See *deuterium.*

heavy water — [Symbol D_2O] Water containing significantly more than the natural proportion (1 in 6500) of heavy hydrogen (deuterium) atoms to ordinary hydrogen atoms. Heavy water is used as a moderator in some reactors because it slows down neutrons effectively and also has a low cross section for absorption of neutrons.

hydrogen — [Symbol H] The lightest element, no. 1 in the atomic series. It has two natural isotopes of atomic weights 1 and 2. The first is ordinary hydrogen, or light hydrogen; the second is deuterium, or heavy hydrogen. A third isotope, tritium, atomic weight 3, is a radioactive form produced in reactors by bombarding lithium 6 with neutrons.

hydrogen bomb — A nuclear weapon that derives its energy largely from fusion.

hyperon One of a class of short-lived elementary particles with a mass greater than that of a proton and less than that of a deuteron. All hyperons are unstable and yield a nucleon as a decay product.

induced radioactivity Radioactivity that is created when substances are bombarded with neutrons, as from a nuclear explosion or in a reactor, or with charged particles produced by accelerators.

intensity The energy or the number of photons or particles of any radiation incident upon a unit area or flowing through a unit of solid material per unit of time. In connection with radioactivity, the number of atoms disintegrating per unit of time.

ionization The process of adding one or more electrons to, or removing one or more electrons from, atoms or molecules, thereby creating ions. High temperatures, electrical discharges, or nuclear radiations can cause ionization.

ionization chamber An instrument that detects and measures ionizing radiation by measuring the electrical current that flows when radiation ionizes gas in a chamber, making the gas a conductor of the electricity.

ionizing radiation Any radiation displacing electrons from atoms or molecules, thereby producing ions. Example: alpha, beta, gamma radiation, shortwave ultraviolet light. Ionizing radiation may produce severe skin or tissue damage.

ion pair A closely associated positive ion and negative ion (usually an electron) having charges of the same magnitude and formed from a neutral atom or molecule by radiation.

isobar One of two or more nuclides having about the same atomic mass but different atomic numbers, hence different chemical properties. Example: ${}^{14}_{6}C$, ${}^{14}_{7}N$, and ${}^{14}_{8}O$ are isobars. See *isotope*.

isotope — One of two or more atoms with the same atomic number (the same chemical element) but with different atomic weights. An equivalent statement is that the nuclei of isotopes have the same number of protons but different numbers of neutrons. Thus, $^{12}_{6}C$, $^{13}_{6}C$, and $^{14}_{6}C$ are isotopes of the element carbon, the subscripts denoting their common atomic numbers, the superscripts denoting the differing mass numbers, or approximate atomic weights. Isotopes usually have very nearly the same chemical properties, but somewhat different physical properties. See *isobar*.

isotope dilution — An analytical technique involving addition of a known amount of an isotopic mixture of abnormal composition to the unknown amount of an element of normal or known isotopic composition.

isotope separation — The process of separating isotopes from one another, or changing their relative abundances, as by gaseous diffusion or electromagnetic separation. All systems are based on the mass differences of the isotopes. Isotope separation is a step in the isotopic enrichment process.

isotopic enrichment — A process by which the relative abundances of the isotopes of a given element are altered, thus producing a form of the element which has been enriched in one particular isotope. Example: enriching natural uranium in the uranium 235 isotope.

kaon — An elementary particle (contraction of K-meson). A heavy meson with a mass about 970 times that of an electron.

kinetic energy — Energy due to motion.

K-meson — See *kaon*.

label — See *tracer, isotopic*.

lanthanide series — The series of elements beginning with lanthanum, element 57, and continuing through lutetium, element

71, which together occupy one position in the periodic table of the elements. These are the rare earths, which all have chemical properties similar to lanthanum. They also are called the lanthanides.

lepton — One of a class of light elementary particles (having small mass). Specifically, an electron, a positron, a neutrino, an antineutrino, a muon, or an antimuon.

lethal dose — A dose of ionizing radiation sufficient to cause death. Median lethal dose (MLD or LD-50) is the dose required to kill within a specified period of time (usually 30 days) half of the individuals in a large group of organisms similarly exposed. The LD-50/30 for man is about 400–450 roentgens.

linear accelerator — A long straight tube (or series of tubes) in which charged particles (ordinarily electrons or protons) gain in energy by the action of oscillating electromagnetic fields.

linear energy transfer — [Acronym LET] A measure of the ability of biological material to absorb ionizing radiation; the radiation energy lost per unit length of path through a biological material. In general, the higher the LET value, the greater is the relative biological effectiveness of the radiation in that material.

luminescence — Emission of light produced by the action of biological or chemical processes or by radiation, or any other cause except high temperature (which produces incandescence).

magnetic bottle — A magnetic field used to confine or contain a plasma in controlled fusion (thermonuclear) experiments.

magnetic mirror — A magnetic field used in controlled fusion experiments to reflect charged particles back into the central region of a magnetic bottle.

Manhattan Project — The War Department program during World War II that produced the first atomic bombs. The term originated in the code-name Manhattan Engineer District, which was used to conceal the nature of the secret work underway. The Atomic Energy Commission, a civilian agency, succeeded the military unit on January 1, 1947.

mass — The quantity of matter in a body. Often used as a synonym for weight, which, strictly speaking, is the force exerted by a body under the influence of gravity.

mass defect — The difference between the atomic mass and the mass number of a nuclide.

mass-energy equation (equivalence relation) — The statement developed by Albert Einstein, physicist, that "the mass of a body is a measure of its energy content," as an extension of his 1905 special theory of relativity. The statement was subsequently verified experimentally by measurements of mass and energy in nuclear reactions. The equation, usually given as $E = mc^2$, shows that when the energy of a body changes by an amount E (no matter what form the energy takes) the mass m of the body will change by an amount equal to E/c^2. (The factor c^2, the square of the speed of light in a vacuum, may be regarded as the conversion factor relating units of mass and energy.) This equation predicted the possibility of releasing enormous amounts of energy (in the atomic bomb) by the conversion of mass to energy.

mass number — [Symbol A] The sum of the neutrons and protons in a nucleus. It is the nearest whole number to an atom's atomic weight. For instance, the mass number of uranium 235 is 235.

mass spectrograph, mass spectrometer — Two related devices for detecting and analyzing isotopes. They separate nuclei that have different charge-to-mass ratios by passing the nuclei through electrical and magnetic fields.

matter — The substance of which a physical object is composed. All materials in the universe have the same inner nature, that is, they are composed of atoms, arranged in different (and often complex) ways; the specific atoms and the specific arrangements identify the various materials.

maximum credible accident The most serious reactor accident that can reasonably be imagined from any adverse combination of equipment malfunction, operating errors, and other foreseeable causes. The term is used to analyze the safety characteristics of a reactor. Reactors are designed to be safe even if a maximum credible accident should occur.

mean life The average time during which an atom, an excited nucleus, a radionuclide, or a particle exists in a particular form.

meson One of a class of medium-mass, short-lived elementary particles with a mass between that of the electron and that of the proton. Examples: pi-mesons (pions) and K-mesons (kaons).

MeV One million (10^6) electron volts.

micromicro- Pico-. See Units and Prefixes, page 403.

moderator A material, such as ordinary water, heavy water, or graphite, used in a reactor to slow down high-velocity neutrons, thus increasing the likelihood of further fission.

molecule A group of atoms held together by chemical forces. The atoms in the molecule may be identical, as in H_2, S_2, and S_8, or different, as in H_2O and CO_2. A molecule is the smallest unit of matter which can exist by itself and retain all its chemical properties.

monitor An instrument that measures the level of ionizing radiation in an area.

multiplication factor (or constant) [Symbol k] The ratio of the number of neutrons present in a reactor in any one neutron generation to that in the immediately preceding generation. Criticality is achieved when this ratio is equal to 1.

mu-meson See *muon.*

muon (Contraction of mu-meson.) An elementary particle, classed as a lepton (not as a meson), with 207 times the mass of an electron. It may have a single positive or negative charge.

mutation A permanent transmissible change in the characteristics of an offspring from those of its parents.

net counting rate Sample counting rate minus background counting rate.

neutrino [Symbol ν (nu)] An electrically neutral elementary particle with a negligible mass. It interacts very weakly with matter and hence is difficult to detect. It is produced in many nuclear reactions, for example, in beta decay, and has high penetrating power; neutrinos from the sun usually pass right through the earth.

neutron [Symbol n] An uncharged elementary particle with a mass slightly greater than that of the proton, and found in the nucleus of every atom heavier than hydrogen. A free neutron is unstable and decays with a half-life of about 13 minutes into an electron, proton, and neutrino. Neutrons sustain the fission chain reaction in a nuclear reactor.

neutron activation analysis Activation analysis in which neutrons are the activating agent.

neutron density The number of neutrons per cubic centimeter in the core of a reactor.

nondestructive testing Testing to detect internal and concealed defects in materials using techniques that do not damage or destroy the items being tested. X rays, isotopic radiation, and ultrasonics are frequently used.

nuclear battery See *radioisotopic generator.*

nuclear energy The energy liberated by a nuclear reaction (fission or fusion) or by radioactive decay.

nuclear power plant Any device, machine, or assembly that converts nuclear energy into some form of useful power, such as mechanical or electrical power. In a nuclear electric power plant, heat produced by a reactor is generally used to make steam to drive a turbine which in turn drives an electric generator.

nuclear reaction A reaction involving a change in an atomic nucleus, such as fission, fusion neutron capture, or radioactive decay, as distinct from a chemical reaction, which is limited to changes in the electron structure surrounding the nucleus.

nuclear reactor A device in which a fission chain reaction can be initiated, maintained, and controlled. Its essential component is a core with fissionable fuel. It usually has a moderator, a reflector, shielding, coolant, and control mechanisms. Sometimes called an atomic furnace, it is the basic machine of nuclear energy.

nuclear rocket A rocket powered by an engine that obtains energy for heating a propellant fluid (such as hydrogen) from a nuclear reactor, rather than from chemical combustion.

nucleon A constituent of an atomic nucleus, that is, a proton or a neutron.

nucleus The small, positively charged core of an atom. It is only about 1/10,000 the diameter of the atom but contains nearly all the atom's mass. All nuclei contain both protons and neutrons, except the nucleus of ordinary hydrogen, which consists of a single proton.

nuclide A general term applicable to all atom forms of the elements. The term is often erroneously used as a synonym for isotope, which properly has a more limited definition. Whereas isotopes are the various forms of a single element (hence are a family of nuclides) and all have the same atomic number and number of protons, nuclides comprise all the isotopic forms of all the elements. A nuclide is distinguished by its atomic number, atomic mass, and energy state.

pair production The transformation of the kinetic energy of a high-energy photon or particle into mass, producing a particle and its antiparticle, such as an electron and a positron.

parent A radionuclide that upon radioactive decay or disintegration yields a specific nuclide (the daughter), either directly or as a later member of a radioactive series.

particle A minute constituent of matter, generally one with a measurable mass. The primary particles involved in radioactivity are alpha particles, beta particles, neutrons, and protons.

periodic table (periodic chart) A table or chart listing all the elements, arranged in order of increasing atomic numbers and grouped by similar physical and chemical characteristics into periods. The table is based on the chemical law that the physical or chemical properties of the elements are periodic (regularly repeated) functions of their atomic weights, first proposed by the Russian chemist Dmitri I. Mendeleev in 1869.

phosphor A luminescent substance; a material capable of emitting light when stimulated by radiation.

photon The carrier of a quantum of electromagnetic energy. Photons have an effective momentum but no mass or electrical charge.

pile Old term for nuclear reactor. This name was used because the first reactor was built by piling up graphite blocks and natural uranium.

pi-meson See *pion.*

pinch effect In controlled fusion experiments, the effect obtained when an electric current, flowing through a column of plasma, produces a magnetic field that confines and compresses the plasma.

pion An elementary particle (contraction of pi-meson). The mass of a charged (positive or negative) pion is about

273 times that of an electron; that of an electrically neutral pion is 264 times that of an electron.

plasma An electrically neutral gaseous mixture of positive and negative ions. Sometimes called the fourth state of matter, since it behaves differently from solids, liquids, and gases. High-temperature plasmas are used in controlled fusion experiments.

Plowshare The Atomic Energy Commission program of research and development on peaceful uses of nuclear explosives. The possible uses include large-scale excavation, such as for canals and harbors, crushing ore bodies, and producing heavy transuranic isotopes. The term is based on a biblical reference: Isaiah 2:4.

plumbology The study of the uranium and thorium-lead decay systems. The name is derived from the Latin name for lead, *plumbum.*

plutonium [Symbol Pu] A heavy, radioactive, man-made, metallic element with atomic number 94. Its most important isotope is fissionable plutonium 239, produced by neutron irradiation of uranium 238. It is used for reactor fuel and in weapons.

pool reactor A reactor in which the fuel elements are suspended in a pool of water that serves as the reflector, moderator, and coolant. Popularly called a swimming pool reactor, it is usually used for research and training.

port An opening in a research reactor through which objects are inserted for irradiation or from which beams of radiation emerge for experimental use.

positron [Symbol β^+ (beta-plus)] An elementary particle with the mass of an electron but charged positively. It is the "antielectron." It is emitted in some radioactive disintegrations and is formed in pair production by the interaction of high-energy gamma rays with matter.

power reactor — A reactor designed to produce useful nuclear power, as distinguished from reactors used primarily for research or for producing radiation or fissionable materials.

primordial — Present at the time of the formation of the earth.

production reactor — A reactor designed primarily for large-scale production of plutonium 239 by neutron irradiation of uranium 238. Also a reactor used primarily for the production of radioactive isotopes.

proton — An elementary particle with a single positive electrical charge and a mass approximately 1837 times that of the electron. The nucleus of an ordinary or light hydrogen atom. Protons are constituents of all nuclei. The atomic number Z of an atom is equal to the number of protons in its nucleus.

pulse — An electrical signal arising from a single event of ionizing radiation.

quantum — Unit quantity of energy according to the quantum theory. It is equal to the product of the frequency of radiation of the energy and 6.6256×10^{-34} joule second. The photon carries a quantum of electromagnetic energy.

quantum theory — The statement according to Max Planck that energy is not emitted or absorbed continuously but in units or quanta. A corollary of this theory is that the energy of radiation is directly proportional to its frequency.

rad — [Acronym for *r*adiation *a*bsorbed *d*ose] The basic unit of absorbed dose of ionizing radiation. A dose of 1 rad means the absorption of 100 ergs of radiation energy per gram of absorbing material.

radiation — The emission and propagation of energy through matter or space by means of electromagnetic disturbances which display both wavelike and particlelike behavior; in this context the "particles" are known as photons.

Also, the energy so propagated. The term has been extended to include streams of fast-moving particles (alpha and beta particles, free neutrons, cosmic radiation, etc.). Nuclear radiation is that emitted from atomic nuclei in various nuclear reactions, including alpha, beta, and gamma radiation and neutrons.

radiation burn — Radiation damage to the skin. Beta burns result from skin contact with or exposure to emitters of beta particles.

radiation chemistry — The branch of chemistry that is concerned with the chemical effects, including decomposition, of energetic radiation or particles on matter. See *radiochemistry.*

radiation damage — A general term for the harmful effects of radiation on matter.

radiation detection instruments — Devices that detect and record the characteristics of ionizing radiation.

radiation dosimetry — The measurement of the amount of radiation delivered to a specific place or the amount of radiation that was absorbed there.

radiation illness — An acute organic disorder that follows exposure to relatively severe doses of ionizing radiation. It is characterized by nausea, vomiting, diarrhea, blood cell changes, and in later stages by hemorrhage and loss of hair.

radiation shielding — Reduction of radiation by interposing a shield of absorbing material between any radioactive source and a person, laboratory area, or radiation-sensitive device. See *shield.*

radiation sterilization — Use of radiation to cause a plant or animal to become sterile, that is, incapable of reproduction. Also the use of radiation to kill all forms of life (especially bacteria) in food, surgical sutures, etc.

radiation therapy — Treatment of disease with any type of radiation. Often called radiotherapy.

radio- A prefix denoting radioactivity or a relationship to it, or a relationship to radiation.

radioactive Exhibiting radioactivity or pertaining to radioactivity.

radioactive chain See *radioactive series.*

radioactive contamination Deposition of radioactive material in any place where it may harm persons, spoil experiments, or make products or equipment unsuitable or unsafe for some specific use. The presence of unwanted radioactive matter. Also radioactive material found on the walls of vessels in used-fuel processing plants, or radioactive material that has leaked into a reactor coolant. Often referred to simply as contamination.

radioactive dating A technique for measuring the age of an object or sample of material by determining the ratios of various radioisotopes or products of radioactive decay it contains. For example, the ratio of carbon 14 to carbon 12 reveals the approximate age of bones, pieces of wood, or other archaeological specimens that contain carbon extracted from the air at the time of their origin.

radioactive series A succession of nuclides, each of which transforms by radioactive disintegration into the next until a stable nuclide results. The first member is called the parent, the intermediate members are called daughters, and the final stable member is called the end product.

radioactive tracer A small quantity of radioactive isotope (either with carrier or carrier-free) used to follow biological, chemical, or other processes, by detection, determination, or localization of the radioactivity.

radioactivity The spontaneous decay or disintegration of an unstable atomic nucleus, usually accompanied by the emission of ionizing radiation.

radiobiology The body of knowledge and the study of the principles, mechanisms, and effects of ionizing radiation on living matter.

radiochemistry The body of knowledge and the study of the chemical properties and reactions of radioactive materials. See *radiation chemistry.*

radioelement An element containing one or more radioactive isotopes; a radioactive element.

radiogenic Of radioactive origin; produced by radioactive transformation.

radiography The use of ionizing radiation for the production of shadow images on a photographic emulsion. Some of the rays (gamma rays or X rays) pass through the subject, while others are partly or completely absorbed by the more opaque parts of the subject and thus cast a shadow on the photographic film. See *autoradiograph.*

radioisotope A radioactive isotope. An unstable isotope of an element that decays or disintegrates spontaneously, emitting radiation. More than 1300 natural and artificial radioisotopes have been identified.

radioisotopic generator A small power generator that converts the heat released during radioactive decay directly into electricity. These generators generally produce only a few watts of electricity and use thermoelectric or thermionic converters. Some also function as electrostatic converters to produce a small voltage. Sometimes called an atomic battery.

radiology The science which deals with the use of all forms of ionizing radiation in the diagnosis and the treatment of disease.

radium [Symbol Ra] A radioactive metallic element with atomic number 88. As found in nature, the most common isotope has an atomic weight of 226. It occurs in minute quantities associated with uranium in pitchblende, carnotite, and other minerals; the uranium decays to radium in a series of alpha and beta emissions. By virtue of being an alpha and gamma emmitter, radium is used as a source of luminescence and as a radiation source in medicine and radiography.

radon [Symbol Rn] A radioactive element, one of the heaviest gases known. Its atomic number is 86, and its atomic weight is 222. It is a daughter of radium in the uranium radioactive series.

rare earths A group of fifteen chemically similar metallic elements, including elements 57 through 71 on the periodic table of the elements, also known as the lanthanide series.

RBE See *relative biological effectiveness.*

recycling The reuse of fissionable material, after it has been recovered by chemical processing from spent or depleted reactor fuel, reenriched, and then refabricated into new fuel elements.

relative biological effectiveness (RBE) A factor used to compare the biological effectiveness of different types of ionizing radiation. It is the inverse ratio of the amount of absorbed radiation, required to produce a given effect, to a standard (or reference) radiation required to produce the same effect.

rem [Acronym for *r*oentgen *e*quivalent *m*an] The unit of dose of any ionizing radiation which produces the same biological effect as a unit of absorbed dose of ordinary X rays. The RBE dose (in rems) = RBE × absorbed dose (in rads).

rep [Acronym for *r*oentgen *e*quivalent *p*hysical] An obsolete unit of absorbed dose of any ionizing radiation, with a magnitude of 93 ergs per gram. It has been superseded by the rad.

research reactor A reactor primarily designed to supply neutrons or other ionizing radiation for experimental purposes. It may also be used for training, materials testing, and production of radioisotopes.

roentgen [Abbreviation r] A unit of exposure to ionizing radiation. It is that amount of gamma or X rays required to produce ions carrying 1 electrostatic unit of electrical charge (either positive or negative) in 1 cubic centimeter of dry air under standard conditions. Named after Wilhelm Roentgen, who discovered X rays in 1895.

roentgen equivalent, man — See *rem.*

roentgen ray — See *X ray.*

safety rod — A standby control rod used to shut down a nuclear reactor rapidly in emergencies.

scanning, radioisotope — A method of determining the location and amount of radioactive isotopes within the body by measurements taken with instruments outside the body; usually the instrument, called a scanner, moves in a regular pattern over the area to be studied, or over the whole body, and makes a visual record.

scattering — A process that changes a particle's trajectory. Scattering is caused by particle collisions with atoms, nuclei, and other particles or by interactions with fields of magnetic force. If the scattered particle's internal energy (as contrasted with its kinetic energy) is unchanged by the collision, elastic scattering prevails; if there is a change in the internal energy, the process is called inelastic scattering.

scintillation counter — An instrument that detects and measures ionizing radiation by counting the light flashes (scintillations) caused by radiation impinging on certain materials (phosphors).

secular equilibrium — The production of a radioactive substance at a rate equal to its decay.

shell — One of a series of concentric spheres, or orbits, at various distances from the nucleus, in which, according to atomic theory, electrons move around the nucleus of an atom. The shells are designated, in the order of increasing distance from the nucleus, as the *k, l, m, n, o, p,* and *q* shells. The number of electrons which each shell can contain is limited. Electrons in each shell have the same energy level and are futher grouped into subshells.

Sherwood — The Atomic Energy Commission program for research in controlled thermonuclear reactions.

shield (shielding) — A body of material used to reduce the passage of radiation. See *radiation shielding*.

slow neutron — See *thermal neutron*.

SNAP — [Acronym for Systems for Nuclear Auxiliary Power.] An Atomic Energy Commisssion program to develop small auxiliary nuclear power sources for specialized space, land, and sea uses. Two approaches are employed: the first uses heat from radioisotope decay to produce electricity directly by thermoelectric or thermionic methods; the second uses heat from small reactors to produce electricity by thermoelectric or thermionic methods or by turning a small turbine and electric generator.

somatic effects of radiation — Effects of radiation limited to the exposed individual, as distinguished from genetic effects (which also affect subsequent, unexposed generations). Large radiation doses can be fatal. Smaller doses may make the individual noticeably ill, may merely produce temporary changes in blood-cell levels detectable only in the laboratory, or may produce no detectable effects whatever. Also called physiological effects of radiation.

special (or restricted) theory of relativity — A theory developed by Albert Einstein in 1905 that is of great importance in atomic and nuclear physics. It is especially useful in studies of objects moving with speeds approaching the speed of light. Two of the results of the theory with specific application in nuclear physics are statements (1) that the mass of an object increases with its velocity and (2) that mass and energy are equivalent.

spectrum — A visual display, a photographic record, or a plot of the distribution of the intensity of a given type of radiation as a function of its wavelength, energy, frequency, momentum, mass, or any related quantity.

spent (depleted) fuel — Nuclear reactor fuel that has been irradiated (used) to the extent that it can no longer effectively sustain a chain reaction.

spontaneous fission — Fission that occurs without an external stimulus. Several heavy isotopes decay mainly in this manner; examples: californium 252 and californium 254. The process occurs occasionally in all fissionable materials, including uranium 235.

stable isotope — An isotope that does not undergo radioactive decay.

stopping power — A measure of the effect of a substance upon the kinetic energy of a charged particle passing through it.

strange particles — A class of very short-lived elementary particles that decay more slowly than they are formed, indicating that the production process and decay process result from different fundamental reactions. They include kaons and hyperons.

strata — Plural of stratum. A sheet or mass of sedimentary rock (formed by deposits of sediments, as from ancient seas) of one kind, usually in layers between beds or layers of other kinds.

subatomic particle — Any of the constituent particles of an atom: electron, neutron, proton, etc.

subcritical assembly — A reactor consisting of a mass of fissionable material and moderator whose effective multiplication factor is less than 1 and that hence cannot sustain a chain reaction. Used primarily for educational purposes.

subcritical mass — An amount of fissionable material insufficient in quantity or of improper geometry to sustain a fission chain reaction.

supercritical mass — A mass of fuel whose effective multiplication factor is greater than 1.

survival curve — Curve obtained by plotting the number or percentage of organisms surviving at a given time against the dose of radiation, or the number surviving at different intervals after a particular dose of radiation.

tails — See *depleted uranium.*

tank reactor — A reactor in which the core is suspended in a closed tank, as distinct from an open pool reactor. These are commonly used as research and test reactors.

target — Material subjected to particle bombardment (as in an accelerator) or irradiation (as in a research reactor) in order to induce a nuclear reaction; also a nuclide that has been bombarded or irradiated.

test reactor — A reactor specially designed to test the behavior of materials and components under the neutron and gramma fluxes and temperature conditions of an operating reactor.

thermal (slow) neutron — A neutron in thermal equilibrium with its surrounding medium. Thermal neutrons are those that have been slowed down by a moderator to an average speed of about 2200 meters per second (at room temperature) from the much higher initial speeds they had when expelled by fission. This velocity is similar to that of gas molecules at ordinary temperatures.

thermonuclear reaction — A reaction in which very high temperatures bring about the fusion of two light nuclei to form the nucleus of a heavier atom, releasing a large amount of energy. In a hydrogen bomb, the high temperature to initiate the thermonuclear reaction is produced by a preliminary fission reaction.

thorium — [Symbol Th] A naturally radioactive element with atomic number 90 and, as found in nature, an atomic weight of approximately 232. The fertile thorium 232 isotope is abundant and can be transmuted to fissionable uranium 233 by neutron irradiation.

thorium series (sequence) — The series of nuclides resulting from the radioactive decay of thorium 232. Many man-made nuclides decay into this sequence. The end product of this sequence in nature is lead 208.

threshold dose — The minimum dose of radiation that will produce a detectable biological effect.

tracer, isotopic — An isotope of an element, a small amount of which may be incorporated into a sample of material (the carrier) in order to follow (trace) the course of that element through a chemical, biological or physical process, and thus also follow the larger sample. The tracer may be radioactive, in which case observations are made by measuring the radioactivity. If the tracer is stable, mass spectrometers, density measurement, or neutron activation analysis may be employed to determine isotopic composition. Tracers also are called labels or tags, and materials are said to be labeled or tagged when radioactive tracers are incorporated in them.

transmutation — The transformation of one element into another by a nuclear reaction or series of reactions. Example: the transmutation of uranium 238 into plutonium 239 by absorption of a neutron.

transplutonium element — An element above plutonium in the periodic table, that is, one with an atomic number greater than 94.

transuranic element (isotope) — An element above uranium in the periodic table, that is, with an atomic number greater than 92. All 11 transuranic elements are produced artificially and are radioactive. They are neptunium, plutonium, americium, curium, berkelium, californium, einsteinium, fermium, mendelevium, nobelium, and lawrencium.

tritium — A radioactive isotope of hydrogen with two neutrons and one proton in the nucleus. It is man-made and is heavier than deuterium (heavy hydrogen). Tritium is used in industrial thickness gauges, and as a label in experiments in chemistry and biology. Its nucleus is a triton.

triton — The nucleus of a tritium (^3H) atom.

unstable isotope — See *radioisotope*.

uranium — [Symbol U] A radioactive element with atomic number 92 and, as found in natural ores, an average atomic weight of approximately 238. The two principal natural isotopes are uranium 235 (0.7% of natural uranium), which is fissionable, and uranium 238 (99.3% of natural uranium), which is fertile. Natural uranium also includes a minute amount of uranium 234. Uranium is the basic raw material of nuclear energy.

uranium series (sequence) — The series of nuclides resulting from the radioactive decay of uranium 238, also known as the uranium-radium series. The end product of the series is lead 206. Many man-made nuclides decay into this sequence.

waste, radioactive — Equipment and materials (from nuclear operations) which are radioactive and for which there is no further use. Wastes are generally classified as high-level (having radioactivity concentrations of hundreds to thousands of curies per gallon or cubic foot), low-level (in the range of 1 microcurie per gallon or cubic foot), or intermediate (between these extremes).

X ray — A penetrating form of electromagnetic radiation emitted either when the inner orbital electrons of an excited atom return to their normal state (these are characteristic X rays), or when a metal target is bombarded with high-speed electrons (these are bremsstrahlung). X rays are always nonnuclear in origin.

Selected Bibliography

BASIC books on atomic energy and closely related subjects

AUERBACH, CHARLOTTE. *Genetics in the Atomic Age.* 2nd ed. New York: Oxford University Press, 1965.

Popular-level, well-written study of genetics and the effects of radiation.

BARNES, D. E., *et al.,* advisory eds. *Newnes Concise Encyclopaedia of Nuclear Energy.* New York: Wiley, 1962.

For scientists and others; includes entries from technical fields with which nuclear energy is interrelated.

BOHR, NIELS. *Atomic Physics and Human Knowledge.* New York: Wiley, 1958.

A collection of addresses and articles which constitutes a valuable contribution to the philosophy of atomic physics.

BOORSE, HENRY A., and MOTZ, LLOYD, eds. *The World of the Atom.* 2 vols. New York: Basic Books, 1966.

Contains the actual text of landmark documents in the history of atomic physics, each preceded by commentary that places it in the context of the discoverer's personal life and in the conditions prevailing in science and in society in his time.

COHEN, BERNARD L. *The Heart of the Atom: The Structure of the Atomic Nucleus.* New York: Doubleday, 1967. (Also in paperback.)

Describes all aspects of this "atomic heart": its structure, motion, radiation, and large-scale application.

DAUGHERTY, CHARLES MICHAEL. *City Under the Ice: The Story of Camp Century.* New York: Macmillan, 1963.

A popular-level, well-illustrated book. Camp Century, an army research station directed toward opening the polar regions for human use, was constructed under the ice 800 miles from the North Pole and used a nuclear reactor to provide power, heat, and light.

DAVIS, GEORGE E. *Radiation and Life.* Ames, Iowa: Iowa State University Press, 1967.

Introduces the student or layman to the principles of atomic physics and those of biology and their interplay, with emphasis on the impact of radiation on human and animal life.

ETTER, LEWIS E. (ed.) *The Science of Ionizing Radiation.* Springfield, Ill.: Charles C. Thomas, 1965.

Readable encyclopedic record that surveys the field from Roentgen's time to the present.

FERMI, LAURA. *Atoms in the Family: My Life with Enrico Fermi.* Chicago: University of Chicago Press, 1954. (Also in paperback.)

By the wife of the physicist who led the group that built the first nuclear reactor.

FORD, KENNETH W. *The World of Elementary Particles.* New York: Blaisdell, 1963. (Also in paperback.)

A brief and simple presentation of this field.

FRANK, PHILIPP. *Einstein: His Life and Times.* New York: Knopf, 1953.

A brilliant biography that reveals the richness of Einstein's life and work and the tremendous impact he made upon physics.

FRISCH, DAVID H., and THORNDIKE, ALAN M. *Elementary Particles.* Princeton, N.J.: Van Nostrand, 1964.

An account of the basic properties of particles and the experimental techniques used to study them.

GAMOW, GEORGE. *The Atom and Its Nucleus.* Englewood Cliffs, N.J.: Prentice-Hall, 1961. (Also in paperback.)

Popular-level discussion of nuclear structure and the applications of nuclear energy.

———. *Thirty Years That Shook Physics: The Story of Quantum Theory.* New York: Doubleday, 1966. (Also in paperback.)

The development of the quantum theory presented in nontechnical language.

GLASSTONE, SAMUEL. *Sourcebook on Atomic Energy.* 3rd ed. Princeton, N.J.: Van Nostrand, 1967.

Standard reference work for both scientists and the public.

GROVES, LESLIE R. *Now It Can Be Told: The Story of the Manhattan Project.* New York: Harper, 1961.

History of the World War II atomic energy effort by its director.

HARPER, DOROTHY. *Isotopes in Action.* New York: Pergamon, 1963.

Isotope use in industry, science, medicine, and agriculture discussed in nontechnical language.

HEWLETT, RICHARD G., and ANDERSON, OSCAR E., JR. *The New World, 1939/1946. History of the United States Atomic Energy Commission,* Vol. I. University Park, Pa.: Pennsylvania State University Press, 1962.

The achievements of the Manhattan Project, the formulation of national and international policy on atomic energy, and the legislative origins of the AEC.

HEWLETT, RICHARD G., and DUNCAN, FRANCIS. *Atomic Shield, 1947/1952.*

History of the United States Atomic Energy Commission, Vol. II. University Park, Pa.: Pennsylvania State University Press, 1969.
Comprehensive history of the development of atomic energy in the United States from the transfer of the government's atomic energy program to the AEC on January 1, 1947, until the end of 1952.

HOGERTON, JOHN F. *The Atomic Energy Deskbook.* New York: Reinhold, 1963.
Encyclopedia for nonspecialists. The entries, arranged alphabetically, range from simple explanation to treatment in depth.

INGLIS, DAVID RITTENHOUSE. *Nuclear Energy: Its Physics and Its Social Challenge.* Reading, Mass.: Addison-Wesley, 1973.
A view of the scientific, political, and economic aspects of nuclear energy by an author who has been active for many years both as a research physicist and a concerned citizen.

IRVING, DAVID. *The German Atomic Bomb: The History of Nuclear Research in Nazi Germany.* New York: Simon and Schuster, 1968.
The German nuclear research program during World War II.

KRAMER, A. W. *Nuclear Propulsion for Merchant Ships.* Washington, D.C.: U.S. Government Printing Office, 1962.
Source book prepared for commercial shippers, port authorities, regulation officials, construction and design engineers, writers, and other interested persons. A substantial portion is devoted to discussions of the first commercial nuclear ship, the NS *Savannah.*

LAMONT, LANSING. *Day of Trinity.* New York: Atheneum, 1965; paperback New American Library.
Lively narrative by a *Time* correspondent focused on the Los Alamos scientists who developed the nuclear device detonated at the Trinity site near Alamogordo, New Mexico, July 16, 1945.

LAURENCE, WILLIAM. *Men and Atoms.* New York: Simon and Schuster, 1959. (Also in paperback.)
History of atomic pioneers and their work, including American development of the nuclear weapon in World War II and subsequent peaceful uses of the atom.

LAWRENCE, JOHN H., MANOWITZ, BERNARD, and LOEB, BENJAMIN S. *Radioisotopes and Radiation.* New York: Dover, 1969.
Surveys the major advances in the use of radioisotopes and radiation in medicine, agriculture, and industry. An AEC presentation volume at the 1964 Geneva Conference. Well illustrated.

MOORE, RUTH. *Niels Bohr: The Man, His Science, and the World They Changed.* New York: Knopf, 1966.
An interesting biography of one of the pioneers in the study of the internal structure of the atom.

MUSSET, PAUL, and LLORET, ANTONIO. *Concise Encyclopedia of the Atom.* Trans-

lated and edited by G. Wylie. Chicago: Follett, 1968. (Also in paperback.)
An admirable guide to terms used in nuclear science, using nontechnical language as much as possible.

SCHONLAND, BASIL. *The Atomists (1805–1933)*. New York: Oxford University Press, 1968.
Atomic theory development from Dalton through Bohr, in a form which can be understood by anyone who has had a high school physics course. Achieves a good balance between popular treatments and highly technical works without slighting the technical aspects.

SCHROEER, DIETRICH. *Physics and Its Fifth Dimension: Society.* Reading, Mass.: Addison-Wesley, 1972.
An intelligent attempt to place physics in a historical and social context, a sizable section of which is devoted to atomic energy and its consequences. Many references are given.

SEABORG, GLENN T. *Man-made Transuranium Elements.* Englewood Cliffs, N.J.: Prentice-Hall, 1963. (Also in paperback.)
The discovery, properties, and applications of elements heavier than uranium; an introduction to the subject.

SHILLING, CHARLES W., ed. *Atomic Energy Encyclopedia in the Life Sciences.* Philadelphia: Saunders, 1964.
Source book designed to be of value to the medical and biological professions and as a quick reference work for researchers, teachers, administrators, and students. The alphabetically arranged entries vary from concise definitions to journal-length articles.

SLATER, JOHN C. *Concepts and Development of Quantum Physics.* New York: Dover, 1969.
Discussion of the development of twentieth-century physics, designed for both scientists and laymen interested in modern physics as a chapter in the history of human thought. Mathematics is kept to a minimum.

SMYTH, HENRY D. *Atomic Energy for Military Purposes.* Princeton, N.J.: Princeton University Press, 1945.
Account of the project that developed the first nuclear weapons and of the considerations that prompted their use.

TOULMIN, STEPHEN, and GOODFIELD, June. *The Architecture of Matter.* New York: Harper and Row, 1963.
Scientific and philosophic concepts concerning the physics, chemistry, and physiology of matter from the beginning of scientific research eloquently presented.

WALLACE, BRUCE, and DOBZHANSKY, THEODOSIUS. *Radiation, Genes, and Man.* New York: Holt, 1959.
Careful, popular-level discussion on the genetic effects of radiation.

WOODBURY, DAVID O. *Fresh Water from Salty Seas.* New York: Dodd, Mead, 1967. (Also in paperback.)

Well-illustrated and interesting account of desalination with a section on nuclear energy applications.

YANG, CHEN NING. *Elementary Particles: A Short History of Some Discoveries in Atomic Physics.* Princeton, N.J.: Princeton University Press, 1961.

Dr. Yang, co-winner of the Nobel Prize with Dr. Tsung Dao Lee for suggesting the experiments that led to the downfall of the conservation of parity principle, provides a general outline for laymen of the history of elementary particle research during the last 60 years.

ZINN, WALTER H., PITTMAN, FRANK K., and HOGERTON, JOHN F. *Nuclear Power, U.S.A.* New York: McGraw-Hill, 1964.

Survey of the United States progress in the development of peaceful uses of atomic power; an AEC presentation volume at the 1964 Geneva Conference. Many photographs.

Index

About the Editors

GRACE MARMOR SPRUCH received her B.A. from Brooklyn College, her M.S. from the University of Pennsylvania, and her Ph.D. from New York University. She has taught at all three and at The Cooper Union in New York City. In addition to research papers in technical journals, she has written extensively for the layman and was co-translator of Robert Jungk's *The Big Machine*. She is presently Associate Professor of Physics at Rutgers University in Newark, New Jersey.

LARRY SPRUCH received his B.A. from Brooklyn College and his Ph.D. from the University of Pennsylvania. A Fellow of the American Physical Society, he has had numerous articles published in technical journals and books and is a corresponding editor of *Comments on Atomic and Molecular Physics*. He is a Professor of Physics at New York University.